# Lecture Notes in Civil Engineering 236

**Lecture Notes in Civil Engineering** (LNCE) publishes the latest developments in Civil Engineering—quickly, informally and in top quality. Though original research reported in proceedings and post-proceedings represents the core of LNCE, edited volumes of exceptionally high quality and interest may also be considered for publication. Volumes published in LNCE embrace all aspects and subfields of, as well as new challenges in, Civil Engineering. Topics in the series include:

- Construction and Structural Mechanics
- Building Materials
- Concrete, Steel and Timber Structures
- Geotechnical Engineering
- Earthquake Engineering
- Coastal Engineering
- Ocean and Offshore Engineering; Ships and Floating Structures
- Hydraulics, Hydrology and Water Resources Engineering
- Environmental Engineering and Sustainability
- Structural Health and Monitoring
- Surveying and Geographical Information Systems
- Indoor Environments
- Transportation and Traffic
- Risk Analysis
- Safety and Security

To submit a proposal or request further information, please contact the appropriate Springer Editor:

- Pierpaolo Riva at pierpaolo.riva@springer.com (Europe and Americas);
- Swati Meherishi at swati.meherishi@springer.com (Asia —except China, Australia, and New Zealand);
- Wayne Hu at wayne.hu@springer.com (China).

**All books in the series now indexed by Scopus and EI Compendex database!**

Humberto Varum · Amadeo Benavent-Climent ·
Fabrizio Mollaioli
Editors

# Energy-Based Seismic Engineering

## Proceedings of IWEBSE 2023

Springer

*Editors*
Humberto Varum
FEUP
CONSTRUCT-LESE, University of Porto
Porto, Portugal

Amadeo Benavent-Climent
Escuela Técnica Superior de Ingenieros Industriales
Technical University of Madrid
Madrid, Spain

Fabrizio Mollaioli
Dipartimento di Ingegneria Strutturale e Geotecnica (DISG)
Sapienza University of Rome
Rome, Italy

ISSN 2366-2557 ISSN 2366-2565 (electronic)
Lecture Notes in Civil Engineering
ISBN 978-3-031-36561-4 ISBN 978-3-031-36562-1 (eBook)
https://doi.org/10.1007/978-3-031-36562-1



This Springer imprint is published by the registered company Springer Nature Switzerland AG
The registered company address is: Gewerbestrasse 11, 6330 Cham, Switzerland

# Preface

Seismic design of structures in most of the developed world relies on a force-based approach where the earthquake-induced forces in the lateral load-resisting system are limited to prescribed values. Characterizing the seismic loading effect in terms of forces lacks rationality since what the earthquake really imparts is energy to the structure and the forces are a product of its resistance. Various international researchers have demonstrated the advantages of an energy-based seismic design methodology and the important role it would play in the future seismic design codes. With the exception of Japan, the energy-based approach did not gain traction until the publication of the Vision 2000 document, which advocated for the development of both displacement-based and energy-based design approaches. One of the advantages of the energy-based approach is that cumulative damage is explicitly addressed. Another advantage is the possibility to account for damage concentration in highly nonlinear systems, for which the use of equivalent single degree of freedom systems is also highly questionable.

The need for energy-based design methodologies for the earthquake design of structures was recognized as early as the mid-1950s, by Housner, and it has been addressed by many researchers (Akiyama, Bertero, etc.). A handful of procedures have been developed for using an energy-based approach in seismic design; however, most of the methods sacrifice one or more important physical characteristics for the sake of simplicity. However, with the widespread availability of seismic structural analysis software, specialized spreadsheets, and mathematical packages, simplicity in design should be reexamined. A numerical procedure is not simpler because its equations have fewer terms or an important aspect of structural response is approximated, or even ignored. Instead, a numerical procedure is simple and well understood when the designer can go from the performance objectives to the design values in an explicit and transparent way. Although, compared to forces and displacements, energy is apparently a more difficult concept for a designer to rationalize, the conservation of energy principle is a law of nature, solid, and familiar to engineers as equilibrium.

A significant amount of research is required to incorporate energy-based methodologies throughout current seismic design codes, particularly for the Performance-Based Seismic Engineering approach. For example, to assess the performance of a structure in terms of energy, engineers must have the necessary analysis tools, including models for hazard, response, damage, and costs. Although there is widespread research on performance energy-based design methodologies, it lacks a common thread for defining key features, such as:

(1) energy-based hazard analysis (intensity measures, IMs, defined in terms of energy),
(2) energy-based engineering demand parameters (EDPs),
(3) energy-based damage measures,
(4) energy-based capacity,
(5) energy-based measures of economic losses.

All of these features require an explicit energy-based formulation of performance criteria, i.e., there is a need to evaluate the energy input into a structure for different levels of earthquake ground motion, the distribution of the energy input throughout the structure so as to control the dissipation demand at each level of the structure, and explicit modeling of the energy absorption capacity of structural elements. Further research questions arise from the aleatory (irreducible and inherent to the characteristics of the phenomenon) and epistemic (reducible, typically modeling error, or lack of data) uncertainty that must to be taken into account for the hazard, demand, damage, and economic measures. All of these steps should be defined not only for newly designed structures, but also for the assessment of existing ones; and not only for conventional structures but also for innovative structures with passive control systems.

Recently, an energy-based research group, called Energy-Based Seismic Engineering Network (EBSENet), was constituted to provide answers toward the solution of all of these issues or a suitable subset therein based on the expertise of the group members. This is being accomplished by the exchange of ideas and sharing of knowledge via workshops, conference calls, and exchange of graduate students where appropriate. These workshops and exchanges will continue permitting the group to improve quickly through the research and potential implementation of the energy-based methodology in seismic design codes and standards.

The 1st International Workshop on Energy-Based Seismic Engineering was held in Madrid on 2021, and it was focused on a broad range of topics related to the Energy-Based Seismic Engineering aiming to bring together experts from several institutions.

The aim of the 2023 workshop is to develop a shared comprehensive vision for Energy-Based Seismic Engineering, tackling new challenges for the design and improvement of new and existing structures, together with the application of new smart technologies.

Amadeo Benavent-Climent
Fabrizio Mollaioli
Humberto Varum

# Contents

# Predicting the Distribution of Hysteretic Energy Demand in Frame Members

Fırat Soner Alıcı[1] and Haluk Sucuoğlu[2](✉)

[1] Department of Civil Engineering, Başkent University, Ankara 06790, Turkey
[2] Department of Civil Engineering, METU, Ankara 06800, Turkey
sucuoglu@metu.edu.tr

**Abstract.** A simple analytical procedure is developed for predicting the distribution of hysteretic energy demand from the building frame into frame members. The frame systems considered are code-designed frames conforming to capacity design principles where the formation of plastic hinges are permitted at member ends. Total hysteretic energy demand is obtained from attenuation relations developed for input energy, by employing response spectrum analysis. Energy dissipated by damping is conservatively ignored in estimating total hysteretic energy demand. Empirical relations are derived first for the ratio of energy dissipated at the maximum displacement cycle to total input energy for elasto-plastic SDOF systems under several ground motions, and expressed in spectral form. Then, hysteretic energy is distributed to the plastic hinges at member ends in accordance with their maximum chord rotations, which are estimated by a modified equivalent linear analysis. Verification of the proposed procedure is demonstrated on a five-story frame subjected to a recorded strong ground motion. The basic advantage of the developed procedure is estimating hysteretic energy dissipation demands in plastic hinges without conducting nonlinear dynamic analysis, which is also a notable advantage for energy-based design.

**Keywords:** Input Energy · Plastic Hinges · Hysteretic Energy Demand · Equivalent Linear Analysis · Chord Rotations · Distribution of Hysteretic Energy

## 1 Introduction

Input energy imparted to structural systems during an earthquake by the earthquake ground motion is a more involved demand parameter compared to the strength and deformation demands. Although force and displacement demands calculated through approximate linear elastic analysis in earthquake resistant design practice form the basis of conventional seismic design, they do not represent the actual seismic actions properly. Furthermore, interaction between forces and displacements are not accounted for in conventional design analysis. Input energy on the other hand represents a comprehensive demand of ground motion from the structure throughout its full duration.

Input energy has a major advantage in seismic design. It is not sensitive to inelastic response. Input energy demand from a linear elastic and an inelastic system with same vibration period and total mass are almost similar. This intrinsic property which


H. Varum et al. (Eds.): IWEBSE 2023, LNCE 236, pp. 1–13, 2023.
https://doi.org/10.1007/978-3-031-36562-1_1

promotes linear elastic analysis was recognized more than half a century ago [1], then applied to the plastic design of steel frames with several simplifying assumptions [2, 3]. However, it did not gain a wide acceptance in international engineering practice because the proposed analytical approaches in transforming energy dissipation demand to structural performance demands were usually complicated, not standard, hence not satisfying the expectations of practical design [4, 5]. Further research studies have revealed additional advantages of input energy such that it is weakly dependent on viscous damping as well as the force reduction factors [6, 7]. Nevertheless, the basic design expectation of demand-to-capacity inequality has not been established in simple energy terms. What is achieved successfully so far is the prediction of input energy demand under variety of seismic excitation characteristics [8–14].

The aim of the study presented herein is to improve the classical seismic design approach with an energy flavor, but not to overrule it. Thus, the initial step of the procedure is to conduct a preliminary design of the building by considering gravity loads and earthquake design forces dictated by contemporary seismic design codes. The main objective of the proposed procedure is to predict the energy dissipation demand from the structural system first, and then to determine the location of plastic hinges throughout the structure and their pertinent characteristics that are required to dissipate the seismic input energy efficiently. In this regard, the energy equilibrium equations for MDOF systems are derived, and then the proposed method is elaborated and verified on a simple case study.

## 2 Energy Equations for MDOF Systems

The equation of motion under a base excitation for a MDOF system has the same form with the equation of motion for a SDOF system, where the scalar displacement related variables are replaced with the vectorial displacement variables. Similarly, the scalar mass, stiffness, and damping property terms are replaced with the associated matrix terms. The equation of motion for a MDOF system is given in Eq. 1, where $\underline{m}$, $\underline{c}$ and $\underline{k}$ are mass, viscous damping and stiffness matrices of the MDOF system respectively. Here, $\underline{u}$ is the relative displacement vector, $\ddot{u}_g$ is the earthquake ground acceleration and $\underline{l}$ is the influence vector for ground acceleration.

$$\underline{m}\,\ddot{\underline{u}} + \underline{c}\,\dot{\underline{u}} + \underline{k}\,\underline{u} = -\underline{ml}\ddot{u}_g \tag{1}$$

The displacement vector $\underline{u}(t)$ can be expressed as a linear combination of the orthogonal modal shape vectors:

$$\underline{u}(t) = \sum_{n=1}^{N} q_n(t)\underline{\phi}_n \tag{2}$$

After substituting $\underline{u}(t)$ in Eq. (2) into Eq. (1), the equation of motion is converted into modal coordinates as shown in Eq. (3).

$$\underline{m}\sum \ddot{q}_n\underline{\phi}_n + \underline{c}\sum \dot{q}_n\underline{\phi}_n + \underline{k}\sum q_n\underline{\phi}_n = -\underline{ml}\ddot{u}_g \tag{3}$$

Pre-multiplying each term in Eq. (3) with $\phi_r{}^T$, and invoking the orthogonality of modes, only those terms with $r = n$ are non-zero. Hence, the equation can be re-arranged as follows,

$$M_n\ddot{q}_n + C_n\dot{q}_n + K_nq_n = -L_n\ddot{u}_g \tag{4}$$

where $M_n = \underline{\phi}_n^T \underline{m}\underline{\phi}_n$ is the modal mass, $C_n = \underline{\phi}_n^T \underline{c}\underline{\phi}_n$ is the modal damping, $K_n = \underline{\phi}_n^T \underline{k}\underline{\phi}_n$ is the modal stiffness, and $L_n = \underline{\phi}_n^T \underline{ml}$ is the modal excitation factor. Dividing all terms by $M_n$ leads to a final normalized form presented by Eq. (5).

$$\ddot{q}_n + 2\xi_n\omega_n\dot{q}_n + \omega_n^2q_n = -\frac{L_n}{M_n}\ddot{u}_g \tag{5}$$

Equation (5) is valid for all modes, $n$ = 1- N. This is equivalent to a SDOF system in modal coordinates $q_n$. Recalling the equation of motion for a SDOF system, i.e. Equation (6), and comparing it with Eq. (5), the only difference between them is the $L_n/M_n$ term applied to the ground excitation in the modal equation of motion, Eq. (5). Therefore, $q_n$ can be expressed in terms of $u$ as $q_n$ (t) = $L_n$ /$M_n$ $u(t)$. This equality will be employed in the derivation of energy equations for MDOF systems.

$$\ddot{u} + 2\xi\omega_n\dot{u} + \omega_n^2u = -\ddot{u}_g \tag{6}$$

Energy equation for a MDOF system is obtained by integrating the equation of motion given in Eq. (4) over the modal displacement term $q_n$ as shown in Eq. (7). Definitions of the integral terms are the same for SDOF energy equation explained above. Since the modal displacement $q_n$ is related to physical displacement $u$ through $L_n/M_n$ $u$, then $dq_n$ in Eq. (7) is also equal to $L_n/M_n$ $du$. After replacing $q_n$ and $dq_n$ terms, the right hand side of the equation representing the modal total input energy becomes as given in Eq. (8).

$$\int M_n\ddot{q}_n dq_n + \int C_n\dot{q}_n dq_n + \int K_nq_n dq_n = -\int L_n\ddot{u}_g dq_n \tag{7}$$

$$E_{I,n} = -\frac{L_n^2}{M_n}\int \ddot{u}_g du = -\frac{L_n^2}{M_n}\int \ddot{u}_g\dot{u}dt \tag{8}$$

Here, $L_n{}^2/M_n$ represents the effective modal mass $M_n{}^*$. Thus, the $n^{th}$ mode total input energy $E_{I,n}$ $(T_n)$ can be obtained by multiplying $M_n{}^*$ with the corresponding input energy spectral ordinate derived for a unit mass, $E_I$ $(T_n)/m$. On the other hand, when the energy spectrum is expressed in terms of energy equivalent velocity $V_{eq}$, $E_{I,n}$ $(T_n)$ can be obtained from Eq. (9).

$$E_{I,n}(T_n) = \frac{1}{2}M_n^*V_{eq}^2(T_n) \tag{9}$$

Total input energy for a MDOF system is equal to the sum of modal energies as given in Eq. (10). A statistical combination is not necessary since all $E_{I,n}$ reach their maximum value at the end of ground motion duration. For determining the minimum required number of modes in this equation, the procedure in the modal combination rule can be utilized where the sum of modal masses associated with the modes considered should be equal to at least 90% of the total mass of the system.

$$E_{I,tot} = \sum_{n=1}^{N} E_{I,n}(T_n) \tag{10}$$

## 3 Energy Dissipation at the Maximum Displacement Cycle

Input energy imparted to a structural system is dissipated simultaneously by viscous and hysteretic damping mechanisms in the inelastic range. The portion of total input energy dissipated by hysteretic damping through inelastic response is a crucial information in energy-based design, because the amplitude and number of hysteretic cycles are directly related to damage in structural systems.

Figure 1 shows the response of an elastoplastic SDOF system under an earthquake ground excitation. The amount of energy dissipated by hysteretic cycles can be calculated simply by integrating the area under the hysteresis curve with respect to deformation, where viscous damping dissipates the remaining energy.

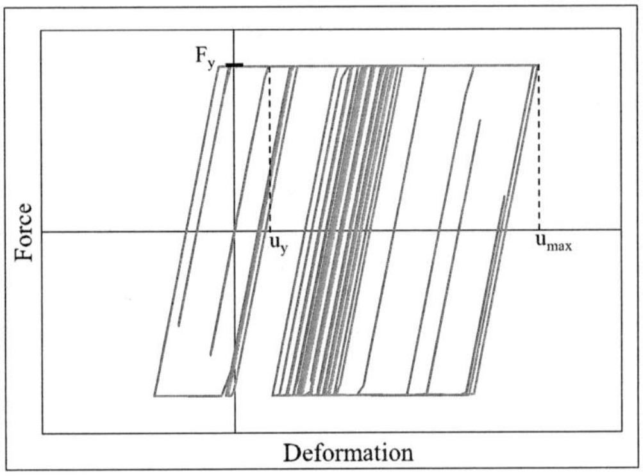


**Fig. 1.** Force-deformation response of an elastoplastic SDOF system

Hysteretic cycles, i.e. inelastic response history during seismic response display significant variability under different earthquake excitations. It is not possible to predict this type of response accurately in the design stage. Instead, more predictable response parameters such as maximum inelastic displacement can be employed as a proxy to predict hysteretic energy dissipation of the system. Figure 1 shows a typical inelastic force – deformation response. Here $F_y$ is the yield force, $u_y$ is the yield displacement corresponding to $F_y$, and $u_{max}$ is the maximum inelastic displacement under the ground excitation. The amount of energy dissipated at the maximum displacement cycle, denoted by $E_O$, can be calculated by using the simple expression given in Eq. (11). In this expression, the area enclosed by $u_y$ and $u_{max}$ is attained when the system reaches its maximum displacement $u_{max}$ during the maximum displacement cycle.

$$E_O = F_y(u_{\max} - u_y) \tag{11}$$

$E_0$ can be predicted by linear elastic analysis if the maximum inelastic displacement can be predicted through the analysis of an equivalent linear system [15, 17]. The relation between $E_0$ and total input energy $E_I$ can be employed for design purposes. The important point here is to determine what portion of the total input energy, i.e. $E_0$ is dissipated at the maximum displacement cycle. In this regard, $E_0/E_I$ ratio is computed in spectral form by employing a near field ground motion database composed of 314 ground motions [15], for different $R_\mu$ values (2, 4 and 6) and 2% viscous damping. The $E_0/E_I$ spectra obtained for each GM in the database and the corresponding mean and mean $\pm$ standard deviations are presented in Fig. 2 for three different $R_\mu$ values.

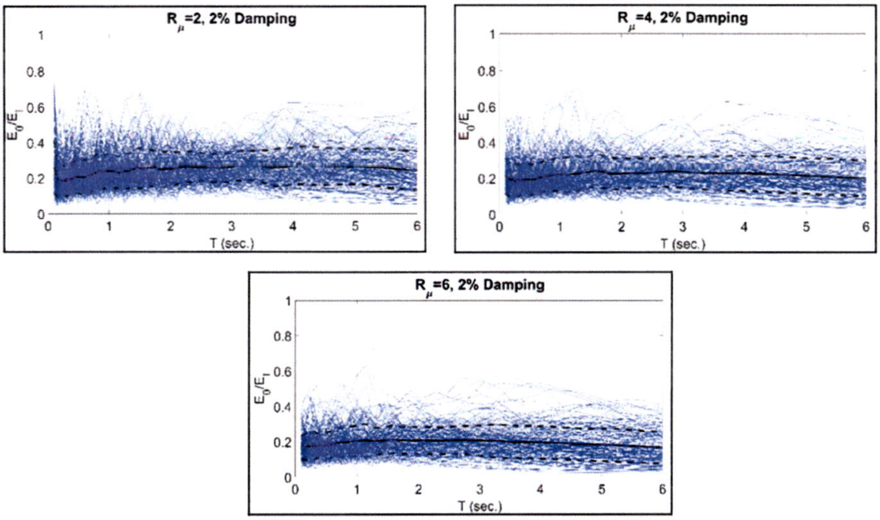


**Fig. 2.** $E_0$ /$E_I$ spectra (2% damped) for three different $R_\mu$ values with the corresponding mean and mean $\pm$ standard deviation values

It can be observed from the Fig. 2 that mean curves are in the range of 0.2 to 0.3, and they display almost a constant variation independent of vibration period $T$. In order to observe the effects of $R_\mu$ more clearly, three mean spectral curves are plotted together, and presented in Fig. 3. As $R_\mu$ increases from 2 to 6, spectral ratios reduce from 0.25 to 0.20 on average, however this difference is acceptably small. Thus, $E_0$ /$E_I$ ratio can be evaluated independently from both $T$ and $R_\mu$ for design purposes. This is a significant advantage in energy-based design compared to the conventional force-based design.

It was stated previously that damping efficiency shows strong dependency on $M_w$ [16]. Accordingly, the effect of $M_w$ on $E_0/E_I$ ratio is also investigated herein. Four different $M_w$ groups ($M_w$ − 1: 5.5–6.0, $M_w$ − 2: 6.0–6.5, $M_w$ − 3: 6.5–7.0, and $M_w$ − 4: > 7.5) are employed for the investigation. Mean $E_0/E_I$ spectra of the four $M_w$ groups obtained for $R_\mu = 4$ are shown in Fig. 4). It can be observed that mean curves do not exhibit significant differences with respect to each other up to the period of 2 s. After $T = 2$ s., lower magnitude groups $M_w$ − 1and $M_w$ − 2 produce smaller values when compared to the $M_w$ − 3 and $M_w$ − 4 groups, and display a decaying trend with period

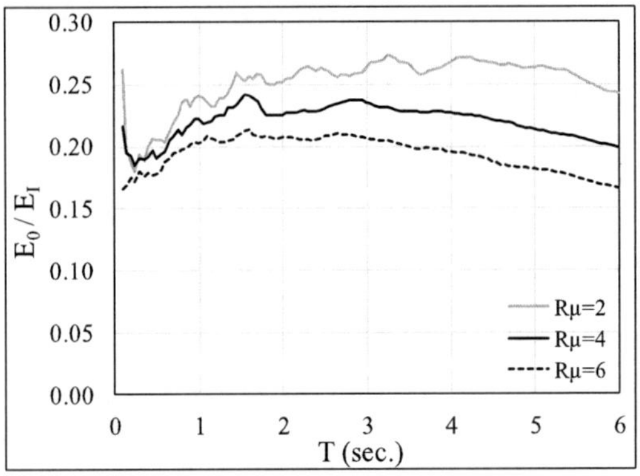


**Fig. 3.** Comparison of mean $E_0/E_I$ spectra (2% damping) obtained for three different $R_\mu$ values: Near fault ground motions

unlike the higher $M_w$ groups. In this period region, the differences between the mean curves are larger.

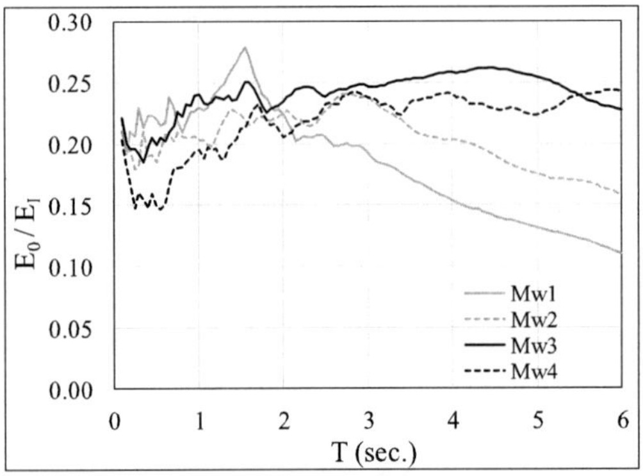


**Fig. 4.** Comparison of mean $E_0/E_I$ spectra (2% damping) of four different $M_w$ groups for $R_\mu =$ 4

## 4 Energy-Based Seismic Assessment

The first step in the assessment procedure is the preliminary design of a structure under vertical service loads and reduced earthquake loads per seismic design code. Then eigenvalue analysis is conducted in order to obtain modal properties of the corresponding system. The next step is to calculate the total input energy ($E_I$) for the MDOF system from input energy spectrum for unit mass, or in terms of equivalent velocity, $V_{eq}$ in order to employ the modal formulation derived above.

Prediction of member deformations is a crucial step in this procedure. Due to its simplicity, response spectrum analysis (RSA) is preferred in the analysis. Response spectrum analysis can be conducted by employing an equivalent linearization method. Classical equal displacement rule with 5% viscous damping, or modified equal displacement rule with equivalent damping spectra [17] can be employed. The important point here is to estimate the member-end deformations of the inelastic (actual) system reasonably well in order to predict the energy dissipation mechanism accurately.

In the following stages of the procedure, energy dissipation at member-ends are evaluated in terms of $E_0$ to set a proper energy dissipation mechanism for the system. In $E_0$ computations, moment – rotation properties of each member-end are taken into account. In this regard, the expression given in Eq. (11) is modified for $M$ and $\theta$ instead of $F$ and $u$ as in Eq. (12) for calculating $E_{0,j}$ for the $j^{th}$ member-end. Hence,

$$E_{0,j} = M_{y,j}(\theta_{\max,j} - \theta_{y,j}) \tag{12}$$

Here, $M_{y,j}$ is the yield moment capacity of the $j^{th}$ member-end, $\theta_{y,j}$ is the associated yield rotation, and $\theta_{max,j}$ is the maximum rotation response obtained from RSA by invoking equal displacement rule. It should be noted that $M_{y,j}$ and $\theta_{y,j}$ are known from the preliminary design phase. Then, the calculated $E_{0,j}$ values for each member-end are converted into total energy that can be dissipated by the $j^{th}$ member-end by employing a simple scaling operation. The scaling value is equal to the inverse of the $E_0/E_I$ ratio, i.e. $(E_0/E_I)^{-1}$. By applying such scaling, the energy dissipated by the $j^{th}$ plastic hinge can be obtained from Eq. (13).

$$E_j = E_{0,j}(E_0/E_I)^{-1} \tag{13}$$

$E_0/E_I$ spectra was presented in Figs. 2, 3 and 4 for the near fault ground motion database above, and it was observed that $E_0/E_I$ ratio is almost constant throughout the entire period region. This brings an important advantage for design calculations, since the $E_0/E_I$ ratio is almost independent from $T$ and $R_\mu$.

$E_j$ values can be calculated for the *NP* member-ends which are expected to undergo plastic rotation during seismic response. The total plastic energy dissipated by these *NP* plastic hinges ($E_{NP}$) is simply calculated from Eq. (14) where the summation of $E_j$'s gives the total dissipated hysteretic energy.

$$E_{NP} = \sum_{j=1}^{NP} E_j \tag{14}$$

*NP* is determined from the internal energy equilibrium given by Eq. (15). That is, the energy dissipated by the *NP* number of plastic hinges should be equal to $E_I$.

$$E_{NP} = E_I \tag{15}$$

Energy-based seismic assessment approach introduced herein suggests that the required number of plastic hinges and their locations, and the energy dissipated by each plastic hinge can be predicted reasonably well by conducting response spectrum analysis on the linear elastic model by using an equivalent linearization approach. Modified equal displacement rule implemented with damping spectra [17] is the preferred procedure here due to its simplicity and improved accuracy.

### 4.1 Verification of Energy-Based Seismic Assessment

The introduced assessment procedure is implemented on a 5-story frame shown in Fig. 5. The design details were presented in [15]. The assessment procedure can be summarized as follows:

1. Preliminary design: Capacity design under design spectrum, reduced with an $R_\mu$ factor. Internal end moments from design are the capacity moments $M_{y,j}$.
2. Obtain modal properties from eigenvalue analysis: $T_n$, $\phi_n$,
3. Employ input energy spectrum for calculating the input energy of mode *n*: $E_{I,n} = 1/2\ M_n{}^*\ V_{eq}{}^2(T_n)$.
4. Compute the total input energy from $E_I = \sum E_{I,n}$.
5. Conduct response spectrum analysis by using modified equal displacement rule with equivalent damping spectra and obtain member-end chord rotations $\theta_{max,j}$. Identify those *NP* hinges where yield rotations $\theta_{y,j}$ are exceeded.
6. For those yielding member-ends *j*, estimate $E_{0,j}$ from Eq. (12).
7. Find the associated $E_0/E_I$ spectral ratio for the 1st mode vibration period.
8. Calculate the energy dissipated by the $j^{th}$ plastic hinge from Eq. (13).
9. Computing the total energy dissipated by the *NP* plastic hinges from Eq. (14).
10. Check internal energy equilibrium expressed by Eq. (15).

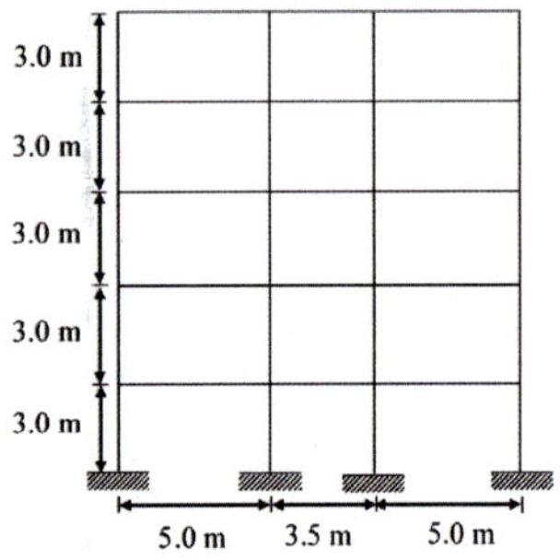


**Fig. 5.** Elevation view of the case study frame

In the following part, one ground motion record is selected from the batch of near-fault $M_w$-3 ground motion database. The ground motion labelled GM46 was recorded during the $M_w$6.69 Northridge (1994) earthquake on site class *D* with the features of reverse faulting, 18 km fault rupture length ($L_{rup}$) and $R_{epi}$ = 4.85 km. 5% damped elastic acceleration spectrum of GM46 is shown in Fig. 6a. The main purpose of the verification is first the prediction of plastic hinge locations under GM46 by using RSA with modified equal displacement rule, then comparing the actual plastic hinges and their hysteretic energy demands under GM46 obtained from NRHA.

$R_\mu$ value is estimated from the design code [18] as R/Ω where R = 8 and Ω = 3 for a ductile frame. It is taken as 3 for brevity. Elastic $V_{eq}$ spectrum of GM46 given in Fig. 6b is utilized for calculating $E_I$ of the frame system. Modal properties of the first three modes and the corresponding modal energies are presented in Table 1. First three

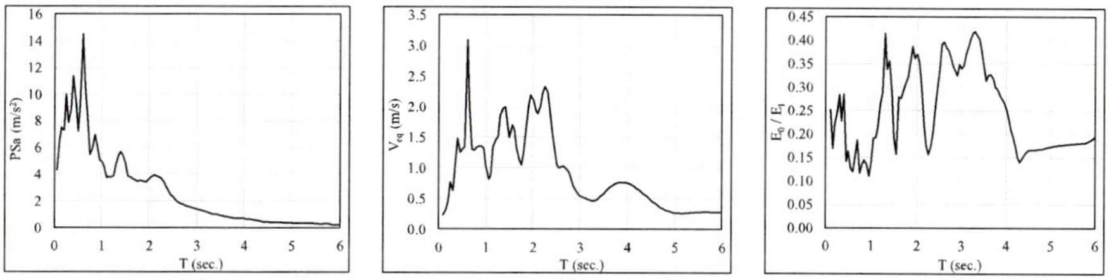

(a) (b) (c)

**Fig. 6.** For GM46, **a** 5% damped PSa spectrum, **b** Elastic equivalent velocity spectrum, **c** $E_0 / E_I$ spectrum for $R_\mu = 3$

**Table 1.** Modal properties and modal input energies of the first three modes

| Mode# | Periods (s) | $M_n^*$ (tons) | $V_{eq}$ ($T_n$) (m/s) | $E_{I,n}$ (kJ) |
|---|---|---|---|---|
| 1 | 0.94 | 208.11 | 1.37 | 194.74 |
| 2 | 0.28 | 28.96 | 0.66 | 6.37 |
| 3 | 0.15 | 12.25 | 0.31 | 0.60 |

modes are taken into consideration, where the sum of effective modal masses is equal to 96% of the total mass ($M_{tot} = 260.74$ tons).

Now, it is required to obtain maximum member-end rotations for calculating $E_{0,j}$. RSA is conducted considering the first three modes with equivalent damping spectrum of the modified equal displacement rule. Maximum responses are obtained from SRSS combination of the associated modal results. In calculating $E_{0,j}$ from Eq. (12), $M_{y,j}$ and $\theta_{y,j}$ of the $j'^{th}$ member-end, such as column-end or beam-end, are obtained from the preliminary design as stated before. $E_{0,j}$ values are tabulated in .

Table 2 for beams and columns separately, where a non-zero value indicates that the corresponding member end exceeds the elastic response range ($\theta_{max,j} > \theta_{y,j}$). In Table 2, labels used for beams are composed of the letter *B* and three numbers expressing that the first number represents the story no, and the last number is used for the bay number from left to right, considering the elevation view of the frame shown in Fig. 5 Similarly, the column labels are composed of the letter *C* and four numbers, where the first number is for the story number, and the last number is for the order from left to right in that story. I-end and J-end in the table are respectively the left end and the right end for beams, bottom end and top end for columns.

In order check the effectiveness of the proposed approximate procedure for calculating member deformations, maximum story displacements and story drift ratios obtained from RSA with modified EDR and NRHA under GM46 are compared in Fig. 7. It can be observed that the obtained maximum story displacements, interstory drifts and beam-end chord rotations obtained from the modified RSA are reasonably close to those obtained from NRHA. RSA with modified EDR gives lower drift ratio and beam-end chord rotation at the first story compared to NRHA, due to the formation of a sway mechanism during inelastic seismic response as a result of plastic hinging at the column bases.

**Table 2.** Computed $E_{0,j}$ for member-ends j

| Member | $E_{0,j}$ (kJ) | |
|---|---|---|
| | I-end | J-end |
| C1001 | 1.88 | 0.37 |
| C1002 | 1.88 | 0.00 |
| C1003 | 1.88 | 0.00 |
| C1004 | 1.88 | 0.37 |
| B111 | 1.22 | 0.91 |
| B112 | 1.14 | 1.14 |
| B113 | 0.91 | 1.22 |
| B211 | 1.42 | 1.12 |
| B212 | 1.34 | 1.34 |
| B213 | 1.12 | 1.42 |
| B311 | 1.07 | 0.81 |
| B312 | 1.02 | 1.02 |
| B313 | 0.81 | 1.07 |
| B411 | 0.46 | 0.32 |
| B412 | 0.52 | 0.52 |
| B413 | 0.32 | 0.46 |

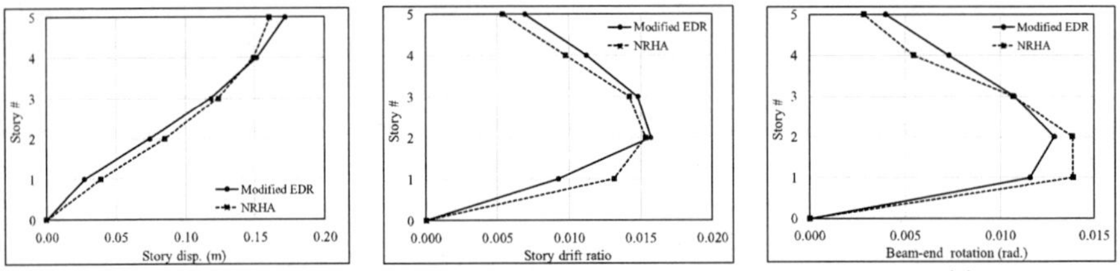


(a) (b) (c)

**Fig. 7.** Comparison of **a** maximum story displacements, **b** maximum story drifts, **c** mean maximum beam-end chord rotations from NRHA and RSA under GM46.

Then, the energy dissipated by the $j^{th}$ plastic hinge (member-end) $E_j$ is computed from Eq. (13). The scaling factor is taken from the $E_0/E_I$ spectrum of GM46 shown in Fig. 6c for the first vibration period $T_1$. The corresponding value of $E_0/E_I$ ($T_1$) is equal to 0.118. $E_j$ values of each plastic hinge obtained for all member-ends are presented in the second and third columns of Table 3.

$E_I$ is computed as 201 kJ from the summation of $E_{I,n}$ values given in Table 1. Then, it is revealed from Table 3 that the total energy dissipated by all plastic hinges (summation of estimated $E_j$) obtained by the proposed procedure is equal to 262 kJ. The difference is 30%, which is the error from the prediction of modified equal displacement rule for a single ground motion. It should be noted that equal displacement rule provides good approximation for the mean values under an ensemble of ground motions.

Hysteretic energy dissipation at plastic hinges estimated by the proposed procedure is compared by the hysteretic energy dissipation obtained from NRHA, which is accepted as the benchmark. NRHA results are presented in the last two columns of Table 3. Total hysteretic energy dissipation calculated by NRHA is 294 kJ, which is 12% larger than the estimated hysteretic energy.

**Table 3.** $E_j$ values of plastic hinges at member-ends

| Member Label | $E_j$ (kJ) PROPOSED | | $E_j$ (kJ) NRHA | |
|---|---|---|---|---|
| | I-end | J-end | I-end | J-end |
| C1001 | 15.97 | 3.15 | 19.83 | 3.50 |
| C1004 | 15.97 | 3.15 | 19.83 | 3.50 |
| C1002 | 15.95 | 0.00 | 19.80 | 0.00 |
| C1003 | 15.95 | 0.00 | 19.80 | 0.00 |
| B111 | 10.38 | 7.70 | 11.42 | 9.24 |
| B113 | 7.70 | 10.38 | 9.24 | 12.98 |
| B112 | 9.69 | 9.69 | 12.11 | 11.24 |
| B211 | 12.05 | 9.49 | 11.81 | 10.44 |
| B213 | 9.49 | 12.05 | 10.72 | 13.62 |
| B212 | 11.32 | 11.32 | 12.79 | 11.89 |
| B311 | 9.03 | 6.88 | 8.31 | 6.88 |
| B313 | 6.88 | 9.03 | 6.88 | 9.21 |
| B312 | 8.61 | 8.61 | 8.87 | 8.35 |
| B412 | 4.39 | 4.39 | 4.04 | 4.39 |
| B411 | 3.92 | 2.70 | 3.92 | 2.75 |
| B413 | 2.70 | 3.92 | 2.78 | 3.80 |

Energy dissipated at each story by the beam-end and column-end plastic hinges are compared in Fig. 8. Zero'th story represents energy dissipated by the first story columns. Columns in the upper stories (2–5) do not yield in conformance with capacity design.

Based on these results, it can be observed that first story columns and the first three story beams dissipate most of the seismic input energy imparted under GM46 record.

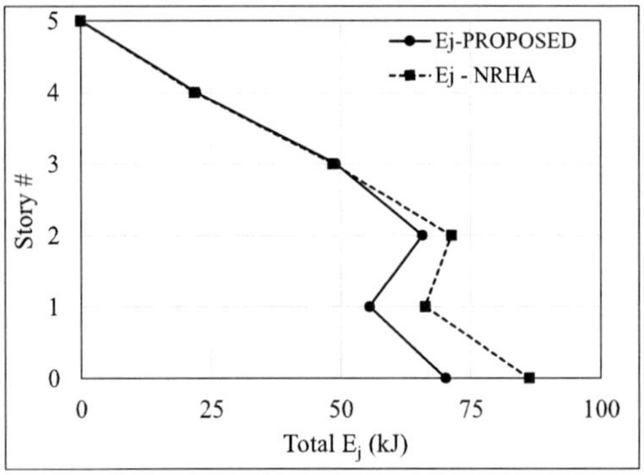


**Fig. 8.** Comparison of energy dissipated at each story from NRHA and RSA with modified EDR.

# 5 Conclusions

Based on the findings from the energy-based assessment procedure developed in this study, the proposed assessment method predicts the required number of plastic hinges and locations, and the energy dissipation mechanism properly under a defined ground excitation by employing only RSA with the modified EDR. Therefore, application of the procedure brings practicality for seismic assessment by using only a linear elastic model instead of conducting nonlinear dynamic analysis. Moreover, seismic performance of a structural system can be evaluated quite accurately from the energy dissipation performance point of view.

Verification under a single ground motion is of course insufficient. A more reliable verification should be conducted under a large set of ground motions which are scaled to either the target design acceleration spectrum, or the target $V_{eq}$ spectrum. This is the scope of our follow-up studies.

# References

1. Housner, G.W.: Limit design of structures to resist earthquakes. In: Proceedings, 1st World Conference on Earthquake Engineering, pp. 186–198. IAEE, Berkeley, California (1956)
2. Akiyama, H.: Earthquake Resistant Limit-State Design for Buildings. University of Tokyo Press, Japan (1985)
3. Akiyama, H.: Earthquake resistant design based on the energy concept. In: Proceedings, 9th World Conference on Earthquake Engineering, pp. 905–910. IAEE, Tokyo, Japan (1988)
4. Bruneau, M., Wang, N.: Normalized energy-based methods to predict the seismic ductile response of SDOF structures. Eng. Struct. **18**(1), 13–28 (1996)
5. Chou, C.C., Uang, C.M.: A procedure for evaluating seismic energy demand of framed structures. Earthquake Eng. Struct. Dynam. **32**, 229–244 (2003)
6. Chapman, M.C.: On the use of elastic input energy for seismic hazard analysis. Earthq. Spectra **15**(4), 607–635 (1999)
7. Sucuoğlu, H., Nurtuğ, A.: Earthquake ground motion characteristics and seismic energy dissipation. Earthquake Eng. Struct. Dynam. **24**, 1195–1213 (1995)
8. Decanini, L.D., Mollaioli, F.: Formulation of elastic earthquake input energy spectra. Earthquake Eng. Struct. Dynam. **27**, 1503–1522 (1998)

9. Okur, A., Erberik, M.A.: Adaptation of energy principles in seismic design of Turkish RC frame structures part I: input energy spectrum. In: Proceedings of the 15th World Conference on Earthquake Engineering, IAEE. Lisbon, Portugal (2012)
10. Alıcı, F.S., Sucuoğlu, H.: Prediction of input energy spectrum: attenuation models and velocity spectrum scaling. Earthquake Eng. Struct. Dynam. **45**, 2137–2161 (2016)
11. Benavent-Climent, A., Pujades, L.G., Lopez-Almansa, F.: Design energy input spectra for moderate seismicity regions. Earthquake Eng. Struct. Dynam. **31**, 1151–1172 (2002)
12. Chou, C.C., Uang, C.M.: Establishing absorbed energy spectra – an attenuation approach. Earthquake Eng. Struct. Dynam. **29**, 1441–1455 (2000)
13. Manfredi, G.: Evaluation of seismic energy demand. Earthquake Eng. Struct. Dynam. **30**, 485–499 (2001)
14. Zahrah, T.F., Hall, W.J.: Earthquake energy absorption in SDOF structures. J. Struct. Eng. **110**(8), 1757–1772 (1984)
15. Alıcı, F.S.: Energy based seismic assessment and design. PhD Thesis, Department of Civil Engineering, Middle East Technical University, Ankara (2019)
16. Alıcı, F.S., Sucuoğlu, H.: Efficiency of viscous damping in seismic energy dissipation and response reduction. In: Benavent-Climent, A., Mollaioli, F. (eds.) IWEBSE 2021. LNCE, vol. 155, pp. 265–276. Springer, Cham (2021). https://doi.org/10.1007/978-3-030-73932-4_17
17. Sucuoğlu, H., Alıcı, F.S.: Damping spectra for estimating inelastic deformations from modal response spectrum analysis. Earthquake Eng. Struct. Dynam. **50**, 436–454 (2021)
18. ASCE. 2016. Minimum design loads for buildings and other structures. ASCE/SEI Standard 7–16. NJ, USA (2016)

Check for updates

# Equivalent Number of Cycles Formulation for a Base-Isolated Building

Kenji Fujii(✉)

Chiba Institute of Technology, Chiba, Japan
kenji.fujii@p.chibakoudai.jp

**Abstract.** In the seismic design of a base-isolated building structure, the properties of an isolation layer are determined to (i) control the peak horizontal displacement within the predetermined limit and (ii) have a sufficient energy absorption capacity. The equivalent number of cycles is an important parameter that relates the hysteresis dissipated energy and the peak displacement of the dampers. The total input energy is a seismic intensity parameter that is related to the hysteresis dissipated energy demand. The maximum momentary input energy is another seismic intensity parameter that is related to the peak displacement. In this study, the relationship between the equivalent number of cycles and these two seismic intensity parameters is formulated for a base-isolated building whose isolation layer consists of a flexible linear isolator and a hysteresis damper. A numerical analysis of an isotropic one-mass two-degree-of-freedom model representing a base-isolated building is performed considering bidirectional ground motions. Comparisons between the formulated equation and the numerical results show that the accuracy of the formulated equation is satisfactory.

**Keywords:** Base-isolated Buildings · Momentary Energy Input · Peak Response · Cumulative Response · Equivalent Number of Cycles

## 1 Introduction

A base-isolated structure is a structure in which an isolation layer is installed at the foundation level of the building [1]. In earthquake-prone countries, base-isolation has been widely applied to important buildings, especially hospital buildings and government offices.

In the seismic design of a base-isolated building structure, the properties of an isolation layer are determined to (i) control the peak horizontal displacement within the predetermined limit and (ii) have a sufficient energy absorption capacity. The total input energy ($E_I$), or the equivalent velocity of $E_I$ ($V_I$), is a seismic intensity parameter related to the cumulative energy demand. This parameter is introduced in the simplified energy-based seismic design method proposed by Akiyama [2, 3]. In the procedure specified by Akiyama [3], the equivalent number of cycles ($n_{eq}$) is a parameter that relates to the hysteretic dissipated energy and the peak displacement [3]. Meanwhile, Hori and Inoue [4]


H. Varum et al. (Eds.): IWEBSE 2023, LNCE 236, pp. 14–26, 2023.
https://doi.org/10.1007/978-3-031-36562-1_2

proposed the maximum momentary input energy ($\Delta E_{\max}$), or the equivalent velocity of $\Delta E_{\max}$ ($V_{\Delta E}$), as another seismic intensity parameter related to the peak displacement.

Ordaz et al. [5] have shown that $E_I$ can be calculated from complex Fourier series of the input ground acceleration and the dynamic properties of the linear system. In addition, a time-varying function for calculating $\Delta E_{\max}$ has been formulated [6]. These findings imply that both seismic intensity parameters ($E_I$ and $\Delta E_{\max}$) can be calculated from complex Fourier series of the input ground acceleration. Therefore, the following question is addressed in this paper.

- What is the relationship between $n_{eq}$ and the above two seismic intensity parameters ($V_I$ and $V_{\Delta E}$)? Can this relationship be formulated in a simple equation?

In this study, the relationship between $n_{eq}$ and the two seismic intensity parameters ($V_I$ and $V_{\Delta E}$) is formulated for a base-isolated building whose isolation layer consists of a flexible linear isolator and a hysteresis damper. A numerical analysis of an isotropic one-mass two-degree-of-freedom model representing a base-isolated building is then performed considering bidirectional ground motions.

## 2 Definition of the Bidirectional Momentary Input Energy

In this study, a base-isolated building is modeled as an isotropic one-mass two-degree-of-freedom model: the superstructure of the building is assumed to be a rigid body whose mass is $M$. The isolation layer consists of (i) a flexible linear isolator whose stiffness is $k_f$ and (ii) a hysteresis damper whose initial stiffness and yield shear strength are $k_s$ and ${}_sQ_y$, respectively. The initial stiffness of the isolation layer is $k_0$ ($= k_s + k_f = (1/\alpha)k_f$). Figure 1 shows the model considered in this study.

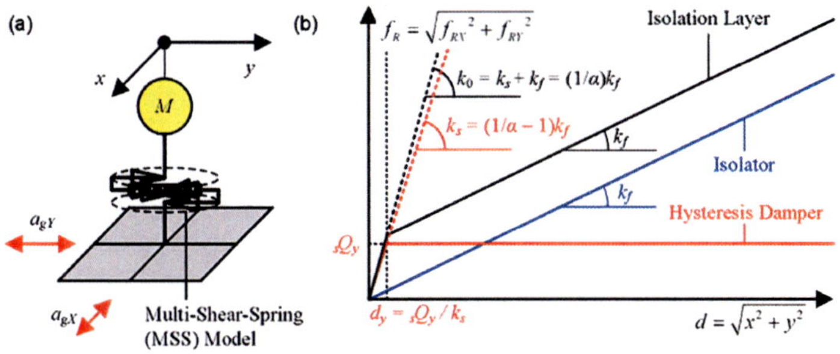


**Fig. 1.** Isotropic one-mass two-degree-of-freedom model: (a) the one-mass two-degree-of-freedom model and (b) envelopes of the restoring force–displacement relationship.

In this figure, $a_{gX}$ and $a_{gY}$ are the X- and Y-components of the horizontal ground acceleration, respectively. In addition, $x$ and $y$ are the X- and Y-components of the displacement, respectively, and $f_{RX}$ and $f_{RY}$ are the X- and Y-components of the restoring force, respectively. The envelope of the restoring force–displacement relationship was assumed to be a bilinear curve, as shown in Fig. 1b. The nonlinear behavior of the isolation layer is modeled as the sum of the isolator and the hysteresis damper: the behavior of the isolator is assumed to be linear elastic, while that of the hysteresis damper is assumed to be elastic–perfectly plastic. The system is assumed to be isotropic, that is, the yield surface of the hysteresis damper is assumed to be circular. The nonlinear behavior of the system was modeled using the Multi-Shear-Spring model proposed by Wada et al. [7]. The viscous damping of the isolation layer was assumed to be zero. The seismic isolation period ($T_f$) and the shear force coefficient of the hysteresis damper (${}_s\alpha_y$) are defined as

$$T_f = 2\pi\sqrt{\frac{M}{k_f}},\ {}_s\alpha_y = \frac{{}_sQ_y}{Mg}. \tag{1}$$

In Eq. (1), $g$ is the acceleration of gravity, which is assumed to be 9.8 m/s$^2$.

The cumulative input energy, the cumulative strain energy, and the kinetic energy at time $t$ ($E_I(t)$, $E_S(t)$, and $E_K(t)$, respectively) are defined as

$$\begin{cases} E_I(t) = -M\displaystyle\int_0^t \left\{\dot{x}(t)a_{gX}(t) + \dot{y}(t)a_{gY}(t)\right\}dt \\ E_S(t) = \displaystyle\int_0^t \{f_{RX}(t)\dot{x}(t) + f_{RY}(t)\dot{y}(t)\}dt \\ E_K(t) = M\displaystyle\int_0^t \{\ddot{x}(t)\dot{x}(t) + \ddot{y}(t)\dot{y}(t)\}dt \end{cases}. \tag{2}$$

Following a previous study [8], the bidirectional momentary input energy ($\Delta E_{BI}$) is defined as

$$\Delta E_{BI} = E_I(t_B) - E_I(t_A) = -M\int_{t_A}^{t_B} \left\{\dot{x}(t)a_{gX}(t) + \dot{y}(t)a_{gY}(t)\right\}dt. \tag{3}$$

In Eq. (3), $t_A$ and $t_B$ ($= t_A + \Delta t$) are the beginning and end times of a half cycle of the structural response, respectively: the times $t_A$ and $t_B$ are defied as the times when the absolute (vector) value of the displacement $d(t)$ is at the local maximum. The absolute value of $d(t)$ is expressed as

$$d(t) = \sqrt{\{x(t)\}^2 + \{y(t)\}^2}. \tag{4}$$

The condition for $d(t)$ at a local maximum is expressed as

$$\begin{cases} \dot{d}(t) = 0 : x(t)\dot{x}(t) + y(t)\dot{y}(t) = 0 \\ \ddot{d}(t) < 0 : x(t)\ddot{x}(t) + y(t)\ddot{y}(t) + \{\dot{x}(t)\}^2 + \{\dot{y}(t)\}^2 < 0 \end{cases}. \tag{5}$$

The maximum momentary input energy for the bidirectional excitation ($\Delta E_{BI,\max}$) is defined as the maximum value of $\Delta E_{BI}$ over the course of the seismic event. Figure 2 shows the nonlinear response of the one-mass two-degree-of-freedom model ($T_f$ = 4.0 s, ${}_s\alpha_y = 9.06 \times 10^{-2}$, $\alpha = k_f/k_0 = 0.01$, and input ground motion, 1999TCU075).

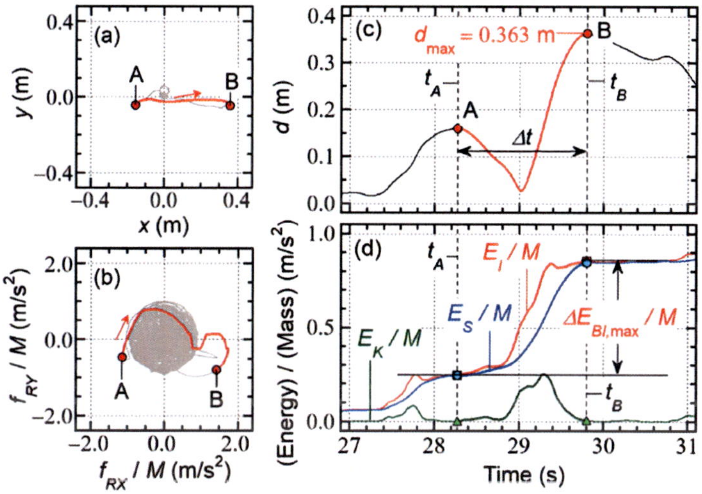


**Fig. 2.** Definition of the maximum momentary input energy for bidirectional excitation: (a) orbit of response displacement; (b) orbit of response restoring force; (c) time-history of the absolute displacement; and (d) time-histories of the input energy, cumulative strain energy, and kinetic energy per unit mass.

The maximum momentary input energy ($\Delta E_{BI,\max}$) is the input energy within the range of $[t_A, t_B]$. Note that (i) the maximum displacement $d_{\max}$ occurs when $t = t_B$, as shown in Fig. 2c, and (ii) the values of the kinetic energy $E_K(t_A)$ and $E_K(t_B)$ are

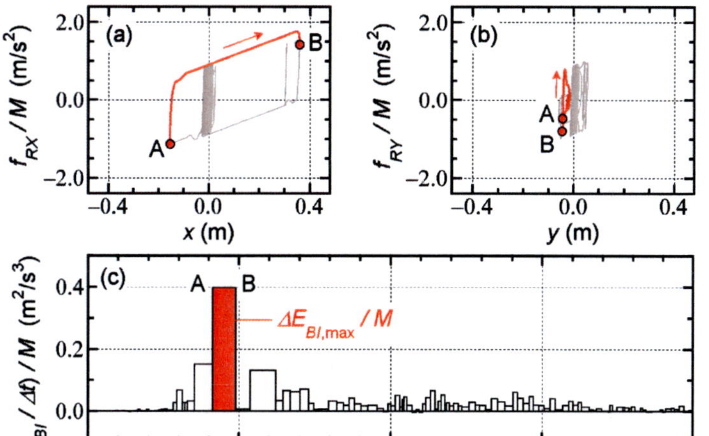


**Fig. 3.** Example of a hysteresis loop and time-history of the momentary input energy: (a) hysteresis loop in the X-direction; (b) hysteresis loop in the Y-direction; and (c) time-history of the momentary input energy per unit mass.

close to zero, as shown in Fig. 2d. Figure 3 shows an example of the hysteresis loop and time-history of the momentary input energy for the case shown in Fig. 2.

As shown in Fig. 3a, the hysteresis loop in the X-direction is similar to that of the unidirectional excitation. However, in the Y-direction, the hysteresis loop is affected by the interaction of the bidirectional strength, as shown in Fig. 3b.

The total input energy ($E_I$) is defined as the energy at the end of a seismic event ($t = t_d$) as

$$E_I = E_I(t_d). \tag{6}$$

The equivalent velocity of $E_I$ and $\Delta E_{BI,\max}$ ($V_I$ and $V_{\Delta E}$, respectively) are defined as

$$V_I = \sqrt{2E_I/M}, V_{\Delta E} = \sqrt{2\Delta E_{BI,\max}/M}. \tag{7}$$

# 3 Equivalent Number of Cycles Formulation

## 3.1 Energy Balance During a Half Cycle of Structural Response

Consider a half cycle of the structural response, AB, shown in Fig. 4. For simplicity, the behavior of the hysteresis damper is assumed to be rigid–perfectly plastic.

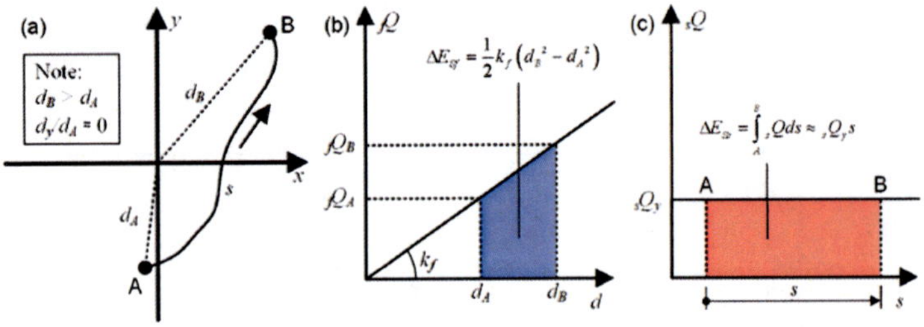


**Fig. 4.** Modelling of the momentary input energy: (a) simplified displacement orbit during a half cycle of the structural response; (b) cumulative strain energy of the isolator; and (c) cumulative strain energy of the hysteresis damper.

The energy balance during half of the structural response, AB, can be expressed as

$$\Delta E_{BI} = \Delta E_K + \Delta E_S = \{E_K(t_B) - E_K(t_A)\} + \{E_S(t_B) - E_S(t_A)\}. \tag{8}$$

It is assumed that the increment of $E_K$ is negligibly small ($\Delta E_K \approx 0$) and that the increment of $E_S$ can be modeled as

$$\Delta E_S = \Delta E_{Sf} + \Delta E_{Ss} = \frac{1}{2}k_f\left(d_B^2 - d_A^2\right) + {}_sQ_y s. \tag{9}$$

The energy balance (Eq. 8) can be rewritten as

$$\Delta E_{BI} = \frac{1}{2}k_f\left(d_B^2 - d_A^2\right) + {}_sQ_y s. \tag{10}$$

Consider the case when $\Delta E_{BI} = \Delta E_{BI,\max}$; it is assumed that the displacements at the times $t_A$ and $t_B$ can be expressed as $d_A = \eta d_{\max}$ $(0 \le \eta \le 1)$ and $d_B = d_{\max}$, respectively. In addition, the length of the path AB, $s$, is assumed to be approximated as

$$s \approx (1 + \eta) d_{\max}. \tag{11}$$

Therefore, the energy balance (Eq. 10) can be rewritten as

$$\Delta E_{BI,\max} = \frac{1}{2} k_f \left(1 - \eta^2\right) d_{\max}^2 + (1 + \eta)\, {}_sQ_y d_{\max}. \tag{12}$$

For convenience, the energy dissipation during the half cycle AB is simplified by calculating the average in the range of $0 \le \eta \le 1$. Therefore,

$$\Delta E_{BI,\max} = \int_0^1 \left\{ \frac{1}{2} k_f \left(1 - \eta^2\right) d_{\max}^2 + (1 + \eta)\, {}_sQ_y d_{\max} \right\} d\eta = \frac{1}{3} k_f d_{\max}^2 + \frac{3}{2}\, {}_sQ_y d_{\max}. \tag{13}$$

The ratio of the maximum force of the flexible linear isolator (${}_fQ_{\max}$) to ${}_sQ_y$, $r_q$, is defined as

$$r_q = {}_fQ_{\max} / {}_sQ_y = k_f d_{\max} / {}_sQ_y. \tag{14}$$

Using $r_q$, Eq. (13) can be rewritten as

$$\Delta E_{BI,\max} = \frac{1}{3}\, {}_fQ_{\max} d_{\max} + \frac{3}{2}\, {}_sQ_y d_{\max} = \frac{3}{2} \left(1 + \frac{2}{9} r_q\right) {}_sQ_y d_{\max}. \tag{15}$$

### 3.2 Equivalent Number of Cycles

Next, the relationship between $n_{eq}$ and the two seismic intensity parameters ($V_I$ and $V_{\Delta E}$) is formulated. Following previous studies [1, 3], the energy balance during an entire seismic event is modeled as

$$E_I = \frac{1}{2} k_f d_{\max}^2 + 4 n_{eq}\, {}_sQ_y d_{\max} = \frac{1}{2}\, {}_fQ_{\max} d_{\max} + 4 n_{eq}\, {}_sQ_y d_{\max}. \tag{16}$$

From Eqs. (14) and (16), $n_{eq}$ can be expressed as

$$n_{eq} = \frac{E_I}{4\, {}_sQ_y d_{\max}} - \frac{1}{8} r_q. \tag{17}$$

Using Eq. (15), Eq. (17) can be rewritten as

$$n_{eq} = \frac{3}{8} \left(1 + \frac{2}{9} r_q\right) \frac{E_I}{\Delta E_{BI,\max}} - \frac{1}{8} r_q = \frac{3}{8} \left(1 + \frac{2}{9} r_q\right) \left(\frac{V_I}{V_{\Delta E}}\right)^2 - \frac{1}{8} r_q. \tag{18}$$

Equation (18) implies that there are two parameters related to the equivalent number of cycles ($n_{eq}$): the ratio of the maximum force of the flexible linear isolator to the yield strength of the hysteresis damper ($r_q = {}_fQ_{\max}/{}_sQ_y$) and the ratio of the equivalent velocities ($V_I/V_{\Delta E}$).

Figure 5 shows the relationship between the ratio $V_I/V_{\Delta E}$ and $n_{eq}$ for various $r_q$. We see that $n_{eq}$ increases as $V_I/V_{\Delta E}$ increases. In addition, when $V_I/V_{\Delta E}$ is larger than $\sqrt{3/2}$, $n_{eq}$ increases as $r_q$ increases.

According to Akiyama [3], the recommended value of $n_{eq}$ for the prediction of the peak displacement is expressed as

$$n_{eq} = \begin{cases} 1 + r_q : 0 \le r_q \le 1 \\ 2 \quad\quad : r_q > 1 \end{cases}. \tag{19}$$

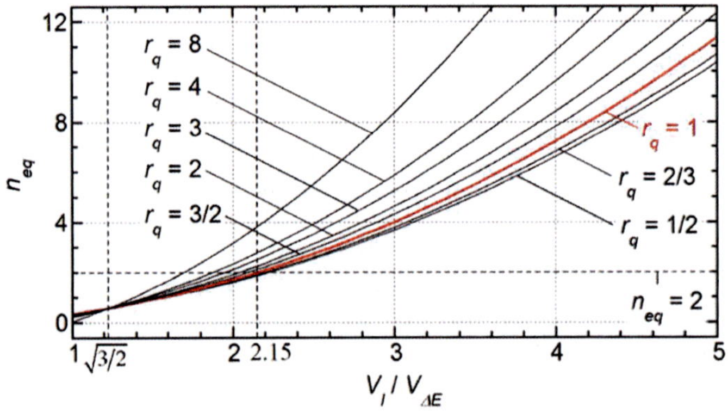


**Fig. 5.** Relationship between $n_{eq}$ and $V_I/V_{\Delta E}$ calculated from Eq. (18).

For $r_q > 1$, $n_{eq}$ is larger than 2.0 when $V_I/V_{\Delta E}$ is larger than 2.15. Therefore, according to Eq. (18), the value $n_{eq} = 2.0$ is conservative for the prediction of the peak displacement when $V_I/V_{\Delta E}$ is larger than 2.15.

Substituting $V_I/V_{\Delta E} = 2.15$ into Eq. (18) gives

$$n_{eq} = \frac{3}{8}\left(1 + \frac{2}{9}r_q\right)(2.15)^2 - \frac{1}{8}r_q = 1.74 + 0.26r_q. \tag{20}$$

Comparing Eqs. (19) and (20), it can be concluded that Eq. (19) is more conservative than Eq. (20) when $V_I/V_{\Delta E}$ is 2.15.

# 4 Numerical Examples

## 4.1 Structure and Ground Motion Data

In the numerical examples, $11 \times 8 \times 2 = 176$ models and 10 recorded ground motions were considered. Therefore, 1760 cases of nonlinear time-history analysis (NTHA) were performed. The seismic isolation period ($T_f$) ranged from 3.0 s to 5.0 s at intervals of 0.20 s. The shear force coefficient of the hysteresis damper (${}_s\alpha_y$) was adjusted to satisfy the predetermined $r_q$: $r_q$ was set to 1/2, 2/3, 1, 3/2, 2, 3, 4, and 8. The stiffness ratio ($\alpha = k_f/k_0$) shown in Fig. 1b was set to 0.01 and 0.10. No viscous damping was considered.

Table 1 lists the 10 recorded ground motions investigated in this study. In this table, $M_J$ indicates the Meteorological Agency Magnitude while $M_W$ indicates the moment magnitude. In addition, distances marked with "*" indicate the closest distance from the rupture plane defined by the Pacific Earthquake Engineering Research Center (PEER) database [9]; other distance values from the Japanese database indicate the epicenter distance.

**Table 1.** List of recorded ground motions investigated in this study.

| ID | Event Date | Magnitude | Distance | Station Name |
|---|---|---|---|---|
| 1968HAC [10] | 1968/05/16 | $M_J = 7.9$ | 177 km [11] | Hachinohe Harbor |
| 1995JKB [12] | 1995/01/17 | $M_J = 7.3$ | 16 km | Kobe JMA Observatory |
| 1999YPT [9] | 1999/08/17 | $M_W = 7.5$ | 4.83 km* | Yarimka |
| 1999TCU [9] | 1999/09/20 | $M_W = 7.6$ | 0.89 km* | TCU075 |
| 2000TTR [13] | 2000/10/06 | $M_J = 7.3$ | 7 km | KiK-NET TTRH02 |
| 2003TOM [13] | 2003/09/26 | $M_J = 8.0$ | 225 km | K-NET HKD129 |
| 2011PRPC [9] | 2011/02/11 | $M_W = 6.2$ | 1.92 km* | Pages Road Pumping Station |
| 2011SND [13] | 2011/03/11 | $M_J = 9.0$ | 170 km | K-NET MYG013 |
| 2016MSK [13] | 2016/04/16 | $M_J = 7.3$ | 7 km | KiK-NET KMMH16 |
| 2018MKW [13] | 2018/09/06 | $M_J = 6.7$ | 14 km | K-NET HKD126 |

In this study, all unscaled ground motion sets were converted from the as-provided datasets to the horizontal major and minor components defined by Arias [14]. In the following analysis, the direction of the horizontal excitation was rotated with respect to the vertical axis such that the major axis of the horizontal excitation coincides with the X-axis.

Figure 6 shows the linear bidirectional $V_I$ and $V_{\Delta E}$ spectra of the recorded ground motion records: viscous damping with 10% critical damping was considered in the calculation.

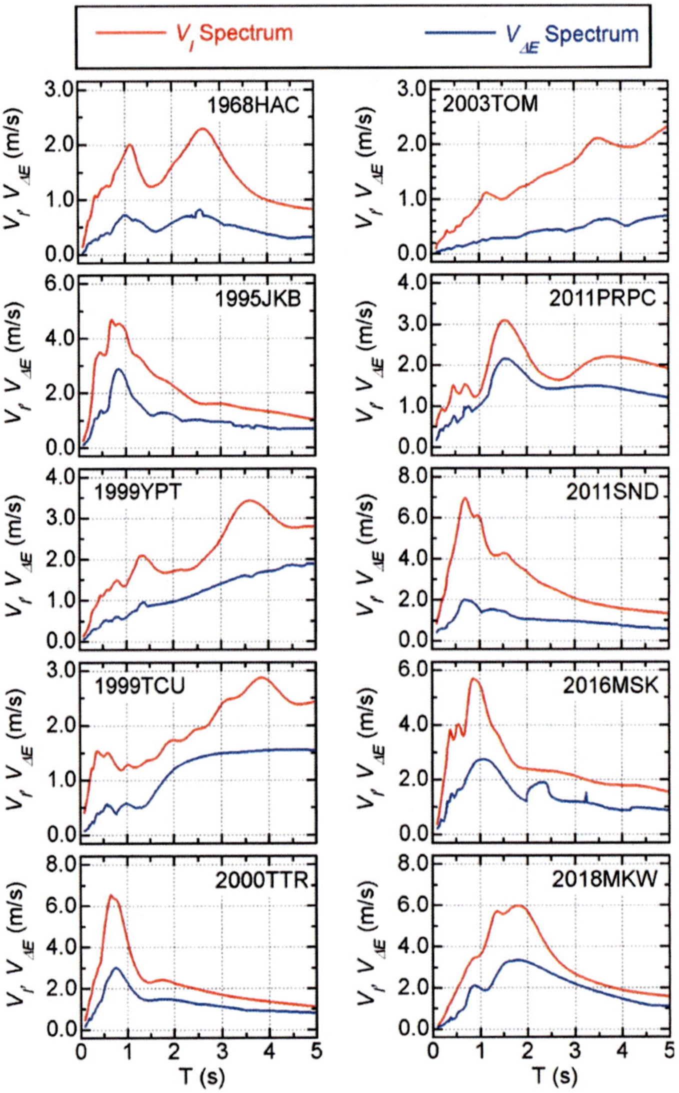


**Fig. 6.** $V_I$ and $V_{\Delta E}$ spectra of the recorded ground motion records.

## 4.2 Analysis Results

Figures 7 and 8 show comparisons of $n_{eq}$ values obtained from NTHA and Eq. (18). In these figures, the $n_{eq}$ value obtained from NTHA was calculated as follows:

$$n_{eq} = \frac{1}{4\,{}_sQ_y d_{\max}}\left(E_I - \frac{1}{2}k_f d_{\max}^2\right). \tag{21}$$

As shown in Fig. 7, the calculated $n_{eq}$ from Eq. (18) agrees well with that obtained from the NTHA results. However, the calculated $n_{eq}$ from Eq. (18) overestimates the

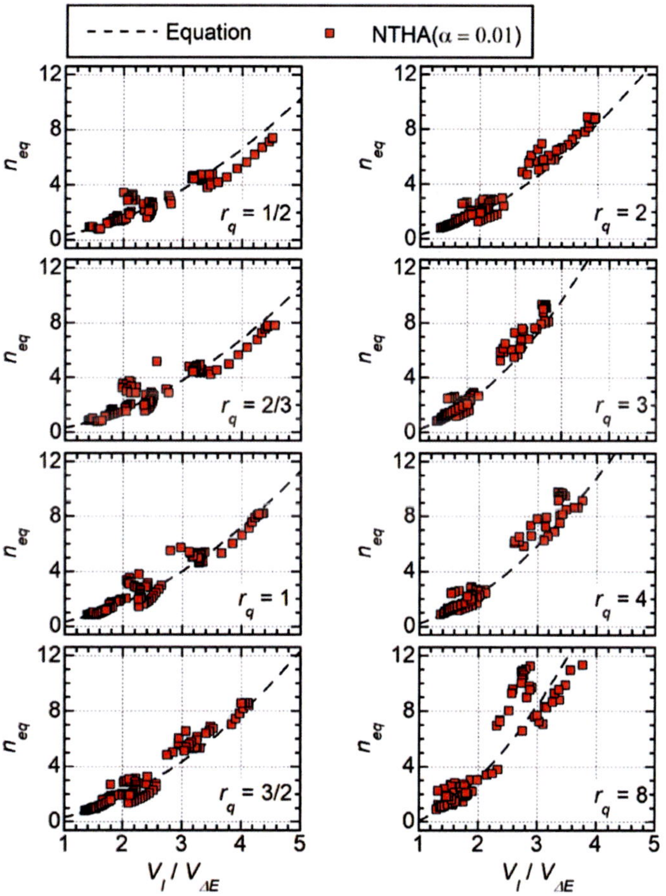


**Fig. 7.** Validation of Eq. (18) for calculating the equivalent number of cycles ($\alpha = 0.01$).

value obtained from the NTHA results for $\alpha = 0.10$ (shown in Fig. 8) when $r_q$ equals 1/2 and 2/3. However, the calculated $n_{eq}$ from Eq. (18) agrees well with that obtained from the NTHA results when $r_q$ is larger than or equal to 1.

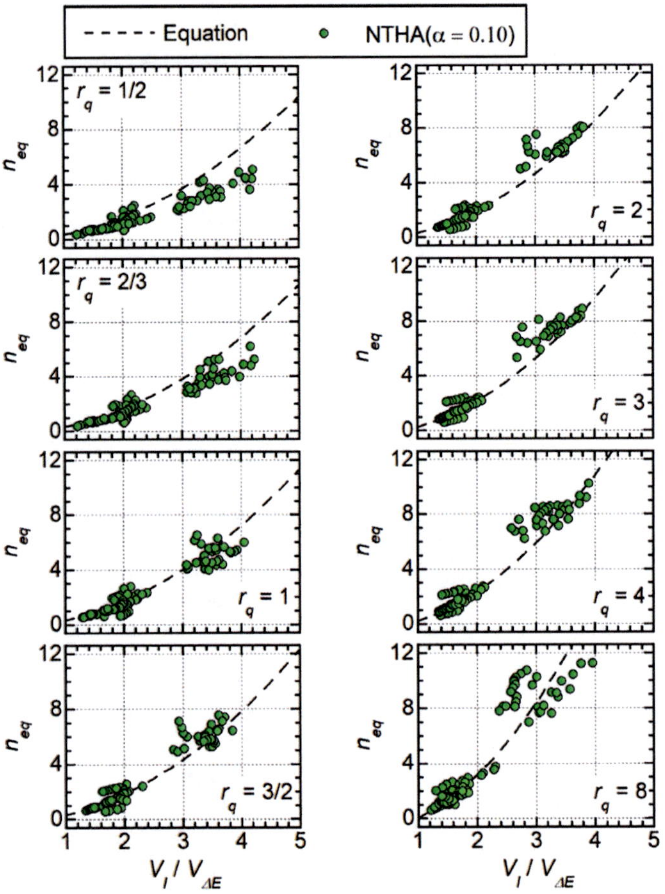


**Fig. 8.** Validation of Eq. (18) for calculating the equivalent number of cycles ($\alpha = 0.10$).

## 5 Discussion and Conclusions

In this study, the relationship between the equivalent number of cycles ($n_{eq}$) and two seismic intensity parameters ($V_I$ and $V_{\Delta E}$) was formulated for a base-isolated building whose isolation layer consists of a flexible linear isolator and a hysteresis damper. The main conclusions of the study are as follows.

(a) The main parameters related to $n_{eq}$ are the ratio of the maximum force of the flexible linear isolator to the yield strength of the hysteresis damper ($r_q = {}_fQ_{\max}/{}_sQ_y$) and the ratio of the equivalent velocities ($V_I/V_{\Delta E}$).
(b) The equation formulated here implies that $n_{eq}$ increases as $V_I/V_{\Delta E}$ increases. In addition, when $V_I/V_{\Delta E}$ is larger than $\sqrt{3/2}$, $n_{eq}$ increases as $r_q$ increases.

(c) The accuracy of the formulated equation is satisfactory when $r_q$ is larger than 1. The calculated $n_{eq}$ from the formulated equation is not conservative when $r_q$ is less than 1.

Note that the ratio of the equivalent velocities $V_I/V_{\Delta E}$ is different from the ratio $V_D/V_E$ discussed by Akiyama [2, 3]: $V_D/V_E$ is the ratio of the equivalent velocities of the cumulative damaging energy (the total input energy minus the cumulative damping energy) and the total input energy. The equivalent velocity $V_D$ is related to the *cumulative* strain energy of the model, not the *peak* displacement. In addition, as discussed in the previous study [8], the ratio $V_I/V_{\Delta E}$ expresses the influence of duration of ground motion set. Therefore, to discuss the relations between the peak and cumulative responses, the ratio $V_I/V_{\Delta E}$ is indispensible.

One of the main achievements of this paper is that the relations the relationship between $n_{eq}$ and the ratio $V_I/V_{\Delta E}$ is formulated in a simple equation. In the Akiyama's theory, $n_{eq}$ was formulated empirically. By introducing the concept of the maximum momentary input energy, more rational expressions of $n_{eq}$ can be made. If $V_I/V_{\Delta E}$ is close to 1 (ground motion with short duration and/or pulsive ground motion: most of the energy input occurs during the short time) $n_{eq}$ becomes small, while $V_I/V_{\Delta E}$ is larger (ground motion with longer duration) $n_{eq}$ becomes larger. Because the both peak and cumulative responses are important, the relations between $n_{eq}$ and two seismic intensity parameters ($V_I$ and $V_{\Delta E}$) must be carefully assessed. The formulation shown in this paper would be the answer to this question.

**Acknowledgments.** Ground motions used in this study were taken from the websites of the Japan Meteorological Agency (https://www.data.jma.go.jp/svd/eqev/data/kyoshin/index.htm, last accessed on 14 December 2019), the National Research Institute for Earth Science and Disaster Resilience (NIED) (http://www.kyoshin.bosai.go.jp/kyoshin/, last accessed on 12 June 2022), and the Pacific Earthquake Engineering Research Center (PEER) (https://ngawest2.berkeley.edu/, last accessed on 16 July 2022). We thank Martha Evonuk, PhD, from Edanz (https://jp.edanz.com/ac), for editing a draft of this manuscript.

## References

1. Architectural Institute of Japan (AIJ).: Design Recommendations for Seismically Isolated Buildings. Architectural Institute of Japan, Tokyo (2016)
2. Akiyama, H.: Earthquake-Resistant Limit-State Design of Buildings. University of Tokyo Press, Tokyo (1985)
3. Akiyama, H.: Earthquake-Resistant Design Method for Buildings Based on Energy Balance. Gihodo Shuppan, Tokyo (1999). (in Japanese)
4. Hori, N., Inoue, N.: Damaging properties of ground motion and prediction of maximum response of structures based on momentary energy input. Earthquake Eng. Struct. Dynam. **31**, 1657–1679 (2002)
5. Ordaz, M., Huerta, B., Reinoso, E.: Exact computation of input-energy spectra from fourier amplitude spectra. Earthquake Eng. Struct. Dynam. **32**, 597–605 (2003)
6. Fujii, K., Kanno, H., Nishida, T.: Formulation of the time-varying function of momentary energy input to a SDOF system by fourier series. J. Japan Assoc. Earthquake Eng. **19**(5), 247–266 (2019). (in Japanese)

7. Wada, A., Hirose, K.: Elasto-plastic dynamic behaviors of the building frames subjected to bi-directional earthquake motions. J. Struct. Construct. Eng. **399**, 37–47 (1989). (in Japanese)
8. Fujii, K.: Bidirectional seismic energy input to an isotropic nonlinear one-mass two-degree-of-freedom system. Buildings **11**, 143 (2021)
9. PEER Ground Motion Database. https://ngawest2.berkeley.edu/. Accessed 30 Jan 2023
10. Midorikawa, S., Miura, H.: Re-digitization of strong motion accelerogram at hachinohe harbor during the 1968 Tokachi-oki, Japan Earthquake. J. Japan Assoc. Earthquake Eng. **10**(2), 12–21 (2010). (in Japanese)
11. Nagai, R., Kikuchi, M., Yamanaka, Y.: Comparative Study on the Source of Recurrent Large Earthquakes in Sanriku-oki Region: the 1968 Tokachi-oki Earthquake and the 1994 Sanriku-oki Earthquake. J. Seismol. Soc. Japan. 2nd ser. **54**, 267–280 (2001). (in Japanese)
12. Japan Meteorological Agency Homepage (Kyoshin Kansoku Data). https://www.data.jma.go.jp/eqev/data/kyoshin/jishin/index.html. Accessed 30 Jan 2023
13. Strong-motion Seismograph Networks (K-NET, KiK-net). https://www.kyoshin.bosai.go.jp/kyoshin/. Accessed 30 Jan 2023
14. Arias, A.: A Measure of Seismic Intensity, Seismic Design for Nuclear Power Plant, pp. 438–483. MIT Press. Cambridge, Mass (1970)

# Demands and Capacities Prediction Equations of R/C Structures Based on Energy

H. Leyva[1](✉), E. Bojórquez[2], J. Bojórquez[2], L. Palemón[3], and M. Barraza[1]

[1] Universidad Autónoma de Baja California, Ensenada, México
herian.leyva@uabc.edu.mx
[2] Universidad Autónoma de Sinaloa, Culiacán, México
[3] Universidad Autónoma del Carmen, Cd. del Carmen, México

**Abstract.** Previous studies have shown that seismic events recorded in soft soil of Mexico City generate up to four times higher energy demands compared to stiff soil. This zone represents a large urban area, where several reinforced concrete buildings are located, therefore, the use of advanced design methods such as energy-based design approaches is necessary. These are based on sizing the structures in order to supply it with an energy dissipation capacity greater or equal than the energy demands. For this reason, this study is composed by two parts. The first consists of the development of hysteretic energy demands prediction equations of low and mid-rise reinforced concrete frames as a function of the ground motion intensity. For this aim, incremental dynamic analyzes of four buildings under the action of thirty narrow-band seismic records are carried out and through statistical studies, prediction equations are formulated. The second part consists of the proposal of a procedure to calculate the cyclic and energy dissipation capacity of reinforced concrete elements. Therefore, equations and a simple method based on the Park and Ang damage index was developed. Figures of cyclic and energy dissipation capacity of concrete beams are presented and it is observed that these capacities are a function of the maximum displacement considered and the desired damage of the section (minor, moderate or severe). The results are promising and by means of this local energy designs are possible to carry out, sizing and reinforcing concrete sections in terms of energy.

**Keywords:** Prediction equations · damage index · cyclic capacity · hysteretic energy

## 1 Introduction

Mexico is located in a high seismic activity due to the interaction of five tectonic plates. In the last five years, sixteen earthquakes with a magnitude greater than 6 have been reported [1], most of which have an epicenter in Oaxaca, Guerrero and Michoacán states, directly affection Mexico City, place that has a zone with unique characteristics. This area is known as lake-zone, where the ground is very soft with periods ranging between 1 and 3 s, in addition, it amplifies ground motions. Seismic events registered in that place


H. Varum et al. (Eds.): IWEBSE 2023, LNCE 236, pp. 27–42, 2023.
https://doi.org/10.1007/978-3-031-36562-1_3

are known as narrow-band records and some of them reach durations greater than five minutes. Likewise, previous studies have shown that they can demand up to four times more energy compared to stiff soil records [2], evidencing the need to modify seismic design practices, especially in structures that are highly affected by the occurrence of several reversible cyclical demands such as reinforced concrete (RC) frames. This type of structural system can reach severe damage and even collapse due to the stiffness and strength degradation caused by cumulative inelastic deformations [3–5], which may be excessive in soft soil. Therefore, the development of advanced design methodologies for this zone that improve and consider these demands on seismic performance is essential, which can be achieve through is energy-based design methods (EBD). In the last decades, its research has increased significantly, since it promises structures with high safety levels because it is directly related to cumulative inelastic deformations and damage. Many EBD methods have been proposed and are based on sizing and reinforcing the structural sections to achieve an energy dissipation capacity higher than the energy demands at expected damage levels: light, moderate or severe. For this, it is necessary to perform studies that determine the energy demands for any structure in critical zones. In addition, develop equations or a method that defines the energy dissipation capacity of reinforced concrete elements based on its main properties such as yield strength (f'c), yield strength of reinforcing steel (fy), longitudinal reinforcement ratio ($p_t$), confinement steel ratio ($p_w$), yield moment (My), ultimate displacement ($\delta_u$), among others.

Due to the above, the main objective of this work is to develop hysteretic energy demands prediction equations for the zone where the most violent seismic record in Mexico has been recorded, caused by the 1985 earthquake ($M_w$ 8.1). These equations must be linked to the design seismic intensity in order to be able to use them in conjunction with the current regulation: Normas Técnicas Complementarias para el Diseño por Sismo (NTCS-2017). For this, four reinforced concrete frames of 4, 6, 8 and 10- story were subjected to incremental dynamic analysis against thirty narrow-band seismic records. Moreover, a simple procedure for the calculation of the cyclical and hysteretic energy dissipation capacity of reinforced concrete elements is also proposed. This method is associated with an objective deformation and is based on the Park and Ang damage index [4]. Through this, it is possible to create a safe and efficient energy-based design method, in addition to raising awareness about the importance of improving the appropriate seismic design in Mexico City. Below, a brief description of some developed EBD methods is presented.

## 2 Energy-Based Design Methods

Several researches about the proposal of this type of methods are based on improving the seismic performance of reinforced concrete frames designed according to old regulations, to reinforce structures that suffered severe damage by soft first story failure mechanism through the application of energy dissipation systems, avoiding damage to the main structure, complex, expensive and time-consuming rehabilitations [6–8]. These methodologies use several parameters to define the seismic demands. For example, the equivalent velocity or pseudo-spectral velocity [6, 7], plastic energy based on lateral load distributions [9], input and hysteretic energy ratios [10], and hysteretic energy and cumulative ductility spectra [11, 12]. One of the main characteristics of EBD approaches is the proposal of equations to size the cross sections of the structural elements at each story based on the ductility demands, energy and/or lateral resistance. Most of these studies compare the designs of their methods with nonlinear time history analysis where energy and displacement demands are below the limits established by the regulations avoiding this type of complex and time-consuming analysis to achieve safe structures. As can be seen, they have extensive advantages over other design methodologies, so their application in areas of high seismic activity is recommended. For this, in the following sections the hysteretic energy demands in RC frames are studied.

## 3 Structural Models

In order to achieve the stated objectives of this study, four three-dimensional RC frames of 4, 6, 8 and 10-story (RC4, RC6, RC8 and RC10) with the same plant dimensions as shown in Fig. 1 and inter-story heights of 3.5 m were designed. In addition, the same cross sections of beams and columns every two consecutive floors were considered.

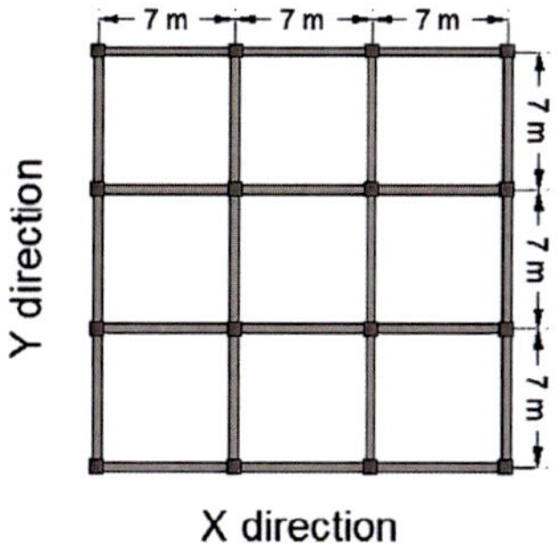


**Fig. 1.** Plant view of buildings.

To perform the seismic design, the spectral modal dynamic analysis according the design spectrum shown in Fig. 2, a seismic behavior or ductility of 4, maximum inter-story drift of 0.03 and the considerations of the current regulations of Mexico City were used (NTCS-2017). Table 1 shows the dimensions (centimeters) of the cross sections of each designed model, while Table 2 shows its main global characteristics: natural period of vibration ($T_1$), yield base shear ($V_y$), yield displacement ($d_y$) and seismic coefficient ($C_y$).

## 4 Incremental Dynamic Analysis

To obtain the seismic demands in hysteretic energy ($E_H$) terms of RC frames, incremental dynamic analyzes (IDA) [13] on the four designed models presented in the previous section are performed. For this purpose, thirty narrow-band seismic records commonly used in various studies are used [14, 15], with intensity intervals corresponding to the spectral acceleration at natural period of vibration of the structure [$Sa(T_1)$] from 0.1 to 2.0 g with 0.1 g increments.

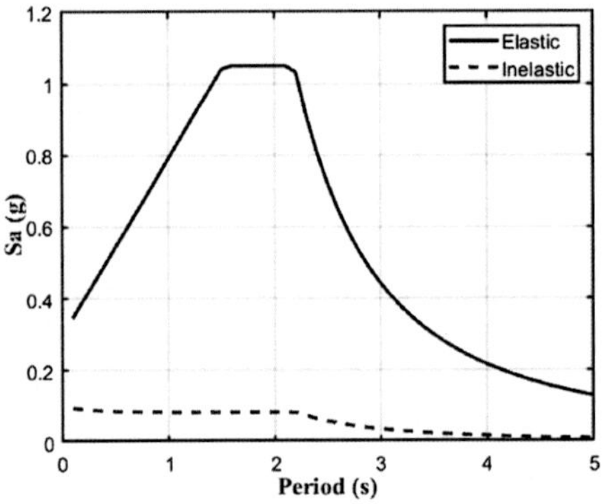


**Fig. 2.** Elastic and inelastic design spectra used.

**Table 1.** Cross sections dimensions (cm × cm) of the designed models.

| Section | Model | | | |
|---|---|---|---|---|
| | RC4 | RC6 | RC8 | RC10 |
| Column1 | 70 × 70 | 85 × 85 | 95 × 95 | 105 × 105 |
| Column2 | 55 × 55 | 75 × 75 | 90 × 90 | 100 × 100 |
| Column3 | | 65 × 65 | 80 × 80 | 90 × 90 |
| Column4 | | | 70 × 70 | 80 × 80 |
| Column5 | | | | 70 × 70 |
| Beam1 | 45 × 85 | 50 × 105 | 65 × 110 | 60 × 125 |
| Beam2 | 35 × 70 | 45 × 95 | 50 × 105 | 55 × 110 |
| Beam3 | | 35 × 70 | 40 × 95 | 45 × 100 |
| Beam4 | | | 35 × 70 | 40 × 80 |
| Beam5 | | | | 35 × 70 |

**Table 2.** Main characteristics of the designed models.

| Property | Model | | | |
|---|---|---|---|---|
| | RC4 | RC6 | RC8 | RC10 |
| $T_1$ (s) | 0.561 | 0.622 | 0.721 | 0.871 |
| $V_y$ (Ton) | 550 | 850 | 1,125 | 1,450 |
| $d_y$ (m) | 0.03 | 0.044 | 0.055 | 0.09 |
| Cy | 0.2816 | 0.2588 | 0.2358 | 0.2376 |

The results obtained are presented in Fig. 3, where the blue dots indicate the hysteretic energy for each records and intensity, while the solid red line corresponds to the median values plus the standard deviation (M+σ). In Fig. 4 the M+σ of each model are shown, where it is observed that the energy demand increases as the height of the buildings increases. Analyzing the IDAs, it is notable that the difference in demands occurs from the moment when the models reach its $V_y$, that is, from intensities of 0.3 g. In order to obtain a better comparison of demands, the $E_H$ of each model at their design intensities are examined. The RC4, RC6, RC8 and RC10 models dissipate 0.956, 24.81, 54.63 and 116.36 MN-m respectively, wide differences for increase of two story heights. This is because in taller buildings there is a greater participation of structural elements in the energy dissipation. In the same way, these elements have higher yield moments and displacements, generating noticeable increase in demands.

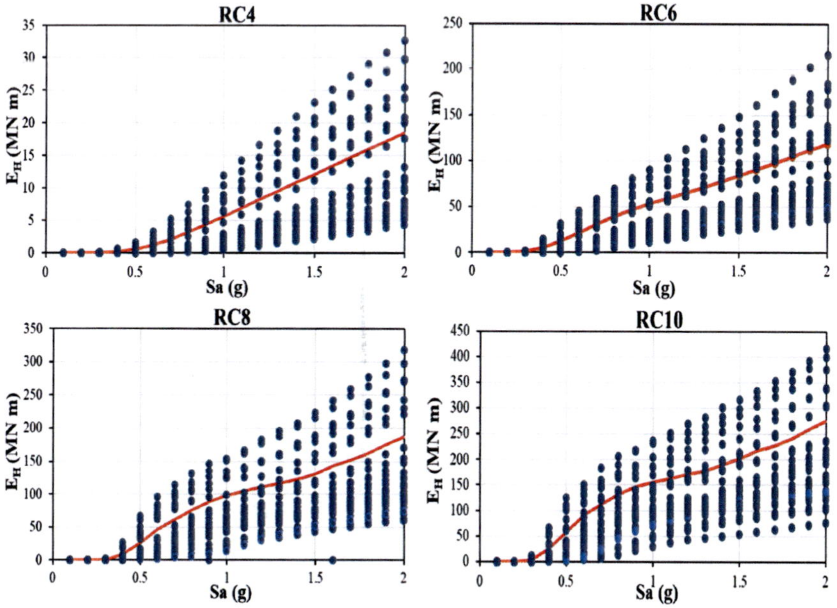


**Fig. 3.** $E_H$ incremental dynamic analysis results.

## 5 $E_H$ Prediction Equations

The first objective of the study is the development of hysteretic energy demands prediction equations for the design or review of low and medium height RC frames placed in the lake zone of Mexico City. For this, the four curves obtained in the previous section were analyzed. From said analyzes it was concluded that by means of polynomial equation very similar demands to the real ones are obtained, that is, with minimum errors. The equation is of second order [Eq. (1d)], therefore, it depends on three constant values that for simplicity are named as *a*, *b* and *c*. These constant were separately examined so that

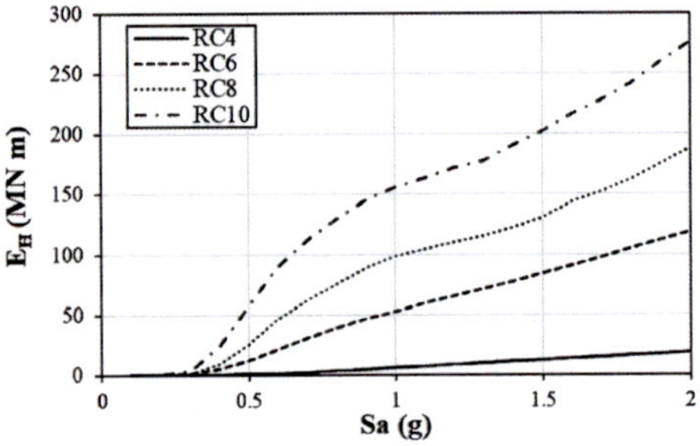


**Fig. 4.** M+σ curves of the four studied models.

they can be obtained as a function of the natural period of vibration of the structure ($T_1$), determining that, in the same way, the most accurate prediction is through polynomial equations: third degree for *a*, and second degree for *b* and *c* [Eqs. (1a) to (1c)].

$$a = 3,239T_1^3 - 7,201T_1^2 + 5,143T_1 - 1,189 \quad (1a)$$

$$b = -718T_1^2 + 1,707T_1 - 726 \quad (1b)$$

$$c = 287T_1^2 - 526T_1 + 206 \quad (1c)$$

$$E_H = ag^2 + bg + c \quad (1d)$$

Figure 5 compares the demands obtained through the proposed equations with the demands of the incremental dynamic analyzes (M+σ), a similarity between those related to the smallest models studied (RC4 and RC6) is observed. In the case of the highest models, the demands obtained by the proposed equations are above to the real ones in intensities lower than 0.7 g; slightly below between 0.7 and 1.0 g; and for the rest intensities are above. This behavior is due to the shape of the real demands, however, the results are accurate.

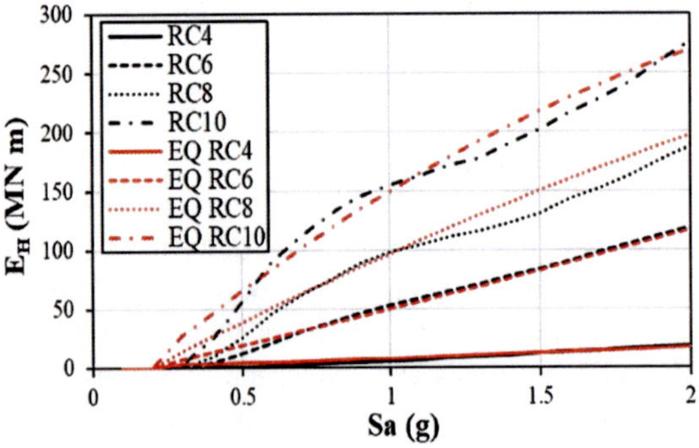


**Fig. 5.** Comparison between M+σ curves and prediction equations

By means of this equations it is possible to obtain the expected $E_H$ demands for low and medium rise structures whose $T_1$ are between 0.56 and 0.9 s, and are also projected or planted in the considered zone. Figure 6, graphically presents the $E_H$ demands for different structures and intensities. Among the shorter periods considered, it is observed that the demands increase considerably, but in periods of 0.7s onwards, the increase is smoother.

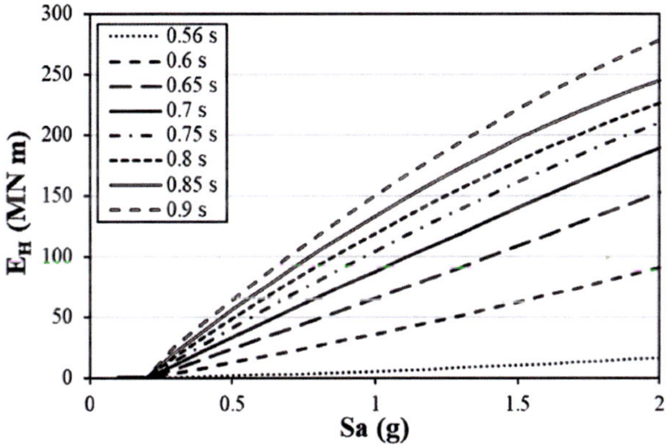


**Fig. 6.** $E_H$ prediction demands for different structures through the proposed equations.

# 6 Cyclical and Energy Capacity of RC Elements

Once the $E_H$ demands of a structure have been calculated, it is necessary to determine if the structure is capable of dissipating them, comparing said demand with its energy dissipation capacity ($C_{EH}$). Consequently, various studies propose equations to calculate the participation in the energy dissipation of each story, in such a way that the comparison is made locally. Therefore, the dissipation capacity of each structural element must be known, which can be obtained through laboratory test studies. On the other hand, investigations to create damage indices of reinforced concrete elements have been carried out, stating that cumulative inelastic deformation is the main factor that defines the failure mechanism of ductile concrete sections [4, 5]. Mehanny and Deirlein [5] use the plastic rotation ($\theta_p$) associated with each hysteretic cycle, while Park and Ang [4] created the β factor shown in Eq. (2) to take into account the effect of accumulated demands, which depends on the steel reinforcement. Additionally, the Park and Ang damage index, shown in Eq. (3), also considers the total dissipated energy and maximum displacement, for which its use is more appropriate. Like any damage index, it takes values between 0 and 1 to define the damage level, whose damage intervals are shown in Table 3.

$$\beta = \left(-0.165 + 0.0315\frac{l}{d} + 0.131p_t\right)0.84^{p_w} \tag{2}$$

$$I_{DPA} = \frac{\delta_{max}}{\delta_u} + \beta\frac{E_H}{F_y\delta_u} \tag{3}$$

**Table 3.** Park and Ang damage intervals.

| $I_{DPA}$ | Damage State |
|---|---|
| 0 – 0.1 | No damage |
| 0.1 – 0.25 | Light damage |
| 0.25 – 0.4 | Moderate damage |
| 0.4 – 1.0 | Severe damage |
| >1.0 | Failure |

This damage index has been widely accepted and studied, so it can be used to obtain the $C_{EH}$ of any reinforced concrete section, associated with a damage level and a maximum expected displacement. The proposed procedure for this aim consists of the following steps:

1. Calculate the longitudinal steel ratio ($p_t$), confinement ratio ($p_w$), shear span ratio (l/d) and β factor of the concrete element.
2. Compute the moment-curvature diagram of the section to obtain its main ductile properties: ultimate deformation ($\delta_u$), yield strength ($F_y$) and yield displacement ($\delta_y$).
3. Calculate the energy dissipation capacity of the section associated with an expected ductility ($C_{EH\mu}$) through its $\delta_y$, using Eqs. (4) and (5), where the second one was obtained clearing $E_H$ from Eq. (3) and equaling the damage to 1.

$$\delta_{max} = \mu\delta_y \tag{4}$$

$$C_{EH\mu} = \left(\frac{1 - \frac{\delta_{max}}{\delta_u}(F_y\delta_u)}{\beta}\right) \tag{5}$$

4. Next, the amount of $E_H$ that the section dissipates in one hysteretic behavior cycle ($E_{H\mu\ cycle}$) with the maximum deformation considered in the previous point is calculated. For this, the Takeda model is used [16].
5. Then, a damage index per cycle associated with the expected ductility is calculated through Eq. (6).

$$I_{DPAcycle} = \frac{E_{H\mu cycle}}{C_{EH\mu}} \tag{6}$$

6. Since the collapse of the section is obtained when the damage index is equal to 1, the cyclic capacity is obtained by the Eq. (7).

$$C_c = \frac{1}{I_{DPAcycle}} \tag{7}$$

To validate the proposed procedure, the results are analyzed and compared with laboratory tests on reinforced concrete columns carries out by various researchers, summarized in Taylor et al. [17]. Figure 7 shows the main characteristics of element GILL79S1, as well as its hysteretic behavior when tested by Gill et al. [18]. Table 4 presents the analytical results of section damage, the main capacities of the section ($C_{EH\mu}$ and $C_{C\mu}$) and the damage occurred in each cycle, which is achieved by the ratio between $E_{H\mu\ cycle}$ and its $C_{EH\mu}$. It is observed that, in both cases, the section collapse shortly before concluding the eight cycle and there is a difference of 1.8% in the dissipated energy.

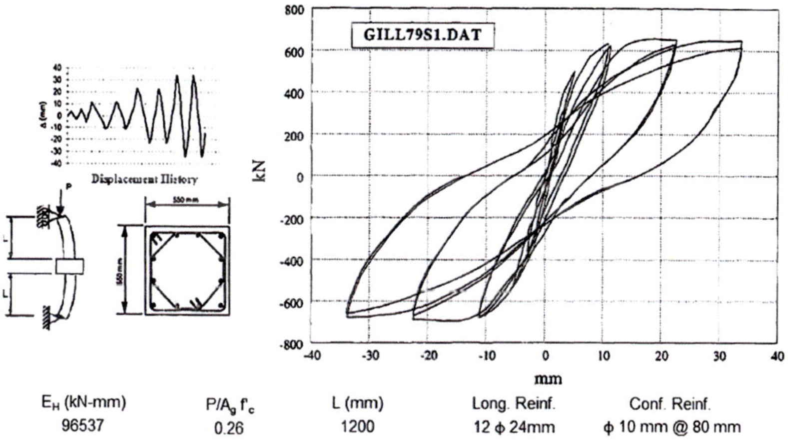


**Fig. 7.** Experimental results of the specimen GILL79S1.

**Table 4.** Analytical results of the specimen GILL79S1.

| Cycle | Ductility | $C_{EH\mu}$ | $E_{H\mu\ cycle}$ | $C_c$ | Damage |
|---|---|---|---|---|---|
| 1 | 1 | INF | 0 | INF | 0 |
| 2 | 1 | INF | 0 | INF | 0 |
| 3 | 2 | 242917 | 7582 | 32.043 | 0.0312 |
| 4 | 2 | 242917 | 7582 | 32.043 | 0.0312 |
| 5 | 4 | 166455 | 15163 | 7.675 | 0.1302 |
| 6 | 4 | 166455 | 15163 | 7.675 | 0.1302 |
| 7 | 6.5 | 62833 | 24640 | 2.551 | 0.3919 |
| 8 | 6.5 | 62833 | 24640 | 2.551 | 0.3331 |
| | | | **94770** | | **1.0478** |

This comparison procedure was carried out on a total of 5 specimens. The results are presented in Table 5 where it is observed that the collapse occurs almost at the same time, demonstrating its precision and validation its application for the calculation of $C_{EH\mu}$ and $C_{c\mu}$ of reinforced concrete elements. In the following chapter, this procedure is used to calculate both capacities of beams and columns sections. It has been determined that reinforcement detailing in concrete sections improves element deformation capacity and energy dissipation [19], however, in this study the same $p_w$ for the same cross sections was considered, to study only the influence of the increase in dimensions, target $\mu$ and $p_t$.

**Table 5.** Comparison between experimental and analytical results of the specimens studied

| Specimen | $E_H$ test (kN-mm) | $E_H$ method (kN-mm) | Damage method | $E_H$ ratio | Damage ratio |
|---|---|---|---|---|---|
| ANG81U3 | 49376 | 50518 | 1.0311 | 2.31% | 3.11% |
| ANG81U4 | 61095 | 59855 | 1.0045 | 2.03% | 0.45% |
| GILL79S1 | 96537 | 94770 | 1.0482 | 1.83% | 4.82% |
| S1RP | 17108 | 17156 | 1.0158 | 0.28% | 1.58% |
| SOES86U2 | 44004 | 41206 | 1.0105 | 6.36% | 1.05% |

## 6.1 Beams

The beams are the main elements responsible for dissipating energy in RC frames without collapsing, which depends on the ductility reached during the hysteretic behavior. It is important to mention that in this study only sections with the same top and bottom longitudinal reinforcement were considered. In the moment-curvature graphs, it was observed that the ultimate ductility of the sections depends mainly on its $p_t$, as this property increases, the d affect its $C_{EH\mu}$ and $C_{c\mu}$. These capabilities are discussed separately.

$C_{EH\mu}$

To begin with, a beam section of 40x90 cm with $p_t$ from minimum to that corresponding to balanced failure according to the NTCS-2017 regulation were studied. For this, target $\mu$ of 2, 4, 6, 8 and 10 were considered, whose deformations depends on the yield displacement of the section. The results are presented by a surface graph (Fig. 8), which corresponds to a set of beam sections with the aforementioned dimensions. It is observed that between $p_t$ percentages from the minimum to 0.36%, the $C_{EH\mu}$ increases, then it is maintained up to a $p_t$ of 0.4%, but for higher percentages, the capacity decreases considerably.

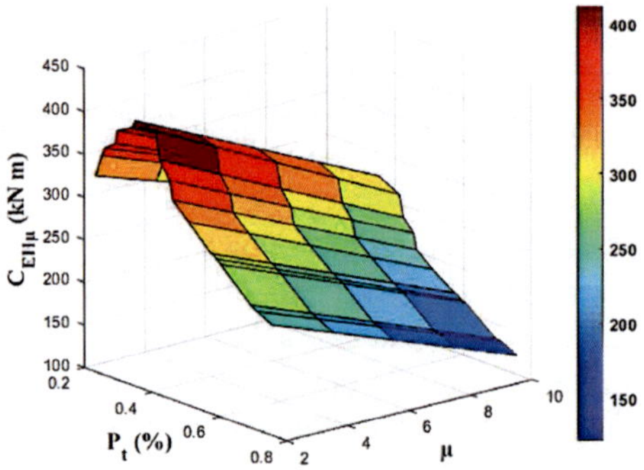


**Fig. 8.** $C_{EH\mu}$ of 40x90 cm beams with several $p_t$.

Analyzing the results, it was concluded that in low-reinforced sections, the ductility barely decreases, but the yield moment increases, which favors energy dissipation capacity. As for sections with longitudinal reinforcement ratios greater than 0.4%, their deformation capacity is significantly affected, and even with very large $M_y$, their $C_{EH\mu}$ is reduced by up to 56% compared to the lower reinforced beam. In the case of the ductility influence, the results are as expected, as the deformation demands increase, the section is capable to dissipate less $E_H$.

Next, the influence of effective depth of cross section ($d$) on $C_{EH\mu}$ is analyzed. For this, sections with a base of 40 cm were analyzed, varying their $d$ from 60 to 95 cm whose $p_t$ values are close to 0.41%. The results are presented in Fig. 9 and like the previous results, as the ductility increase, the capacity decreases. Furthermore, since they are section with the same $p_t$ but different dimensions, those with greater $d$ have best ductile capacity, which benefits the $C_{EH\mu}$. Likewise, it is observed that when the section is designed for a target ductility of 2, the section with $d$ equal to 60 cm is capable of dissipate 102.28 kN-m before collapsing, while the section of 95 cm dissipated 472.63 kN-m, that is, 4.7 time more energy having 1.58 times greater $d$.

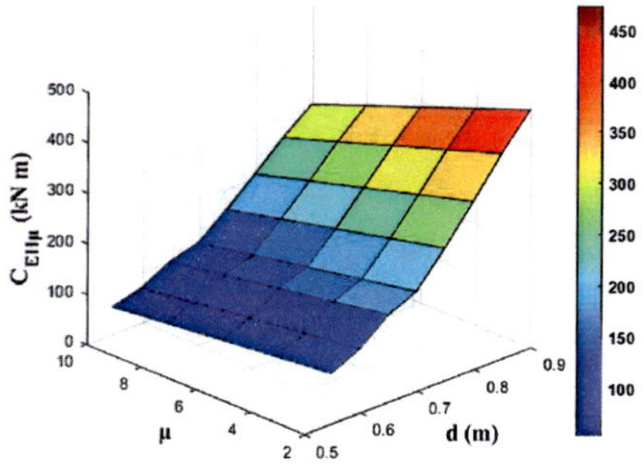


**Fig. 9.** $C_{EH\mu}$ of beams with same base dimension (40 cm), different d and an approximate pt of 0.41%.

Finally, three different beams sections with several $p_t$ are compared to verify if the behavior is similar in all sections (Fig. 10). The results confirm that the behavior is similar depending in $p_t$, a small increase in $p_t$ less than 0.4% and a significant reduction for higher ratios; and the $C_{EH\mu}$ decreases with increasing ductility. In addition, the increase in the cross section increase the $C_{EH\mu}$.

***$C_{c\mu}$***

In the case of $C_{c\mu}$ the same observations were obtained. Figure 11 shows the results for the 40x90 cm beam for several $p_t$ percentages and target $\mu$. As both increase, the cyclical capacity of the section decreases, however, in this case the decrease is abrupt. For example, considering the section with less $p_t$, it can achieve up to 270 hysteretic cycles with a ductility of 2, but it is reduced to 38 cycles with a ductility of 4, and 15 for the maximum target ductility studied. That is, from target $\mu$ from 2 to 10, the $C_{c\mu}$ is reduced by 94%, which explains the decrease in $C_{EH\mu}$.

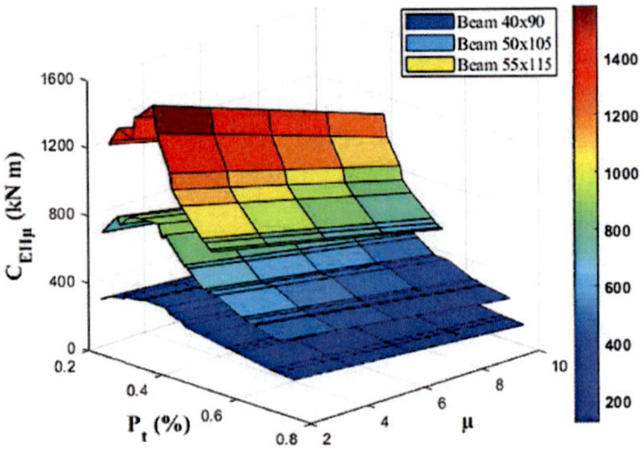


**Fig. 10.** Comparison of $C_{EH\mu}$ of three beam with several $p_t$.

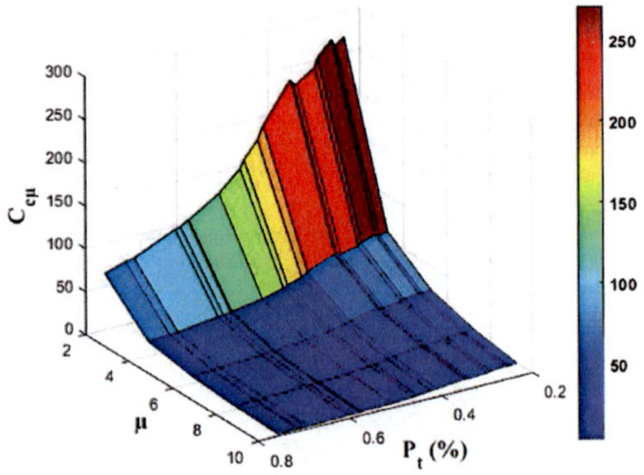


**Fig. 11.** $C_{c\mu}$ of 40x90 beams with different $p_t$.

Finally, in order to have a better view of the influence of the dimensions in both capacities studied, Table 6 is presented, where three different sections with very similar $p_t$ are compared.

**Table 6.** Comparison of cyclic and energy dissipation capacity of three beams

| Dimension | $p_t$ (%) | $C_{EH\mu}$ (kN-m) | | | | $Cc\mu$ | | | |
|---|---|---|---|---|---|---|---|---|---|
| | | μ = 2 | μ = 4 | μ = 6 | μ = 8 | μ = 2 | μ = 4 | μ = 6 | μ = 8 |
| 40 × 90 | 0.294 | 376 | 346 | 317 | 287 | 253 | 69 | 35 | 21 |
| 50 × 105 | 0.288 | 810 | 746 | 689 | 629 | 377 | 103 | 53 | 32 |
| 55 × 115 | 0.285 | 1378 | 1285 | 1191 | 1098 | 536 | 148 | 76 | 47 |

### 6.2 Columns

Columns are the main structural elements that support the weight of the structure, therefore, their failure must be controlled as they must be the last elements to reach non-linear behavior. Despite this, recent studies have shown that most of the energy dissipated in the floor story is through the columns [7], therefore it is very important to consider its participation in the energy dissipation and to design them in such a way that they no suffer severe damage. In the moment-curvature diagrams of these sections, maximum ductilities close to 15 was observed for those with minimum $p_t$, so these should not exceed a ductility of 5 during the occurrence of strong seismic events. Next, the $C_{EH\mu}$ and $C_{C\mu}$ of this type of sections are analyzed.

$C_{EH\mu}$

In this case, a 70 × 70 cm square column is examined with $p_t$ that varies from 1 to 4%, minimum and maximum longitudinal steel ratios according to NTCS-2017. Results are shown in Fig. 12. Unlike beam sections, it is observed that the $C_{EH\mu}$ decreases as the $p_t$ increases, although the maximum reduction is not so sharp, for example, considering the same target $\mu$, the $C_{EH\mu}$ of the column with maximum $p_t$ is reduced by a bit more than 50% compared to the less reinforced section. Increase ductility generates the same impact. This behavior is observed for all analyzed columns.

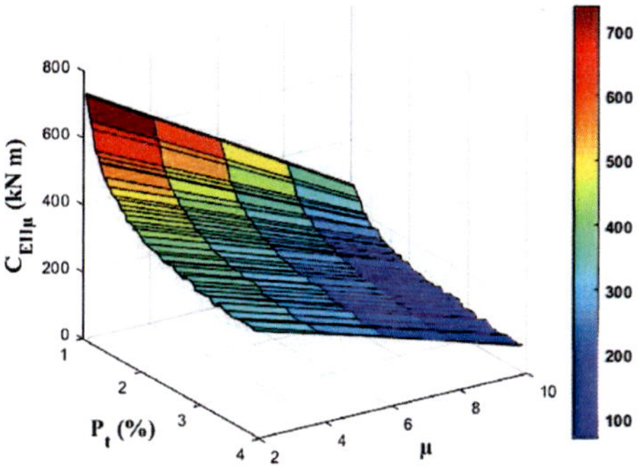


**Fig. 12.** $C_{EH\mu}$ of several 70 × 70 columns with different $p_t$.

In order to analyze the influence of the increase in cross sections, Fig. 13 compares the $C_{EH\mu}$ of two reinforced concrete columns. It can be seen that the larger column dissipates more amount of $E_H$, however, the increase in $p_t$ and target $\mu$ affects them more. In the case of target ductility of two, the 85 × 85 column reduce its $C_{EH\mu}$ up to 72%, while in smaller section it is reduced by 55%.

$C_{c\mu}$

In this chapter, the cyclic capacity of reinforced concrete columns is studied. The analysis was carried out on 70 × 70 columns with several longitudinal reinforcements ratios reaching the five target ductilities considered. The results are presented in Fig. 14 where it is shown that in columns sections, the $C_{c\mu}$ is mainly affected by the ductility. The column with the less $p_t$, reduces its cyclic capacity by more than 75% from target

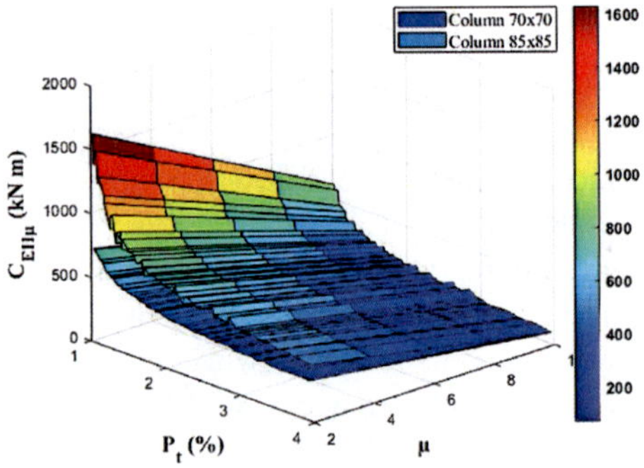


**Fig. 13.** Comparison of $C_{EH\mu}$ of two columns with several $p_t$.

ductility of 2 to 4, and 97% to a ductility of 10. In addition, sections with $p_t$ greater than 2% reduce their $C_{c\mu}$ by more than 70%.

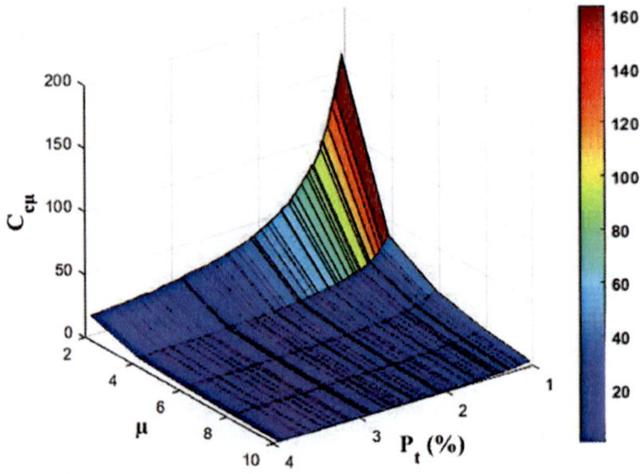


**Fig. 14.** $C_{c\mu}$ of 70 × 70 columns with several $p_t$.

Table 7 compares three columns sections of different cross sections sizes with very close $p_t$, where it is concluded that the increase in size positively affects the studied capacities and the reduction in percentages of their capacities due to the increase in ductility are very close to each other.

It is important to mention that the capacities calculated for all structural elements corresponds to that related to collapse, that is, when its damage index has a value of 1. To obtain structures with lower level of damage, it is recommended to perform the energy comparison using the Eq. (8).

$$\frac{E_{HD}}{C_{EH\mu_c}} \leq 0.25 \tag{8}$$

where $E_{HD}$ is the hysteretic energy demand and 0.25 corresponds to the light damage limit value for reinforced concrete sections according to Park and Ang damage index.

**Table 7.** Comparison of cyclic and energy dissipation capacity of three columns

| Dimension | $p_t$ (%) | $C_{EH\mu}$ (kN-m) | | | | $Cc\mu$ | | | |
|---|---|---|---|---|---|---|---|---|---|
| | | $\mu = 2$ | $\mu = 4$ | $\mu = 6$ | $\mu = 8$ | $\mu = 2$ | $\mu = 4$ | $\mu = 6$ | $\mu = 8$ |
| 70 × 70 | 1.047 | 738 | 621 | 504 | 386 | 164 | 41 | 18 | 10 |
| 85 × 85 | 1.041 | 1393 | 1169 | 945 | 720 | 226 | 56 | 25 | 13 |
| 105 × 105 | 1.055 | 3212 | 2715 | 2219 | 1723 | 316 | 79 | 36 | 19 |

## 7 Conclusions

In this article, it is emphasized that all energy-based design methods are based on the comparison between the energy demands and dissipation capacity of the structures. Therefore, equations to determine the seismic demands in terms of hysteretic energy for reinforced concrete buildings designed for the Mexico City lake zone are developed. In addition, a simple procedure to calculate the energy dissipation capacity and cyclical capacity of reinforced concrete elements based on their longitudinal steel ratio and a target ductility is proposed.

The calculation of the energy demands is obtained by means of a quadratic equation formed by three variables and is based on the spectral pseudo-acceleration designed intensity in units of gravity. The three variables are computed through other equations that depend on the natural period of vibration of the structure ($T_1$). This procedure is only valid for low and medium rise reinforced concrete frames with $T_1$ between 0.56 and 0.9 s.

The proposed method for the determination of cyclical ($C_{C\mu}$) and energy dissipation capacities ($C_{EH\mu}$) of reinforced concrete structural elements, as well as its comparison with laboratory test are shown in Sect. 6. The results show that increasing the cross sections considerably increases its capacity. The increase in $p_t$ shows a slight increase in the $C_{EH\mu}$ between reinforcements from the minimum to 0.36%, and at higher percentages of $p_t$ is reduced abruptly. In the case of columns, increasing $p_t$ always causes a decrease in said capacity. On the other hand, as expected, the increase in displacement demands also reduces $C_{C\mu}$ and $C_{EH\mu}$, especially in columns. Based on these observations, it is concluded that for energy-based design of reinforced concrete frames is recommended to use beam sections with higher d to base ratio greater than 1.7 and low $p_t$, between the minimum and 0.5%; while in the case of the columns, use sections with $p_t$ less tan 2%.

Based on the results of this study, it is possible to develop an energy-based design method which includes the following steps: first, the global hysteretic energy demands of the structure are calculated according the developed Eqs. (1a) to (1d); then, energy demands are distributed to each inter-story and structural element through expressions proposed by several authors; and finally, $E_{HD}$ are compared with $C_{EH\mu}$ by means of Eq. (8) to define if the design is satisfactory.

## References


1. México. Universidad Nacional Autónoma de México, Instituto de Geofísica, Servicio Sismológico Nacional. Catálogo de sismos. http://www2.ssn.unam.mx:8080/catalogo/
2. Terán-Gilmore, A., Jirsa, J.: A damage model for practical seismic design that accounts for low cycle fatigue. Earthq. Spectra **21**(3), 803–832 (2005)
3. Bai, Y., Guan, S., Lin, X., Mou, B.: Seismic collapse analysis of high-rise reinforced concrete frames under long-period ground motions. Struct. Design Tall Sepc. Build. **28**(1) (2019)
4. Park, Y., Ang, A.: Mechanistic seismic damage model for reinforced concrete. J. Struct. Eng. **111**(4), 722–739 (1985)
5. Mehanny, S., Deierlein, G.: Seismic damage and collapse assessment of composite moment frames. J. Struct. Eng. **127**(9), 1045–1053 (2001)
6. Khampanit, A., Leelataviwat, S., Kochanin, J., Warnitchai, P.: Energy-based seismic strengthening design of non-ductile reinforced concrete frames using buckling-restrained braces. Eng. Struct. **81**, 110–122 (2014)
7. Benavent-Climent, A., Mota-Páez, S.: Earthquake retrofitting of R/C frames with soft story using hysteretic dampers: energy-based design method and evaluation. Eng. Struct. **137**, 19–32 (2017)
8. Terenzi, G.: Energy-based design criterion of dissipative bracing systems for the seismic retrofit of frame structures. Appl. Sci. **8**(2) (2018)
9. Merter, O., Ucar, T.: Energy-based design base shear for Rc frames considering global failure mechanism and reduced hysteretic behavior. Struct. Eng. Mech. **63**(1), 23–35 (2017)
10. Ye, L., Cheng, G., Qu, Z.: Study on energy-based seismic design method and the application for steel braced frame structures. J. Build. Struct. **33**(11), 36–45 (2012)
11. Choi, H., Kim, H.: Energy-based seismic design of buckling-restrained braced frames using hysteretic energy spectrum. Eng. Struct. **28**, 304–311 (2006)
12. Qiu, C., Qi, J., Chen, C.: Energy-based seismic design methodology of SMABFs using hysteretic energy spectrum. J. Struct. Eng. **146**(2) (2020)
13. Vamvatsikos, D., Cornell, C.: Incremental dynamic analysis. Earthquake Eng. Struct. Dyn. **31**, 491–514 (2002)
14. Bojórquez, E., Baca, V., Bojórquez, J., Reyes-Salazar, A., Chávez, R., Barraza, M.: A simplified procedure to estimate peak drift demands for mid-rise steel and R/C frames under narrow-band motions in terms of the spectral-shape-based intensity measure $I_{NP}$. Eng. Struct. **150**, 334–345 (2017)
15. Baca, V., Bojórquez, J., Bojórquez, E., Leyva, H., Reyes-Salazar, A., Ruiz, S.: Enhanced seismic structural reliability on reinforced concrete buildings by using buckling restrained braces. Shock Vibrat. **2021** (2021)
16. Takeda, T., Sozen, M., Nielsen, N.: Reinforced concrete response to simulated earthquakes. J. Struct. Div. **96**(12), 2557–2573 (1970)
17. Taylor, A., Kuo, C., Wellenius, K., Chung, D.: A summary of cyclic lateral loads tests on rectangular reinforced concrete columns. Building and Fire Research Laboratory, NISTIR 5984, Gaithersburg, Maryland (1997)
18. Gill, W., Park, R., Priestley, M.: Ductility of rectangular reinforced concrete columns with axial load. Report 79-1, Department of Civil Engineering, University of Canterbury, Christchurch, New Zealand (1979)
19. Liel, A., Haselton, C., Deierlein, G.: Seismic collapse safety of reinforced concrete buildings: II. Comparative assessment of non-ductile and ductile moment frame. J. Struct. Eng. **137**(3), 492–502 (2011)

Check for updates

# Proposal of Design Energy Equivalent Velocity Spectrum Using Long-Period Ground Motion Records of Magnitude 7 to 7.5 Earthquakes

Ju-Chan Kim[1](✉) and Sang-Hoon Oh[2]

[1] Seismic Research and Test Center at Pusan National University, Busan, Republic of Korea
ju-chan@pusan.ac.kr
[2] Pusan National University, Busan, Republic of Korea

**Abstract.** Typically, the duration of strong motion in long-period ground motions is longer than that of short-period ground motions. This is because long-period ground motions are primarily caused by large-scale earthquakes. When the duration of strong motion increases, the energy input to a building may increase. However, the acceleration response spectrum used in seismic design does not consider the duration of strong motion. On the contrary, the energy equivalent velocity ($V_E$) spectrum can consider the duration of strong motion as $V_E$ represents the total energy input to a building. This paper proposes a design spectrum using long-period ground motions generated by large-scale earthquakes (7–7.5 magnitude). The proposed design spectrum includes acceleration, velocity, and $V_E$ spectrum. This study considered 22 large-scale earthquakes that occurred in the seismic zone of Japanese archipelago. Herein, the long-period ground motion refers to the ground motion with $T_g$ (predominant period of acceleration response spectrum) of 1 s or longer. This is in consideration of the fact that the $T_S$ (natural period at which acceleration response spectrum begins to decrease) of design spectrum proposed in previous studies is 1 s or less. The $T_S$ of the seismic design code (KDS 41 17 00) is approximately 0.3 s < $T_S$ <0.83 s. The acceleration and velocity spectrum of long-period ground motions are proposed using a line enveloping 84th percentile of the normalized response spectrum. The $V_E$ spectrum of long-period ground motion is proposed based on the $V_E$–$S_V$ relationship and the $V_E$ spectrum of previous studies.

**Keywords:** Long-period Ground Motion · Energy-based Seismic Design · Energy Spectrum · Duration of Strong Motion · Artificial Seismic Wave

## 1 Introduction

Currently, building seismic design is primarily conducted using two methods: limit state design or performance-based design method. The former method aims to design a building with a resistance capacity greater than the load acting on it, using a stochastic statistical approach to calculate the coefficient of safety. This method considers seismic load as an additional load to the building. However, the performance-based design aims


H. Varum et al. (Eds.): IWEBSE 2023, LNCE 236, pp. 43–56, 2023.
https://doi.org/10.1007/978-3-031-36562-1_4

to achieve the target performance level required for a building through nonlinear analysis. Both methods have been validated as rational design approaches. However, since both evaluate the seismic performance of a building with the maximum value of displacement and acceleration generated by the interaction between ground motion and the building, they may underestimate the damage caused by long-period ground motions. This is because the duration of strong motions caused by long-period ground motions may appear longer than that of short-period ground motions. This problem can be addressed relatively simply by designing the internal and external forces of the building as a unit of energy.

The energy-based seismic design for buildings originated from Housner's (1956) idea, which established the relationship between the total energy input to the building ($E$) and the energy dissipated by the hysteretic behavior of the building ($E_P$). Later, Akiyama (1985) further developed this concept by organizing the seismic energy and damage distribution input to the building and systematizing the energy-based seismic design method. In this method, the seismic energy input to a building is calculated using the energy equivalent velocity ($V_E$) spectrum. This spectrum provides a reliable evaluation of the energy input to a building, as verified by various researchers including Oh et al. (2019).

The Building Research Institute (BRI) in Japan has proposed a design spectrum for long-period ground motion in response to long-period ground motion generated in the Nankai Trough. However, the proposed design spectrum is based on the earthquake records observed in the target area, and its shape varies according to the response spectrum of the ground motion observed in each area. However, this approach can be confusing for designers, who may find it difficult to predict the ground motion that will affect a building. Therefore, it is desirable to have a more intuitive and concise design spectrum to reduce confusion.

This study aimed to propose the design $V_E$ spectrum for long-period ground motion. The proposal of the design $V_E$ spectrum is based on the $V_E$-$S_V$ relationship from a previous study. Therefore, the design spectrum of the acceleration ($S_A$) and velocity ($S_V$) of the long-period ground motion is first proposed. Both are developed via statistical analysis of ground motion observed through large-scale earthquakes (magnitude 7 to 7.5). In this study, long-period ground motion refers to ground motion in which the $T_g$ (predominant period of ground motion) observed in large-scale earthquakes is 1 s or longer. This range was chosen due to the high frequency of long-period ground motion in large-scale earthquakes and the fact that the $T_S$ proposed in previous studies is 1 s or shorter (AIK 2019; ASCE 2022).

## 2 Literature Review

### 2.1 Energy Equivalent Velocity Spectrum

In energy-based seismic design, the seismic energy input to a building is calculated using the energy equivalent velocity ($V_E$) spectrum. The $V_E$ represents the total input energy of ground motion to a building, and Akiyama (1985) reported that the $V_E$ spectrum of a single degree of freedom (SDOF) system with a damping ratio of 10% is meaningful as a design spectrum. The equation for $V_E$ is shown in Eq. (1), where $M$ denotes the total mass of the building and $E$ denotes the total energy input to the building.

$$V_E = \sqrt{\frac{2E}{M}} \tag{1}$$

The total energy input to a building by ground motion is a stable value that depends mainly on the total mass and the natural period of the building (Housner 1956). Akiyama (1985) showed that the $V_E$ spectrum is a stable value for the SDOF system's hysteresis characteristics and degree of plasticity. Oh (2014) verified that the $V_E$ spectrum reliably evaluates the seismic energy input to a multi-layered structure by examining the total input energy based on the mass distribution, horizontal load distribution, stiffness distribution, and degree of plasticity of each layer of the multi-degree-of-freedom (MDOF) system.

Akiyama (1985) put forward design $V_E$ spectra based on site class for seismic design. Later, Oh & Ryu (2007) in South Korea proposed their own design $V_E$ spectrum that takes into consideration the site class. The design $V_E$ spectra proposed by Oh & Ryu (2007) based on site class is depicted in Fig. 1. The site classification used in this proposal is according to the one presented in ASCE7-16.

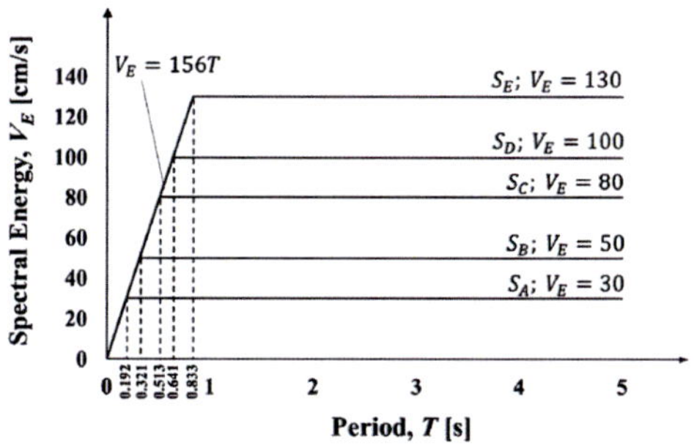


**Fig. 1.** Design energy equivalent velocity spectra based on site class (Oh & Ryu 2007).

## 2.2 Relationship Between Energy and Velocity Spectrum

Akiyama & Kitamura (2006) proposed a $V_E$–$S_V$ relationship based on the damping ratio and the duration of strong motion. In the duration of strong motion, the time between Arias intensity progress from 5% to 95% was used. This relationship is reported in Eq. (2), where $\zeta$ denotes the damping ratio and $t_{95-5}$ denotes the time between Arias intensity progress from 5% to 95%.

$$\frac{S_V}{V_E} = \frac{1}{\sqrt{f}\sqrt{1 + 12\pi\zeta}} \tag{2}$$

$$f = 1 \; if \; t_{95-5} < 50$$

$$f = 1 + 0.017(t_{95-5} - 50)\ if\ t_{95-5} > 50$$

When strong motion lasts 50 s or longer, the magnitude of $V_E$ for $S_V$ increases as the duration of strong motion increases. In general, the duration of strong motion is longer in long-period ground motion than that in short-period ground motion. Therefore, if the velocity responses of long-period and short-period ground motions are the same in the same natural period ($T$) region, the input energy of long-period ground motion can be evaluated as being greater than that of short-period ground motion. In addition, when applying the $V_E$–$S_V$ relationship to the design spectrum, the $T_S$ and $T_G$ of the design spectrum are the same value.

## 3 Characteristics of Large-Scale Earthquakes

### 3.1 General

The study used data from the data provided by the National Research Institute for Earth Science and Disaster Resilience (NIED) for the ground motion observed during large-scale earthquakes. The large-scale earthquakes selected in this study included 22 earthquakes that occurred in the seismic zone of the Japanese Archipelago since 2000. Table 1 provides the general characteristics of these selected large-scale earthquakes. The study used 7562 pieces of ground motion observation data, each consisting of the acceleration time history for three axes (NS, EW, UD), and the acceleration time history for the horizontal direction (NS, EW) was used for this analysis. Most large-scale earthquakes that have occurred since 2000 are primarily concentrated in the Tohoku region, which may have been due to the foreshocks and aftershocks of the Great Tohoku Earthquake (M9.1) that occurred on March 11, 2011.

### 3.2 Relationship Between Epicentral Distance and $T_g$

The relationship between epicentral distance and the $T_g$ for large-scale earthquakes is depicted in Fig. 2. The figure shows the statistical value of $T_g$, calculated in sections of 50 km epicentral distance, as a line graph for the epicentral distance. Figure 2 shows a line connecting the $\mu + \sigma$ and 84th percentile values of the $T_g$ according to the epicentral distance in the scatter plot. As per the epicentral distance–$T_g$ relationship, $T_g$ increased until the epicentral distance of approximately 400 km, after which it remained constant till approximately 800 km. At approximately 900 km, $T_g$ showed a large deviation, which declined thereafter. The Kumamoto earthquake (Table 1, No. 19) was identified as the one that showed a large deviation in $T_g$ at an epicentral distance of approximately 900 km. Based on the epicentral distance–$T_g$ relationship, the epicentral distance in which $T_g$ exceeds 1 s was found to be approximately 250–300 km.

The $T_g$ value of a single ground motion reflects the properties of the station's ground conditions. However, there is no apparent correlation between the ground characteristics of the observation station and the epicentral distance. Therefore, it can be seen that $T_g$ increases up to the epicentral distance of approximately 400 km regardless of the ground characteristics in the relationship between epicentral distance and $T_g$ for the large-scale

**Table 1.** General characteristics of the large-scale earthquakes.

| No | Origin time (JST) | Lat | Long | M | Number of stations |
|---|---|---|---|---|---|
| 1 | 2000/10/06-13:30:00 | 35.278 | 133.345 | 7.3 | 303 |
| 2 | 2003/05/26-18:24:00 | 38.808 | 141.678 | 7.0 | 411 |
| 3 | 2003/09/26-06:08:00 | 41.710 | 143.691 | 7.1 | 287 |
| 4 | 2004/09/05-23:57:00 | 33.146 | 137.139 | 7.4 | 430 |
| 5 | 2004/11/29-03:32:00 | 42.946 | 145.274 | 7.1 | 214 |
| 6 | 2005/03/20-10:53:00 | 33.738 | 130.175 | 7.0 | 299 |
| 7 | 2005/08/16-11:46:00 | 38.150 | 142.278 | 7.2 | 455 |
| 8 | 2008/05/08-01:45:00 | 36.227 | 141.607 | 7.0 | 340 |
| 9 | 2008/06/14-08:43:00 | 39.028 | 140.880 | 7.2 | 330 |
| 10 | 2008/09/11-09:21:00 | 41.775 | 144.150 | 7.1 | 221 |
| 11 | 2011/03/11-15:09:00 | 39.838 | 142.780 | 7.4 | 257 |
| 12 | 2011/04/07-23:32:00 | 38.203 | 141.920 | 7.1 | 506 |
| 13 | 2011/04/11-17:16:00 | 36.945 | 140.672 | 7.0 | 376 |
| 14 | 2011/07/10-09:57:00 | 38.032 | 143.507 | 7.3 | 429 |
| 15 | 2012/01/01-14:28:00 | 31.427 | 138.565 | 7.0 | 308 |
| 16 | 2012/12/07-17:18:00 | 38.018 | 143.867 | 7.3 | 560 |
| 17 | 2013/04/19-12:06:00 | 45.300 | 150.957 | 7.0 | 238 |
| 18 | 2015/11/14-05:51:00 | 30.942 | 128.590 | 7.1 | 95 |
| 19 | 2016/04/16-01:25:00 | 32.753 | 130.762 | 7.3 | 370 |
| 20 | 2016/11/22-05:59:00 | 37.353 | 141.603 | 7.4 | 371 |
| 21 | 2020/02/13-19:34:00 | 45.055 | 149.162 | 7.2 | 216 |
| 22 | 2021/02/13-23:08:00 | 37.728 | 141.698 | 7.3 | 546 |

earthquake. This can be due to the attenuation of the short-period component waveform acting more than that of the long-period component waveform as the ground motion generated in the fault propagates through the medium.

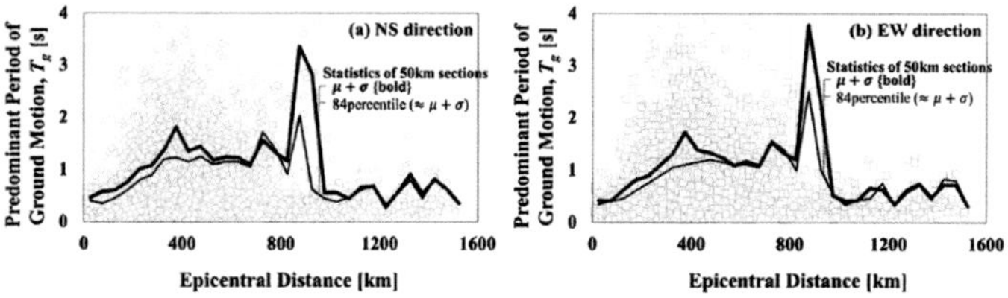


**Fig. 2.** Relationship between epicentral distance and $T_g$ for large-scale earthquakes: (a) NS direction; (b) EW direction.

### 3.3 Relationship Between Epicentral Distance and $t_m$

The relationship between epicentral distance and duration of strong motion ($t_m$) for the large-scale earthquakes is presented in Fig. 3. Here, $t_m$ is the time between Arias intensity progress from 5% to 75% (Salmon et al. 1992). The figure shows the statistical value of $t_m$, calculated in 50 km epicentral distance sections, as a line graph for the epicentral distance. Figure 3 shows a line connecting the $\mu + \sigma$ and 84th percentile values of the $t_m$ according to the epicentral distance in the scatter plot. The trend shows that $t_m$ increases until an epicentral distance of approximately 400 km. After this value, $t_m$ gently decreases as the epicentral distance increases.

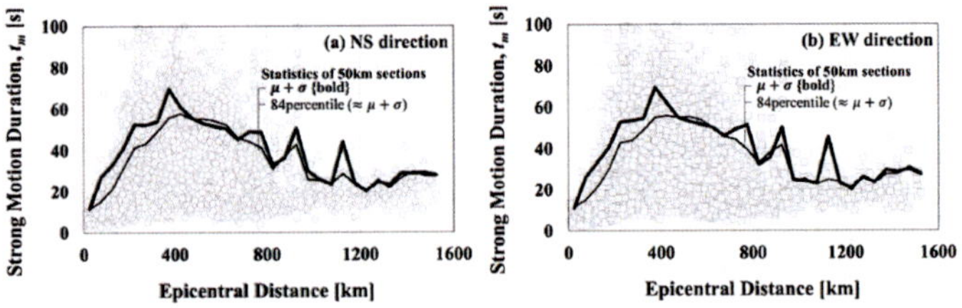


**Fig. 3.** Relationship between epicentral distance and $t_m$ for large-scale earthquakes: (a) NS direction; (b) EW direction.

### 3.4 Relationship Between Epicentral Distance and PGA

The relationship between epicentral distance and peak ground acceleration (PGA) relationship for the large-scale earthquakes is shown in Fig. 4, with the PGA of long-period ($T_g \geq 1$) and short-period ($T_g < 1$) ground motions classified. The statistical values of PGA, calculated in 50 km epicentral distance sections, are shown as a line graph for the epicentral distance. Figure 4 shows a line connecting the $\mu + \sigma$ and 84th percentile values of the PGA according to the epicentral distance in the scatter plot. There are 1189 long-period ground motions in the NS direction and 1107 long-period ground motions in the EW direction. Long-period ground motions are mainly observed at epicentral distances of 250 km or more; however, some can be observed at distances less than 100 km. This is similar to the result in which $T_g$ exceeds 1 s after an epicentral distance of approximately 250 km in the epicentral distance–$T_g$ relationship. The PGA of long-period ground motions tended to be smaller than that of short-period ground motions. At an epicentral distance of 250 km, the PGA of long-period ground motions is approximately 69% smaller than that of short-period ground motions.

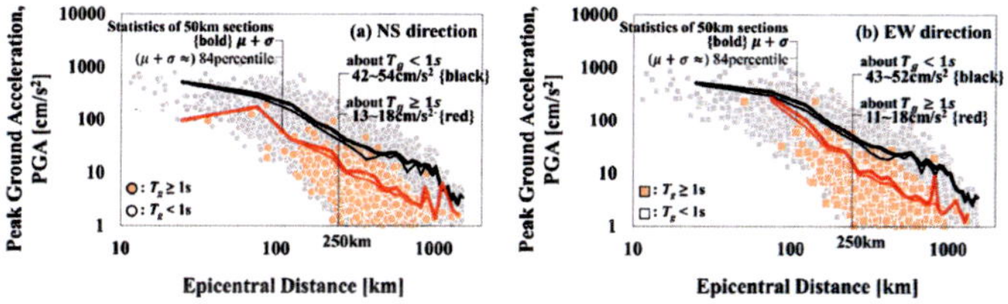

**Fig. 4.** Relationship between epicentral distance and PGA for large-scale earthquakes: (a) NS direction; (b) EW direction.

# 4 Spectrum of Long-Period Ground Motions

## 4.1 Acceleration and Velocity Response Spectrum

The normalized acceleration and velocity response spectra of the long-period ground motion are shown in Figs. 5 and 6. The normalized acceleration response spectra shown in Fig. 5 are a statistical value obtained by dividing the acceleration response spectrum of each long-period ground motion by the 0-period acceleration (≈PGA). The normalized velocity response spectra shown in Fig. 6 were calculated by dividing the normalized acceleration response spectrum in Fig. 5 by the natural angular frequency ($2\pi/T$). In the standards, the amplification ratio of the constant-acceleration segment for the PGA (hereinafter referred to as the PGA amplification ratio) is proposed as 2.5 (ASCE 2022).

In the case of the normalized acceleration response spectrum of the long-period ground motion (hereinafter referred to as ${}_NS_{A,long}$), the constant-acceleration segment was approximately 1.0 to 1.5 s. The 84th percentile and $\mu + \sigma$ of ${}_NS_{A,long}$ showed the same value until a period of approximately 5 s. Beyond this point, the $\mu + \sigma$ of ${}_NS_{A,long}$ exceeded its 84th percentile due to the large deviation of ${}_NS_{A,long}$ in the long-period region of 5 s or more. The PGA amplification ratio of the 84th percentile of ${}_NS_{A,long}$ was approximately 3.3 in the NS direction and approximately 3.4 in the EW direction. This is approximately 1.4 times higher than the PGA amplification ratio of the design acceleration spectrum.

The constant-velocity segment of the design velocity spectrum proposed in the standards is formed between $T_S$ and $T_L$ (AIK 2019; ASCE 2022). In the normalized velocity response spectrum of long-period ground motion (hereinafter referred to as ${}_NS_{V,long}$), the slope decreased from a period of approximately 1 s. The 50th percentile of ${}_NS_{V,long}$ decreased after 2 s, and the average of ${}_NS_{V,long}$ was higher than the 50th percentile after 2 s. The 84th percentile of ${}_NS_{V,long}$ showed the constant-velocity segment between approximately 3–6 s. In the 8–10 s period, the value was approximately 10% lower than the ${}_NS_{V,long}$ value in the 3–6 s period. The $\mu + \sigma$ of ${}_NS_{V,long}$ showed a value similar to the 84th percentile until approximately 5 s, but it became approximately 20% higher in the 3–8 s period than in the 6 to 10 s period.

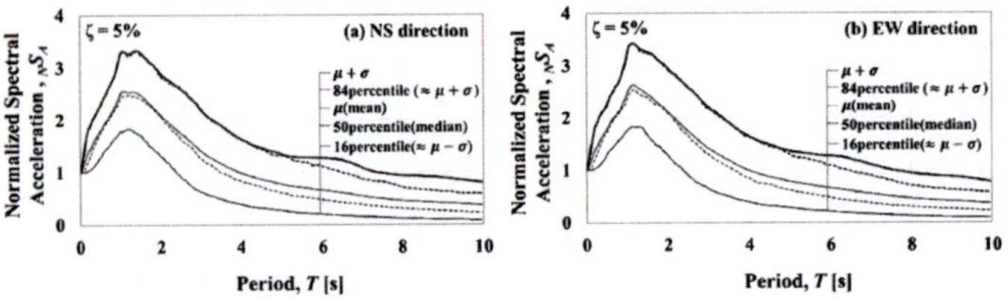


**Fig. 5.** Normalized acceleration response spectrum for long-period ground motions: (a) NS direction; (b) EW direction.

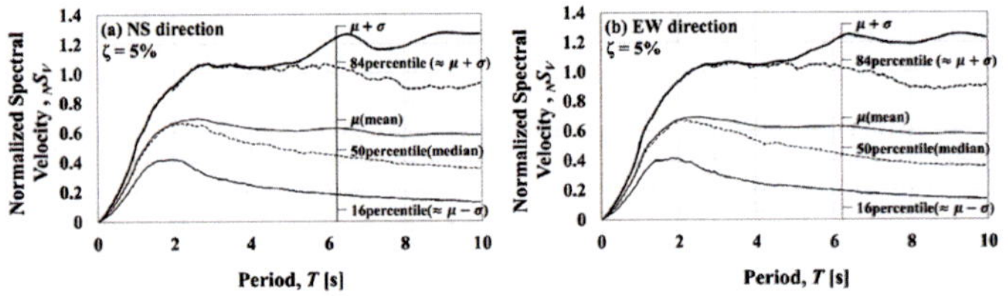


**Fig. 6.** Normalized velocity response spectrum for long-period ground motions: (a) NS direction; (b) EW direction.

## 4.2 Proposal of Design Spectrum

In this paper, the design spectrum for long-period ground motion based on its response spectrum is proposed. As in the literature, it was developed based on the line that envelops the 84th percentile of the normalized response spectrum of the long-period ground motion (Newmark & Hall 1982).

In proposing the design spectrum for long-period ground motion, it is difficult to determine the exact value of the inflection point dividing the constant-acceleration segment and the constant-velocity segment using only the data from this study. This is because the shape of the design spectrum may appear differently depending on the response spectrum sample when the design spectrum is proposed. Therefore, in this study, the inflection point was approximately determined when the design spectrum for long-period ground motion was proposed.

The proposal of normalized design spectrum for long-period ground motions is shown in Fig. 7. In the normalized design spectrum, the PGA amplification ratio is suggested to be 3.5, considering that the PGA amplification ratios in the NS and EW directions are approximately 3.3 and 3.4 respectively. It is suggested that $T_0$ be 1 s and $T_S$ be 2 s. When $T_S$ is 1.9 s, the design velocity spectrum for long-period ground motions more closely envelops the 84th percentile of ${}_N S_{V,long}$. However, considering the tendency of ${}_N S_{V,long}$ $\mu + \sigma$ to increase in the long-period region, $T_S$ is proposed to be 2 s. In addition, the design velocity spectrum for long-period ground motions is proposed to have a constant value from $T_S$. The normalized design acceleration and velocity spectra for long-period ground motions proposed in this study are expressed in Eqs. (3) and (4).

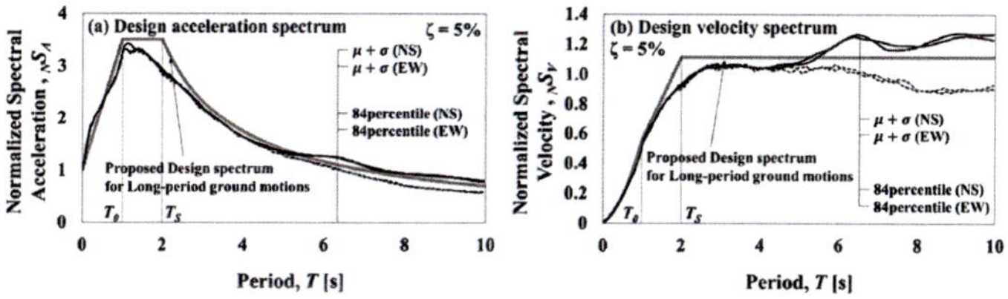


**Fig. 7.** Proposal of normalized design spectrum for long-period ground motions: (a) Design acceleration spectrum; (b) Design velocity spectrum.

$$_{N}S_{A,long} = \begin{cases} 1 + 2.5T, & 0 \le T < 1 \\ 3.5, & 1 \le T < 2 \\ 7/T, & 2 \le T \end{cases} \tag{3}$$

$$_{N}S_{V,long} = \begin{cases} (T + 2.5T^2)/2\pi, & 0 \le T < 1 \\ 3.5T/2\pi, & 1 \le T < 2 \\ 7/2\pi, & 2 \le T \end{cases} \tag{4}$$

Considering that the $T_S$ of the ground motion increases as the site class becomes weaker in the design spectrum, the design spectrum for long-period motions is proposed by matching the constant-velocity segment of the design spectrum that for the seismic zone I and the site class $S_5$ (= $S_E$) in KDS 41 17 00 (AIK 2019). In KDS 41 17 00 (AIK 2019), the constant-velocity segment of the design spectrum for seismic zone I and site class $S_5$ is proposed to be approximately 60 cm/s. Accordingly, the PGA of the design acceleration spectrum for long-period ground motions is calculated to be approximately 54 cm/s$^2$. In the relationship between epicentral distance and PGA, the long-period ground motion shows PGA values of approximately 11–18 cm/s$^2$ ($\mu + \sigma$) at the 250 km epicentral distance. The PGA of the design acceleration spectrum for long-period ground motions appears at an epicentral distance of approximately 100 km in the relationship between epicentral distance and PGA. The proposal of design spectrum for long-period ground motions is shown in Fig. 8.

In this study, when determining the design spectrum for long-period ground motions, it is proposed to match the constant-velocity segment of the design spectrum proposed in the standard. This is because, even if the PGA of long-period ground motion is smaller than the PGA of short-period ground motion, the $S_V$ of long-period ground motion may appear larger in the long-period region than the $S_V$ of short-period ground motion. In general, since the PGA of long-period ground motion appears small, if the design spectrum for long-period ground motions is determined by scaling the PGA value, there is a risk of underestimating long-period ground motions. In addition, considering that in this study, long-period ground motions of 100 cm/s$^2$ or more are partially observed in the 100–250 km section, the PGA value of the design spectrum for long-period ground motions is evaluated to be at an appropriate level.

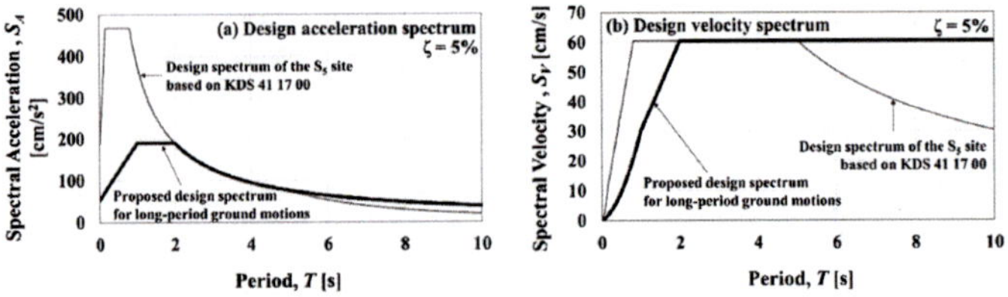


**Fig. 8.** Proposal of design spectrum for long-period ground motions: (a) Design acceleration spectrum; (b) Design velocity spectrum.

### 4.3 Energy Equivalent Velocity Spectrum

**Artificial Waves Corresponding to Design Spectra.** In this section, artificial waves corresponding to the design spectrum are calculated with reference to the standards (ASCE 2000, MPSS 2017). Table 2 shows the duration ($t_r$, $t_m$, $t_d$), which is a major variable in creating artificial waves. The duration of short-period artificial waves is a value proposed in the standards (MPSS 2017). The duration of long-period artificial waves is proposed by referring to the epicentral distance–$t_m$ relationship of large-scale earthquakes.

**Table 2.** Duration of artificial waves corresponding to design spectra.

| | Magnitude | Rise time ($t_r$) | Duration of strong motion ($t_m$) | Decay time ($t_d$) |
|---|---|---|---|---|
| Short-period artificial wave | 7.0–7.5 | 2.0 s | 12.5 s | 13.5 s |
| Long-period artificial wave | | 9.0 s | 60.0 s | 43.0 s |

Figure 9 shows the response spectra of artificial waves corresponding to design spectra. In the entire period range, the calculated response spectrum of each artificial wave does not exceed 1.3 times the design spectrum. Figure 10 shows the normalized Fourier spectra of artificial waves corresponding to design spectra. This Fourier spectrum was normalized by PGA to compare the frequency characteristics of each artificial wave. Long-period artificial waves were found to be predominant to short-period artificial waves in low-frequency components below 1 Hz. The PGA of the long-period artificial waves was approximately 3.5 times smaller than that of the short-period artificial waves. However, the long-period artificial waves showed the same acceleration and velocity response as short-period artificial waves in the 2–5 s period range, exhibiting relatively higher acceleration and velocity response in the period range beyond 5 s.

**Proposal of Design Energy Equivalent Velocity Spectrum.** Here, the deign $V_E$ spectrum for long-period ground motions is proposed as an inductive reasoning considering the design $V_E$ spectrum proposed by Oh & Ryu (2007), the short-period and long-period artificial waves, and the $V_E$–$S_V$ relationship in previous studies.

To calculate the $V_E$ spectrum using the $V_E$–$S_V$ relationship of Akiyama & Kitamura (2006), the duration is required; it is the time between Arias intensity progress from 5%

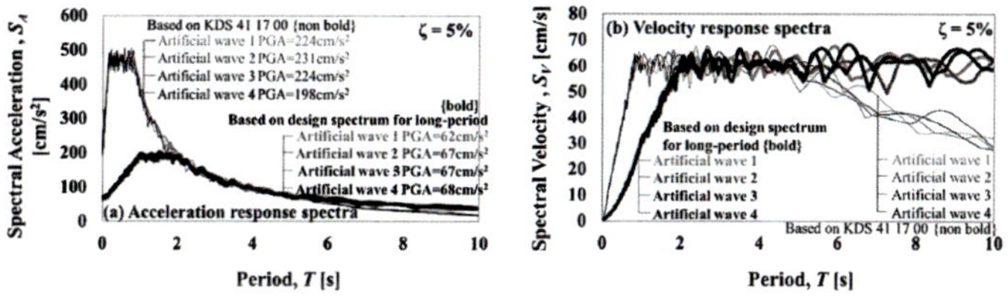


**Fig. 9.** Response spectra of artificial waves corresponding to design spectra: (a) Acceleration response spectra; (b) Velocity response spectra.

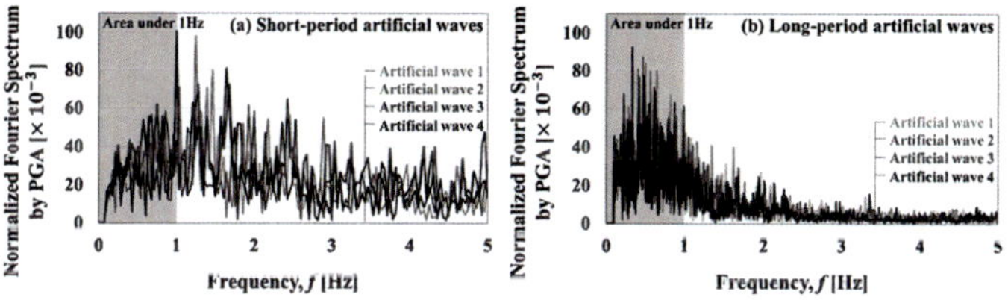


**Fig. 10.** Normalized Fourier spectra of artificial waves corresponding to design spectra: (a) Short-period artificial waves; (b) Long-period artificial waves.

to 95% ($t_{95-5}$). The average $t_{95-5}$ for short-period artificial waves was approximately 18.82 s, while the average $t_{95-5}$ for long-period artificial waves was approximately 79.69 s. In the $V_E$–$S_V$ relationship, the duration is applied from 50 s over, while the duration for the long-period artificial wave was considered to be 80 s.

In the previous study of Oh & Ryu (2007), site class $S_E$ have the same average shear wave velocity range ($\overline{v}_s < 180\text{m/s}$) as site class $S_5$. Therefore, in this study, the design $V_E$ spectrum for long-period ground motions was inferred by comparing the design $V_E$ spectrum proposed by Oh & Ryu (2007) with the $V_E$ response spectrum of artificial waves. The inference of the design $V_E$ spectrum for long-period ground motions is shown in Fig. 11.

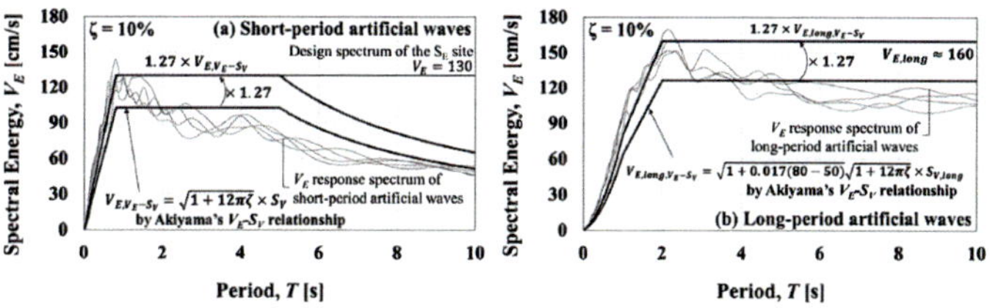


**Fig. 11.** Inference of design energy equivalent velocity spectrum for long-period ground motions: (a) Short-period artificial waves; (b) Long-period artificial waves.

The design $V_E$ spectrum for long-period ground motions (hereinafter referred to as $V_{E,long}$) is estimated to be calculated by multiplying the velocity response spectrum ($S_V$) of the long-period ground motions by the $V_E$-$S_V$ relationship and 1.27. $V_{E,long}$ showed a tendency to envelop the $V_E$ response spectrum of the long-period artificial wave. In this study, $V_{E,long}$ is proposed in a constant (160 cm/s) bilinear form from a period of 2s to a period of 10s. The proposal of design $V_E$ spectrum for long-period ground motions is shown in Fig. 12.

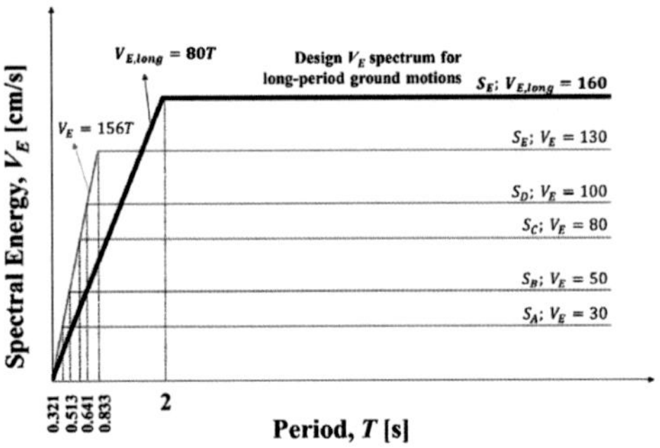


**Fig. 12.** Proposal of design energy equivalent velocity spectrum for long-period ground motions.

## 5 Conclusions

In this study, the design energy equivalent velocity ($V_E$) spectrum for long-period ground motions was proposed using ground motions observed in large-scale earthquakes (7–7.5 magnitude). The main results of this study are summarized as follows.

From the epicentral distance–$T_g$ relationship for the large-scale earthquakes, $T_g$ showed an increasing trend up to an epicentral distance of approximately 400 km. This was because the attenuation of the short-period component of ground motions more than the long-period component attenuation when the ground motion generated in the fault propagated through the medium.

The $t_m$ for large-scale earthquakes up to the epicentral distance of approximately 400 km showed a tendency to increase from approximately 10 s to 60 s. After 400 km, it showed a gentle decrease as the epicentral distance increased. Considering that most long-period ground motions caused by large-scale earthquakes occur at epicentral distances of 250 km or more, the $t_m$ for long-period ground motions needs to be longer than the 12.5 s proposed in the standards. Therefore, in this study, the $t_m$ for long-period ground motions was proposed to be 60 s.

It is proposed that the design acceleration spectrum for long-period ground motions decreases in proportion to $1/T$ in the period range beyond 2 s. However, it is proposed that the design acceleration spectrum of ASCE 7-22 decreases in proportion to $1/T^2$ in the period range beyond $T_L$ (long-period transition period). Therefore, when designing a long-period structure with a longer period than $T_L$ with the design acceleration spectrum for long-period ground motions proposed in this study, the seismic load acting on the structure is evaluated higher than ASCE7-22. It is proposed that the design $V_E$ spectrum for long-period ground motions increases by 23% compared to the design $V_E$ spectrum for site class $S_E$ ($\overline{v}_s < 180\text{m/s}$) in the period range beyond 2 s. This is because the $t_m$ is reflected in the design $V_E$ spectrum.

It is difficult to apply the design spectrum for long-period ground motions to building design as most buildings are affected by short-period ground motions. In addition, when the design spectrum for long-period ground motions is applied to the design of a low-rise building, there is a concern that the seismic energy input to the building may be underestimated. Therefore, when designing a building, the design spectrum envelope for long-period ground motions and short-period motions is suitable as a method to consider long-period ground motions. However, in order to use the design spectrum envelope as the design spectrum to consider long-period ground motions, it is necessary to examine the damage distribution of buildings caused by each ground motion. This is because the behavior of buildings due to each ground motion can appear differently.

**Acknowledgements.** This work was supported by the National Research Foundation of Korea (NRF) grant funded by the Korea government. (MSIT) (2022R1A2C2007001).

## References

Akiyama, H.: Earthquake-Resistant Limit-State Design for Buildings. University of Tokyo Press (1985)

Akiyama, H., Kitamura, H.: Relationship between energy spectra and velocity response spectra. J. Struct. Construct. Eng. (Trans. AIJ) **608**, 37–43 (2006)

American Society of Civil Engineers.: Seismic analysis of safety-related nuclear structures and commentary. ASCE 4-98 (2000)

American Society of Civil Engineers.: Minimum design loads and associated criteria for buildings and other structures. ASCE 7-16 (2017)

American Society of Civil Engineers.: Minimum design loads and associated criteria for buildings and other structures. ASCE 7-22 (2022)

Architectural Institute of Korea.: KDS 41 17 00. Seismic Building Design Code (2019)

Arias, A.: A measure of earthquake intensity. Seismic design for nuclear plants, pp. 438–483 (1970)

Chopra, A.K.: Dynamics of Structures: Theory and Applications to Earthquake Engineering, 4th edn. Prentice-Hall, Upper Saddle River, NJ (2011)

Housner, G. W.: Limit design of structures to resist earthquakes. In: Proceedings of WCEE, vol. 5, pp. 1–13 (1956)

Ministry of Public Safety and Security, Common Application of Seismic Design Criteria (2017)

Newmark, N.M., Hall, W.J.: Earthquake Spectra and Design. Earthquake Engineering Research Institute, Berkeley, CA (1982)

NIED: NIED K-NET, KiK-net (2019)

Novikova, E.I., Trifunac, M.D.: Duration of strong ground motion in terms of earthquake magnitude, epicentral distance, site conditions and site geometry. Earthquake Eng. Struct. Dynam. **23**(9), 1023–1043 (1994)

Oh, S.H., Ryu, H.S.: The proposal of energy equivalent velocity spectra for energy input by earthquake. In: Conference of Korean Society of Steel Construction, pp. 187–187 (2007)

Oh, S.H.: The proposal of design spectrum for performance-based design and vibration control structural system. J. Architect. Inst. Korea Struct. Construct. **30**(12), 3–11 (2014)

Oh, S.-H., Shin, S.-H.: A proposal for a seismic design process for a passive control structural system based on the energy equilibrium equation. Int. J. Steel Struct. **17**(4), 1285–1316 (2017). https://doi.org/10.1007/s13296-017-1203-z

Oh, S.H., Shin, S.H., Bagheri, B.: Stability evaluation of the acceleration and energy response spectra. Soil Dyn. Earthq. Eng. **123**, 124–143 (2019)

Salmon, M.W., Short, S.A., Kennedy, R.P.: Strong motion duration and earthquake magnitude relationships (No. UCRL-CR-117769). Lawrence Livermore National Lab.(LLNL), Livermore, CA (1992)

Seed, H.B., Ugas, C., Lysmer, J.: Site-dependent spectra for earthquake-resistant design. Bull. Seismol. Soc. Am. **66**(1), 221–243 (1976)

# Engineering Demand Parameters for Cumulative Damage Estimation in URM and RC Buildings

P. García de Quevedo Iñarritu(✉), N. Šipčić, and P. Bazzurro

IUSS, Pavia, Italy
pablo.garcia@iusspavia.it

**Abstract.** Seismic sequences are often a source of damage in unreinforced masonry (URM) and reinforced concrete (RC) buildings, which is underestimated by the standard earthquake assessment framework that is based solely on mainshock seismicity. Even when clustered seismicity is considered, buildings are assumed to be in intact conditions once any event in the sequence happens. Simply put, damage accumulation is overlooked. Furthermore, even when the damage accumulation is taken into account, engineering demand parameters (EDPs) often utilized to estimate the damage are not tailored for capturing the increasing levels of damage that some structural members may experience throughout the subsequent events. Typically these metrics, which are drift-based (for structural components) and acceleration-based (for nonstructural components and contents), focus on peak values and, therefore, are not able to capture the evolution of damage in cases when the ground motions caused by aftershocks may cause a lower demand in the already damaged structure. This work explores a method for measuring cumulative damage effects on structural members that considers both drift and energy-related terms via a modified version of the well known Park and Ang Damage Index. The proposed index is recalibrated and tested using experimental data that include different levels of damage up to collapse for components of URM buildings, such as piers and spandrels, and of RC buildings, such as columns. The resulting new EDP is found to be effective in the identification of different damage states and, therefore, a good candidate for the development of fragility and vulnerability curves to be used in the context of clustered seismicity risk assessment.

**Keywords:** Damage accumulation · Assessment · EDP · URM · RC

## 1 Introduction

Following a severe earthquake, a building that has sustained damage and remains unrepaired will be less fit to cope with the ground shaking of potential additional seismic events. Typically, after severe earthquakes, the ability to conduct thorough investigations and take appropriate interventions to ensure structural safety and functionality is limited. Due to either financial constraints or to the short interval between subsequent earthquakes. As a result, the building's vulnerability to additional damage or collapse may increase.


H. Varum et al. (Eds.): IWEBSE 2023, LNCE 236, pp. 57–71, 2023.
https://doi.org/10.1007/978-3-031-36562-1_5

Large magnitude earthquakes (denoted hereinafter as MSs), may, in the next days, weeks, or months, trigger one or more significant aftershocks (denoted hereinafter generally as ASs, regardless of whether they break part of the same MS rupture or not). In some cases, these aftershocks can result in extensive damage to the infrastructure (e.g., Moon et al. 2014). Recent sequences, such as the 2010–2011 Darfield-Christchurch (Shcherbakov et al. 2012) in New Zealand, the 2011 Van in Turkey (Di Sarno et al. 2013), the Great East Japan Earthquake (Goda et al. 2013), the 1997–98 Umbria-Marche (ref), the 2012 Emilia (Penna et al., 2013), and the 2016–17 earthquake sequences (Sextos et al. 2018) in central Italy, as well as the recent Zagreb-Petrinja 2020–2021 earthquakes (Atalić et al 2021) in Croatia, have shown the devastating cumulative damage effects of repeated shocks. These events were especially damaging to URM and old RC structures as indicated by Kam et al., 2011; Ingham and Griffith, 2012 and Sextos et al., 2018. These structures were already damaged in the MS event (or in some cases, by significant foreshocks, as in the case of the 2016–17 sequence in Central Italy) and suffered a considerable increase in damage or even collapse in the following events of the sequence.

The focus of performance-based earthquake engineering (PBEE) (Cornell and Krawinkler, 2000), however, is on the quantification of seismic risk caused by MS events only with the implicit assumption that that the impact on structures due to ASs is less important if not negligible. A straightforward application of PBEE, therefore, may result in lower damage and loss estimates and, in turn, to potentially misinformed decisions in, for example, emergency planning, insurance premiums evaluation and so on. To address this issue, we contend, that the PBEE methodology should be modified to account for the impact of clustered seismicity on both hazard and vulnerability assessments.

Underestimation in seismic hazard arising from the removal of spatially and temporally dependent events from the seismic catalogs (i.e., via declustering) and the consequent assumption that the events occur as a Poissonian process has been investigated in several studies (e.g., Yeo and Cornell, 2009, Boyd 2012, Iervolino et al., 2014, Šipčić et al, 2022). The impact of this simplification on vulnerability assessment has been investigated in previous studies that investigated damage accumulation (e.g., Mouyiannou et al. 2014, Grimaz and Malisan 2016, Rinaldin and Amadio, 2018 for URM and Iervolino et al. 2020, Furtado et al. 2018, Goda and Taylor 2012 for RC structures) and derived damage state-dependent building fragility curves. However, most of these works have used either simplified methods such as modeling structures as SDOF systems or monitoring their response via global drift-based engineering demand parameters (EDPs).

These approaches have led to conclusions that appear to contradict the observations, as the analytically predicted change in the damage state of the structure has been observed to occur only when the intensity of the AS ground motion is higher than that of the MS. This is not what happens in reality especially to buildings that already suffered a moderate to an extensive level of damage. This analytical prediction of relatively low impact of aftershock shaking on the progressive damage can be attributed to the use of either simplified structural models that are unable to capture damage accumulation and/or improper global damage EDPs that do not reliably reflect the deterioration of the structural elements. Garcia de Quevedo *et al.* 2021 have briefly addressed this issue by suggesting that a component-based approach using also energy-related metrics may be a better fit for purpose than peak drift-based metrics, whose limitations were also recognized by Baraschino et al. 2023. A recent study by Shen et al. 2022, devoted to the development of damage state-dependent fragility curves for URM buildings found that a global energy-based damage index was better suited than the drift-based peak EDP measures.

Our proposed approach to assess the progression of damage in individual structural elements is to employ a component-based method that goes beyond the use of maximum drift alone, which is unable to capture incremental damage in a seismic sequence. To achieve this, we will use an energy-based, monotonically-increasing damage metric, which provides a better representation of the cumulative damage effects of repeated ground motions. Our work focuses on the development of such EDP for both URM and RC members.

## 2 EDP for URM Piers

### 2.1 Cumulative Damage on URM Elements

One important issue to investigate for assessing the effects of cumulative damage is what happens within the cyclic loading of masonry walls, beyond only assessing the drift limits on a monotonic capacity curve or envelope. This investigation helps find ways to understand how damage increases in real cases, which is through the formation and growth of cracks and a general loss of capacity that can appear due to cyclic damage. For this to occur, there is no need for the pier to undergo a displacement larger than the peak one experienced during the mainshock. Information related to this process is scarce since most of the damage metrics and limit states for masonry are derived from drift values only, with no information about the loading history. Little information is available on different metrics such as dissipated energy or the number of cycles exceeding a certain displacement. However, Wilding et al. (2017) studied the impact of loading protocols on the response of masonry structures, using a set of micromodel numerical simulations which were calibrated with experimental data. Several walls with different shear span ratios (boundary conditions) and axial loads were tested with the same aspect ratio and evaluated with different loading protocols (monotonic and cyclic with variations). Some important findings of this study revealed that walls with a predominant shear failure mode are those impacted the most by the loading history. These impacts cause a reduction of the ultimate and peak drifts, as well as an increase in the energy dissipated in cyclic cases compared to that of the monotonic ones. The authors also mention that lower axial load ratios cause walls to be more sensitive to the specific loading history. On the contrary, flexure-controlled walls are not as influenced by the loading history because damage is concentrated at the toes, and as soon as the corner crushing occurs the wall fails. In shear walls, however, the whole wall area can be affected. With this, they conclude that the influence of load history can be related to how widespread the damage is within the wall. Additionally, the work by Beyer and Mergos (2015) also includes a discussion of the influence of loading history on the response of URM walls based on a comparison of results of a few works from the literature with experimental tests that are useful for assessing these effects together with some numerical and experimental tests carried out by her. These comparisons showed changes in effective stiffness, maximum force, and ultimate displacement and, more specifically, big reduction in ultimate displacement and gains in stiffness. These findings are very much in line with Wilding et al.'s (2017) observations that stem from their own experimental campaign of cyclic vs. monotonic tests for walls with different failure mechanisms. They mention that for low levels of damage, the differences in URM walls behavior are negligible. Differences only start once when the pier has experienced irreversible damage, such as failure of compression zone (flexure) or concentration of damage on a diagonal crack (shear), a condition that only happens close when the pier is loaded in plane with aforce close to the maximum capacity.

From all these considerations, we can conclude that the most critical piers for which to model cumulative damage are those that eventually exhibit a shear or hybrid failure, where the damage is spread throughout the element, and especially after significant cracks have formed. Given the preliminary nature of this work, and the relative higher importance of piers over spandrels on the overall damage state of a building, the calibration and assessment will be done for piers only, and shall be extended for spandrels in the future.

## 2.2 Calibration of the EDP

In the literature, very few EDPs that capture effects of cumulative damage in masonry elements are available. One such example is the Park and Ang damage index (Young-Ji Park et al., 1985) that was initially calibrated for flexure-dominated RC members and utlized to describe the response of entire RC structures. It is defined as the sum of a displacement-based term and a hysteretic energy term (weighted by the coefficient β), and takes on values ranging from 0 (undamaged state) to 1 (total collapse). Kunnath et al. (1992) later modified the proposed RC damage index as follows:

$$DI = \frac{d_m - d_y}{d_{um} - d_y} + \frac{\beta E_H}{E_{mon}} \tag{1}$$

where $d_m$ represents the maximum displacement reached during the analysis, $d_{um}$ represents the ultimate displacement under monotonic loading, $E_H$ represents the dissipated energy, $E_{mon}$ represents the equivalent strain energy, and $\beta$ is a parameter calibrated based on the experimental data. To adequately calibrate the EDP for URM piers, we propose here to make changes to the index given in Eq. (1). The first modification is to balance energy and drift terms, which in the previously proposed index were heavily dominated by the former term (i.e., coefficient $\beta$ was small). We also aim to achieve a maximum value of 1 for both terms by introducing the coefficient γ, which balances the ratio of the monotonic to the full energy dissipated. The second change is to add a *1-β* factor in front of to the displacement term, ensuring that the sum at collapse gives a DI of 1. The modified index proposed here is shown in Eq. (2):

$$DI = (1 - \beta) * \frac{d_m - d_y}{d_{u,m} - d_y} + \frac{\beta(E_H)}{\gamma E_m} \tag{2}$$

The yield point is taken from the bilinear idealization that considers $F_y$ as 60% of the maximum shear strength. For the energy-related terms, $E_H$ is found by integrating the full force-displacement curve of the cyclic response, and $E_m$ is computed as $E_m = F_y(d_{u,m} - 0.5d_y)$.

The database utilized to calibrate the proposed index consists of 135 tests on URM piers that include information about the force-displacement behavior, failure types and, in some cases, a description of the different levels of damage experienced by the specimen during the test. This database is mainly built on that assembled by Vanin et al., 2017, with additional tests extracted from (Graziotti *et al.*, 2019; Salmanpour et al., 2013).To investigate the issue of the value for energy being much larger than 1 and to estimate a value for the $\gamma$ coefficient, it is required to evaluate the ratio of the cyclic energy to the monotonic energy at collapse (computed as: $E_H/E_m$).This operation was repeated for all the experimental tests. Figure 1 shows this ratio plotted versus the cumulative ductility, $\mu_{cum}$, in each case. The value of cumulative ductility is found through adding all the values of ductility at every nonlinear excursion during the load history of the element.This relationship between these two variables allowed us to find in each case the value of $\gamma$ that leads to a value of 1 to the energy term at collapse.

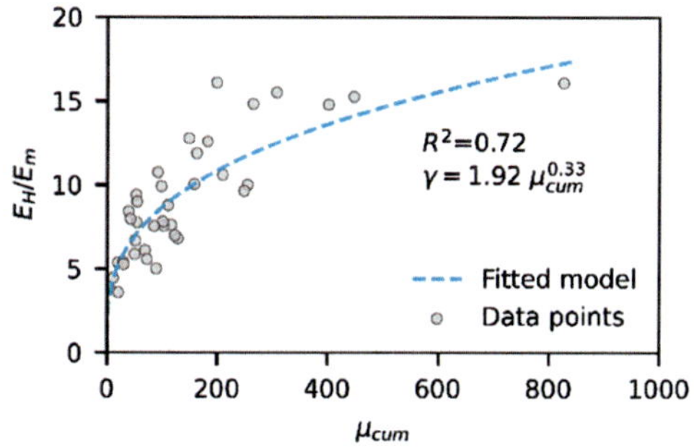


**Fig. 1.** Relationship between the cumulative ductility and the ratio of cyclic to monotonic dissipated hysteretic energy. $R^2$ is the coefficient of determination of the regression model

The value of $\beta$, which determines the weight of the drift and energy terms in Eq. (2), was obtained using all shear and hybrid failing experimental tests (45% of all tests), at the reported point of collapse. To do this, all the relevant parameters of Eq. (2) are taken from the reported values of each experimental test while the values of additional quantities, such as $F_y$ and $d_y$ are derived from the data provided. Next, $\gamma$ was computed for each case using the cumulative ductility. The maximum displacement was taken from what is reported as the collapse on every test, and the dissipated hysteretic energy at the onset of collapse is computed by integrating the force-displacement response up to that point. Finally, the equation is solved for $\beta$, while setting the $DI = 1$. The obtained values, are found in Fig. 2a.

### 2.3 Damage State Definition

Finally, to estimate the thresholds at the onset of each damage state (DS), it was necessary to calibrate Eq. (2) to different levels of physical damage experienced by the piers during the tests. This was done using data extracted from the subset of 20 experimental tests that included a detailed description of the pier damage throughout the test. This allowed us to match the different levels of damage observed to the descriptions of damage shown in FEMA 306 (ATC, 1998), which defines four distinct damage states (slight, moderate and extreme -or collapse-) for different types of failure of URM piers. After the identification of the onset of each of those damage states were identified in tehse 202 tests, we computed all the values of the parameters of Eq. (2) and assigned a DI to each DS. The resulting set of values for each of the DS was then fitted into a normal distribution (see its CDF in Fig. 2b). After this exercise, we defined the ranges of DI that encompass each DS, with the objective of minimizing the overlap and gaps between them. The ranges of proposed DI values for each DS as well as the 16$^{th}$ and 84$^{th}$ percentiles extracted from the distribution fitting are shown in Table 1.

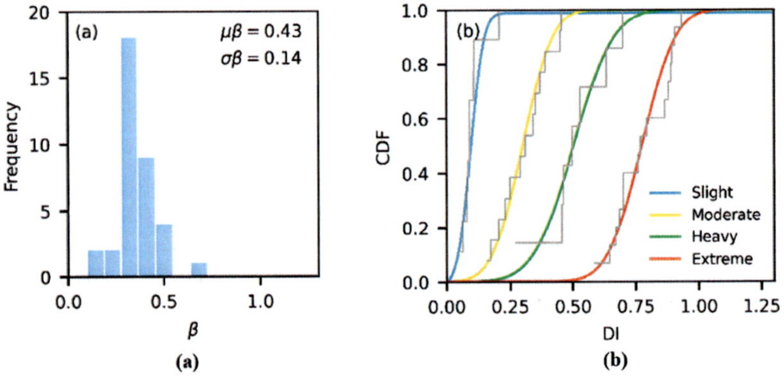


**Fig. 2.** (a) distribution of beta values for all experimental tests and (b) CDF of the different DI corresponding to different DS used for calibration.

**Table 1.** Percentiles and ranges of the DI for the different DS according to FEMA-306.

| DS | 16$^{th}$-84$^{th}$ range | Proposed range |
|---|---|---|
| 0 - No damage | – | <0.050 |
| 1 - Slight | 0.055–0.160 | 0.050–0.175 |
| 2 - Moderate | 0.180–0.400 | 0.175–0.400 |
| 3 - Heavy | 0.410–0.710 | 0.400–0.750 |
| 4 - Extreme (collapse) | 0.740–0.980 | >0.750 |

# 3 EDP for RC Columns

## 3.1 Calibration of the EDP

The first step in developing a damage index to describe the damage observed in RC columns involves collecting available experimental data. To assemble the database, we use two main sources: (1) data gathered by the ACI committee 369 (described in more detail in (Sivaramakrishnan, 2010), as the primary source for the database, and (2) the database developed for the SERIES research project (Peruš et al., 2014). These databases expand on the Pacific Earthquake Engineering Research center (PEER) Structural Database (Berry et al. 2004), with some additional experiments extracted from published reports of different authors. In the current version, only rectangular RC columns are included. The compiled database provides force-displacement histories in comma-separated value (CSV) files of 370 tests that are either obtained directly from the researchers or digitized from force-displacement plots in the available paper format of the source documents.

Similarly as in the case of masonry components, the modified Park and Ang DI (Eq. 1) is used as the starting point. The ultimate displacement under monotonic loading (dum) was determined following the ATC-52 guidelines, taking into consideration the displacement at yield, the maximum displacement under cyclic loading, and the displacement at the capping point under cyclic loading. In instances where the maximum displacement coincided with the capping point, indicating no loss in strength, the maximum displacement under cyclic loading was estimated as 1.1 times the displacement at the capping point under cyclic loading. Admittedly, this might be a limitation, although not critical, of the proposed approach. Park et al. (1987) estimated the value of β by performing regression analysis using 402 RC and 132 steel specimens and suggested a value of 0.05 for RC. Later, researchers proposed higher values of β, with Kunnath et al. (1992) suggesting a value of 0.1, and Cosenza and Manfredi (1998) suggesting a value of 0.15. In this study, the value of β was estimated for each experiment in the assembled database using Eq. (3) below.

$$\beta = \frac{E_{mon}}{\int^{d} E}\left(1 - \frac{d_{max} - d_y}{d_{u,m} - d_y}\right) \tag{3}$$

Out of all the tests collected, we only used 97 experiments related to columns that fail in flexure and have good-quality data (additional 12 experiments were retained for validation). Shear-dominated and flexure-shear dominated columns are outside the scope of this study. The median value of $\beta$ obtained was 0.1, which is in line with the recommendations of Kunnath et al. (1992), and the standard deviation of the DI calculated with the median $\beta$ was 0.32 (COV of 0.3). To reduce the standard deviation, the value of $\beta$ was kept at 0.1, and a new parameter γ was added, resulting in:

$$DI = \frac{d_{max} - d_y}{d_{u,m} - d_y} + \beta \frac{E_p}{\gamma F_y (d_{u,m} - d_y)} \tag{4}$$

From Eq. (4) above the value of γ was estimated for each column in the database. As we observed that the ratio between displacement and energy terms differs for cases

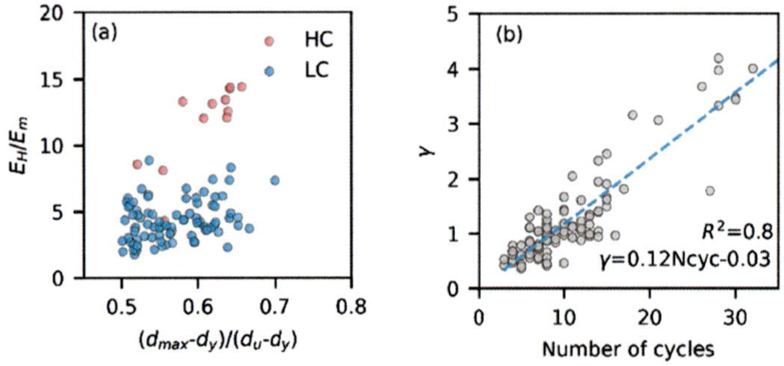


**Fig. 3.** (a) Relationship between the displacement term and the ratio of cyclic to monotonic hysteretic energy (HC: tests with number of cycles $\geq$ 17; LC: tests with number of cycles < 17); (b) relationship between the number of cycles and $\gamma$ parameter. Data points represent the values from 97 experimental tests.

with many cycles (at least 17) and fewer cycles (Fig. 3a), we expressed $\gamma$ as a function of the number of cycles using linear regression (Fig. 3b).

The DI was calculated for each specimen in the database using the $\gamma$ values calculated as illustrated in Fig. 3b, resulting in a standard deviation of 0.12 that is significantly lower than the previous value of 0.32 (Fig. 4).

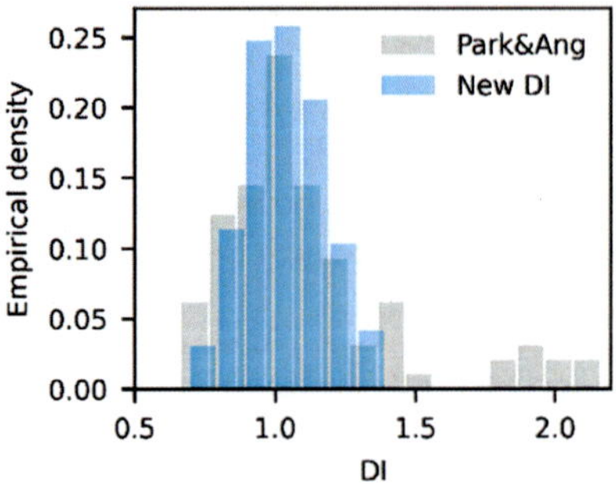


**Fig. 4.** Distribution of Park and Ang DI (grey) and the proposed DI (blue) (Eq. 4).

### 3.2 Damage State Definition

To define the damage states associated with the progression of damage in RC columns that fail in flexural mode, we used the experimental data provided by several researchers (Bearman (2012), Sivaramakrishnan (2010), Moe and Wang (2000), Paal et al. (2014)) along with the reports and photos accompanying these experiments. We defined four damage states, with $DS_1$ corresponding to the onset of slight damage where flexural cracking, longitudinal cracking and shear cracking occur. $DS_2$ is associated with the onset of concrete cover spalling, indicating moderate damage, while $DS_3$ is associated with a heavy damage, marked by significant spalling of concrete with exposed steel rebars. $DS_4$ corresponds to major safety implications, including buckling of longitudinal bars, crushing of the concrete core, rebar fractures and complete failure of the column. To determine the values of the DI that correspond to the thresholds of the proposed damage states, we conducted a detailed examination of all the available experiments that provided a comprehensive description of damage, namely 23 experiments for prediction and an additional 5 for validation. We identified the instances at which the specified damage states were reached and calculated the values of the parameters in Eq. (4), leading to the calculation of the corresponding DI. Using these results, the proposed ranges for each of the four damage states are presented in Table 2. In this case, we did not fit the normal distribution to the data as the dispersion in the obtained values was too low, probably due to the small number of available experiments. One should note that further investigation might be needed to assess if the proposed damage states are applicable in the cases when we have RC beams.

**Table 2.** Ranges of the DI for the different DS

| DS | Proposed range |
|---|---|
| 0 – No damage | <0 |
| 1 – Slight | 0–0.2 |
| 2 – Moderate | 0.2–0.4 |
| 3 – Heavy | 0.4–0.75 |
| 4 – Extreme (collapse) | >0.75 |

# 4 Validation

## 4.1 URM

To validate the newly proposed damage metric on piers experiencing cumulative damage it is necessary to use realistic case with documented values of pier damage caused by known levels of ground motions. This validation would allow comparing the accuracy of different approaches and would pave the way for the development of damage state-dependent fragilities. Unfortunately, to our knowledge, no shaking table test has yet been conducted on structures with the aim of capturing the effect of a realistic seismic sequence. Such tests consist of incremental analysis where each input motion is set up to be larger than the previous one applied.

Here, we compared the analytical prediction of damage states of piers in the building Prototype 2 from Magenes et al. 2010, Magenes et al. 2014, and Senaldi et al., 2019. This building was subjected to shake table tests with increasing levels of ground motion intensity. The damage suffered by the piers in terms of descriptions and measurements of the crack patterns was recorded. Additionally, experimental data on base forces, accelerations, and displacements at various sensor locations are also available.

The Tremuri (Lagomarsino et al, 2013) model (described in Penna et al, 2016), calibrated for these tests was used to compute the different values of the *DI* after each ground motion. The mean values of axial loads and end moments were used to determine the failure mechanism and capacity of the elements.

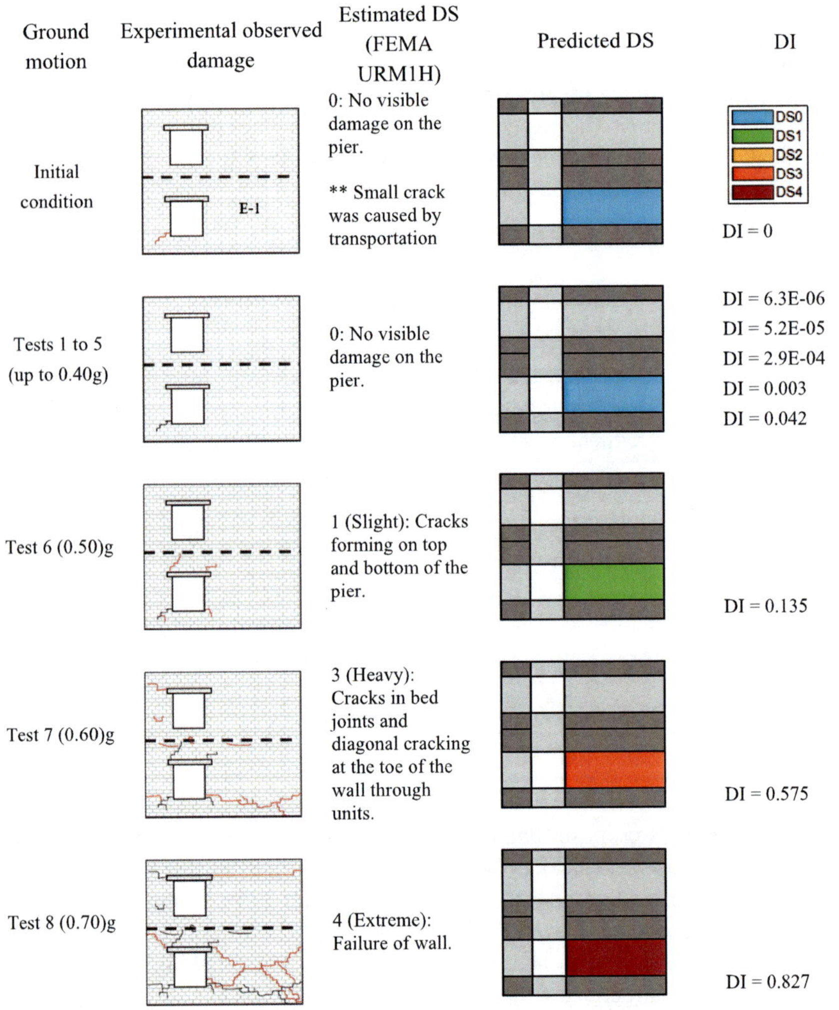


**Fig. 5.** Comparison between experimental and predicted damage for increasing levels of shaking.

For this experiment, only one of the piers at the bottom floor exhibited hybrid failure (pier E-1), while the remaing ones failed in flexure. Therefore,only this specific wall was used for the comparing predicted and recorded results. Figure 5 shows the experimental observed damage, with red lines highliting the new cracks developed by the specific ground motion. Additionally, the third column contains the estimated DS according to FEMA 306, using the most fitting case (URM1H). The two rightmost panels show the value of DI computed after each of the ground motions, as well as the corresponding DS from 0 to 4 as per Table 1. We find a good match between observations and predictions all the way to the highest level of damage.

## 4.2 RC

In the case of RC columns that fail in flexure, we could not find any shake table test with the data necessary for comparing predicted and observed damage states. As a result, as mentioned earlier we decided to validate our findings using the cyclic tests that were not part of the calibration (12 experiments). As an example, we show the results obtained for specimen C1–3 from Mo and Wang (2000) who provided a detailed explanation of the damage progression throughout the experiment. Using that information, we identified the time step during the test when the specific damage occurred and determined the corresponding DI value using Eq. (4). The specimen first started cracking followed by the yielding of longitudinal bars in tension. As the loading increased, concrete spalling started and ties began to get loose. Finally, the rupture of longitudinal bars occurred. These points, which correspond to damage states one to four, respectively, are shown on the hysteresis force-displacement history in Fig. 6a. According to Table 3, the calculated values of the proposed DI predict well the observed damage states. A similar good agreement is observed for the remaining 11 specimens used for validation.

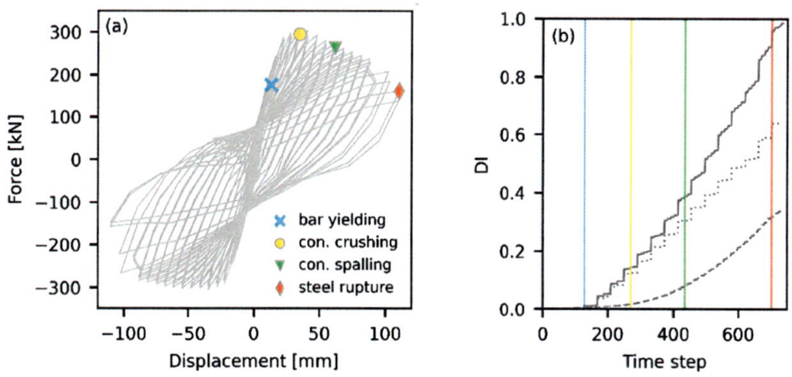


(a) (b)

**Fig. 6.** (a) Hysteretic loops of load-displacement relationship; (b) progression of DI, calculated with Eq. 3, dotted line indicates the displacement part and dashed line the energy part. Results correspond to the specimen C1–3 from Mo and Wang (2000).

**Table 3.** Obtained values of the maximum drift and proposed DI for Specimen C1–3

| | $DS_1$ | $DS_2$ | $DS_3$ | $DS_4$ |
|---|---|---|---|---|
| DI (Eq. 3) | 0.01 | 0.15 | 0.42 | 0.95 |
| Max drift (%) | 1.2 | 2.5 | 4.4 | 7.8 |

## 5 Conclusions

The work presented herein is an initial attempt to enhance the current methods for assessing the effects of seismic sequences on URM and RC buildings within the consolidated PBEE framework. We used a damage index that contains both energy and drift-based terms to evaluate the damage on shear and hybrid failing URM piers and on RC columns failing in flexure. The proposed DIs are monotonically increasing functions that are able to capture the progression of damage resulting from repetitive shaking These proposed EDPs were derived from experimental data, by first calibrating the index values corresponding to the collapse of the member, and then by defining thresholds for the intermediate damage levels based on detailed and in-depth descriptions of the progressive damage states reached by members in experimental tests.

These initial findings are very promising and indicate that the EDPs can accurately assess the damage to the components. However, an additional more extensive set of experimental test is needed that account for the damage suffered by all the elements of a structure. Ideally, such shake table tests should include cases where the structure is subjected to the ground motion of a realistic seismic sequence, rather than to ground motions incrementally increasing in severity, such as those discussed here. Alternatively, this research on damage progression could greatly benefit by collecting damage data suffered by real instrumented buildings after the occurrence of each significant event in a sequence. Any such data set would provide insight into the sensitivity of this metric to capture cumulative damage for more "common" cases and aid in the development of PBEE techniques suitable for assessing risk in presence of clustered seismicit.

## References

Atalić, J., Uroš, M., Šavor Novak, M., Demšić, M., Nastev, M.: The Mw5.4 Zagreb (Croatia) earthquake of March 22, 2020: impacts and response. Bull. Earthq. Eng. **19**(9), 3461–3489 (2021)

Bae, S.: Seismic Performance of Full-Scale Reinforced Concrete Columns, 312 p. Ph.D. Thesis, The University of Texas at Austin, Austin, TX (2005)

Baraschino, R., Baltzopoulos, G., Iervolino, I.: A note on peak inelastic displacement as a proxy for structural damage in seismic sequences. Procedia Struct. Integrity **44**, 75–82 (2023). https://doi.org/10.1016/j.prostr.2023.01.011

Bearman, C.F.: Post-Earthquake Assessment of Reinforced Concrete Frames. University of Washington (2012)

Berry, M., Parrish, M., Eberhard, M.: PEER Structural Performance Database User's Manual (Version 1.0). University of California Berkeley, Berkeley, CA. Retrieved from PEER Structural Performance Database: http://nisee.berkeley.edu/spd (2004)

Beyer, K., Mergos, P.: Sensitivity of drift capacities of urm walls to cumulative damage demands and implications on loading protocols for quasi-static cyclic tests (2015)

Boyd, O.S.: Including foreshocks and aftershocks in time-independent probabilistic seismic-hazard analyses. Bull. Seismol. Soc. Am. **102**(3), 909–917 (2012). https://doi.org/10.1785/0120110008

Cornell, C.A., Krawinkler, H.: Progress and challenges in seismic performance assessment. PEER Center News **3**(2), 1–2 (2000)

Cosenza, E., Manfredi, G.: A seismic design method including damage effect. In: Proceedings of the 11th European Conference on Earthquake Engineering, Paris, 6–11 Sep 1998

Di Sarno, L., Yenidogan, C., Erdik, M.: Field evidence and numerical investigation of the Mw = 7.1 October 23 Van Tabanlı and the M = 5.7 November earthquakes of 2011. Bull. Earthq. Eng. **11**(1), 313–346 (2013)

Furtado, A., Rodrigues, H., Varum, H., Arêde, A.: Mainshock-aftershock damage assessment of infilled RC structures. Eng. Struct. **175**, 645–660 (2018). https://doi.org/10.1016/j.engstruct.2018.08.063

de Quevedo Iñarritu, P.G., Šipčić, N., Kohrangi, M., Bazzurro, P.: Effects of pre-existing damage on fragility of URM and RC frame buildings. In: Benavent-Climent, A., Mollaioli, F. (eds.) IWEBSE 2021. LNCE, vol. 155, pp. 11–28. Springer, Cham (2021). https://doi.org/10.1007/978-3-030-73932-4_2

Goda, K., Taylor, C.A.: Effects of aftershocks on peak ductility demand due to strong ground motion records from shallow crustal earthquakes. Earthq. Eng. Struct. Dynam. **41**(15), 2311–2330 (2012). https://doi.org/10.1002/eqe.2188

Goda, K., et al.: Ground motion characteristics and shaking damage of the 11th March 2011 M w 9.0 Great East Japan earthquake. Bull. Earthq. Eng. **11**, 141–170 (2013)

Graziotti, F., Penna, A., Magenes, G.: A comprehensive in situ and laboratory testing programme supporting seismic risk analysis of URM buildings subjected to induced earthquakes. Bull. Earthq. Eng. **17**(8), 4575–4599 (2018). https://doi.org/10.1007/s10518-018-0478-6

Grimaz, S., Malisan, P.: How could cumulative damage affect the macroseismic assessment? Bull. Earthq. Eng. **15**(6), 2465–2481 (2016). https://doi.org/10.1007/s10518-016-0016-3

Iervolino, I., Chioccarelli, E., Suzuki, A.: Seismic damage accumulation in multiple mainshock–aftershock sequences. Earthq. Eng. Struct. Dyn. **49**(10), 1007–1027 (2020). https://doi.org/10.1002/eqe.3275

Iervolino, I., Giorgio, M., Polidoro, B.: Sequence-based probabilistic seismic hazard analysis. Bull. Seismol. Soc. Am. **104**(2), 1006–1012 (2014). https://doi.org/10.1785/0120130207

Ingham, J., Griffith, M.: The performance of unreinforced masonry buildings in the 2010/2011 canterbury earthquake swarm. In: 15th World Conference on Earthquake Engineering, Lisboa, Portugal (2012)

Kam, W.Y., Pampanin, S., Elwood, K.: Seismic performance of reinforced concrete buildings in the 22 February Christchurch (Lyttelton) earthquake. Bull. New Zealand Soc. Earthq. Eng. **44**(4), 239–278 (2011). https://doi.org/10.5459/bnzsee.44.4.239-278

Kunnath, S.K., Reinhorn, A.M., Lobo R.F.: IDARC Version 3.0: A Program for the Inelastic Damage Analysis of Reinforced Concrete Structures. Technical report NCEER-92-0022. State University of New York at Buffalo, Buffalo, NY (1992)

Lagomarsino, S., Penna, A., Galasco, A., Cattari, S.: TREMURI program: an equivalent frame model for the nonlinear seismic analysis of masonry buildings. Eng. Struct. **56**, 1787–1799 (2013)

Magenes, G., Penna, A., Galasco, A., Magenes, G., Penna, A., Galasco, A.: Seismic Vulnerability Assessment. Retrofit Strategies and Risk Reduction View project Contribution in the preparation of the Italian Norms for Construction View project A full-scale shaking table test on a two-storey stone masonry building. https://www.researchgate.net/publication/257333070 (2010)

Magenes, G., Penna, A., Senaldi, I.E., Rota, M., Galasco, A.: Shaking table test of a strengthened full-scale stone masonry building with flexible diaphragms. Int. J. Architectural Heritage **8**(3), 349–375 (2014). https://doi.org/10.1080/15583058.2013.826299

Mo, Y.L., Wang, S.J.: Seismic behavior of RC columns with various tie configurations. J. Struct. Eng. **126**(10), 1122–1130 (2000). https://doi.org/10.1061/(ASCE)0733-9445(2000)126:10(1122)

Moon, L., et al.: The Demise of the URM Building Stock in Christchurch during the 2010–2011 Canterbury Earthquake Sequence. Earthq. Spectra **30**, 253–276 (2014). https://doi.org/10.1193/022113EQS044M

Mouyiannou, A., Penna, A., Rota, M., Graziotti, F., Magenes, G.: Implications of cumulated seismic damage on the seismic performance of unreinforced masonry buildings. Bull. New Zealand Soc. Earthq. Eng. **47**(2), 157–170 (2014). https://doi.org/10.5459/bnzsee.47.2.157-170

Paal, S.G., Jeon, J.-S., Brilakis, I., DesRoches, R.: Automated damage index estimation of reinforced concrete columns for post-earthquake evaluations. J. Struct. Eng. **141**(9), 04014228 (2014)

Park, Y.J., Ang, A.H.S.: Mechanistic seismic damage model for reinforced concrete. J. Struct. Eng. **111**, 722–739 (1985)

Park, Y.J., Ang, A.H.S., Wen, Y.K.: Damage-limiting aseismic design of buildings. Earthq. Spectra. **3**(1), 26 (1987). https://doi.org/10.1193/1.1585416

Penna, A., Morandi, P., Rota, M., Manzini, C.F., da Porto, F., Magenes, G.: Performance of masonry buildings during the Emilia 2012 earthquake. Bull. Earthq. Eng. **12**(5), 2255–2273 (2013). https://doi.org/10.1007/s10518-013-9496-6

Penna, A., Senaldi, I.E., Galasco, A., Magenes, G.: Numerical simulation of shaking table tests on full-scale stone masonry buildings. Int. J. Architectural Heritage **10**(2–3), 146–163 (2016). https://doi.org/10.1080/15583058.2015.1113338

Perus, I., Biskinis, D., Fajfar, P., Fardis, M. N., Grammatikou, S., Krawinkler, H., Lignos, D.: THE SERIES DATABASE OF RC ELEMENTS. 2nd European Conference on Earthquake Engineering and Seismology. Istanbul (2014). http://www.eaee.org/Media/Default/2ECCES/2ecces_eaee/145.pdf

Rinaldin, G., Amadio, C.: Effects of seismic sequences on masonry structures. Eng. Struct. **166**, 227–239 (2018). https://doi.org/10.1016/j.engstruct.2018.03.092

Salmanpour, A.H., Mojsilovic, N., Schwartz, J.: Deformation capacity of unreinforced masonry walls subjected to in-plane loading: a state-of-the-art review. Int. J. Adv. Struct. Eng. (IJASE) **5**(1), 1–12 (2013). https://doi.org/10.1186/2008-6695-5-22

Senaldi, I., et al.: Experimental seismic response of a half-scale stone masonry building aggregate: effects of retrofit strategies. In: Aguilar, R., Torrealva, D., Moreira, S., Pando, M.A., Ramos, L.F. (eds.) Structural Analysis of Historical Constructions. RB, vol. 18, pp. 1372–1381. Springer, Cham (2019). https://doi.org/10.1007/978-3-319-99441-3_147

Sextos, A., et al.: Local site effects and incremental damage of buildings during the 2016 Central Italy Earthquake sequence. Earthq. Spectra **34**, 1639–1669 (2018). https://doi.org/10.1193/100317EQS194M

Shcherbakov, R., Nguyen, M., Quigley, M.: Statistical analysis of the 2010 MW 7.1 Darfield Earthquake aftershock sequence. New Zealand J. Geol. Geophys. **55**(3), 305–311 (2012). https://doi.org/10.1080/00288306.2012.676556

Shen, J., Zhang, Y., Chen, J.: Vulnerability assessment and collapse simulation of unreinforced masonry structures subjected to sequential ground motions. Bull. Earthq. Eng. **20**, 8151–8177 (2022). https://doi.org/10.1007/s10518-022-01509-6

Šipčić, N., Kohrangi, M., Papadopoulos, A.N., Marzocchi, W., Bazzurro, P.: The Effect of Seismic Sequences in Probabilistic Seismic Hazard Analysis. Bul. Seismol. Soc. Am. XX **1–16**, 1694–1709 (2022). https://doi.org/10.1785/0120210208

Sivaramakrishnan, B.: Non-linear modeling parameters for reinforced concrete columns subjected to seismic loads. (Masters Thesis). University of Texas at Austin, Austin, TX (2010)

Vanin, F., Zaganelli, D., Penna, A., Beyer, K.: Estimates for the stiffness, strength and drift capacity of stone masonry walls based on 123 quasi-static cyclic tests reported in the literature. Bull. Earthq. Eng. **15**(12), 5435–5479 (2017). https://doi.org/10.1007/s10518-017-0188-5

Washington, D.C.: FEMA 306 Evaluation Of Earthquake Damaged Concrete And Masonry Wall Buildings Basic Procedures Manual Applied Technology Council (ATC-43 Project) The Partnership for Response and Recovery (1998)

Wilding, B.V., Dolatshahi, K.M., Beyer, K.: Influence of load history on the force-displacement response of in-plane loaded unreinforced masonry walls. Eng. Struct. **152**, 671–682 (2017). https://doi.org/10.1016/j.engstruct.2017.09.038

Yeo, G.L., Cornell, C.A.: A probabilistic framework for quantification of aftershock ground-motion hazard in California: methodology and parametric study. Earthq. Eng. Struct. Dynam. **38**, 45–60 (2009). https://doi.org/10.1002/eqe.840

# New Energy Based Metrics to Evaluate Building Seismic Capacity

Giulio Augusto Tropea[1(✉)], Giulia Angelucci[1], Davide Bernardini[1], Giuseppe Quaranta[2], and Fabrizio Mollaioli[1]

[1] Department of Structural and Geotechnical Engineering, Sapienza University of Rome, Via Gramsci 53, 00197 Rome, Italy
{augusto.tropea,giulia.angelucci,davide.bernardini, fabrizio.mollaioli}@uniroma1.it

[2] Department of Structural and Geotechnical Engineering, Sapienza University of Rome, Via Eudossiana 18, 00184 Rome, Italy
giuseppe.quaranta@uniroma1.it

**Abstract.** Since the advent of Performance-Based Earthquake Engineering (PBEE), increasingly effective methodologies have been investigated in characterizing capacity and identifying collapse, in particular. The possibility of accurately defining a limit state with a number, by means of damage indices, has always shown greater attractiveness, given the intrinsic capacity to simplify a complex problem. In recent years, methodologies based on an energy approach have shown great reliability in characterizing the collapse. A new metric aimed at characterizing damage is presented, developed starting from the idea that it can be represented in a measurable space taking into account simultaneously the internal energy and the internal power. In other words, what is essential in characterizing the damage is not represented only by the energy dissipation, but also the velocity at which the energy is imparted to the structure. The energy, as a measure of plastic work, with its first derivative (power) were used as parameters to build a damage index that would represent the dissipative state of the system in a measurable space. The theory, which was developed starting from a series of simple oscillators, was verified on a sample of experimental tests.

**Keywords:** Structural Damage Characterization · Energy Dissipation Capacity · Energy-Based Seismic Design · Nonlinear analysis · Seismic performance design

## 1 Introduction

Having a prediction of structural performance is essential both in the design of new structures and in the assessmentof existing ones. This concept is at the basis of Performance-Based Earthquake Engineering (PBEE) [1], which has led to the search for ever more efficient methodologies to characterize the structural dynamic response. The regulations and guidelines suggest design criteria based on the achievement of pre-established performance objectives [2]. In the context of Performance-Based Design (PBD), numerous


H. Varum et al. (Eds.): IWEBSE 2023, LNCE 236, pp. 72–86, 2023.
https://doi.org/10.1007/978-3-031-36562-1_6

damage indices (DIs) [3–7] have been introduced in recent years, aimed at characterizing the response numerically, in particular with the aim of locating the collapse and damage states. These indices have been designed to represent the response globally or locally, and for this reason they can be divided into global or local damage indices [8]. However, none of the proposals present in the literature manages to give a universally acceptable definition of collapse and state of damage. Some neglect the path necessary to reach a given level of performance, others do not consider the effects of cumulative plastic deformation and fatigue effects, which make displacement-based methods unreliable for these types of cases [9–11]. To overcome these limitations, indices based on the degradation of stiffness and strength, which occurs for numerous load reversals, or cumulative approaches aimed at characterizing fatigue have been proposed [12]. Energy approaches have shown greater effectiveness in describing all these aspects simultaneously, based on the energy balance of the system [13–25]. From the studies carried out to verify the reliability of the damage indices as a function of the actual collapse of the buildings, a very critical picture emerged, particularly in the definition of the collapse, due to the subjectivity of the limit state thresholds, the lack of a physical interpretation of the phenomenon and therefore from the consequent excessive dependence on the predictive capabilities of the adopted structural model. The need therefore emerges to define a physical description of the structural response which also takes into account the velocityofchange of the response. The works that have expressed in this direction have shown great solidity, but the current state of the art is still far from allowing an exhaustive representation of the damage and a reliable identification of the collapse that can be universally shared and applicable to different materials and models structural. Therefore, triggered by the enormous potential of energy theories, this work proposes a new generalized metric for the description of the structural response, which avoids approximations or arbitrary interpretations of the physical phenomenon to be characterized. For this purpose, the following work takes into account not only the internal energy, linked to the hysteretic behavior, but also its dissipation rate, i.e. the internal power [26]. Comparisons are then made on numerical predictions and on experimental results.

## 2 Energy Based Seismic Design Basis

An energetic approach has a number of advantages in describing the seismic problem. Energy is a scalar measure, it is not influenced by the direction of the actionand it contains information about the amplitude, frequency, and duration of the signal at the same time. The energetic analysis of the seismic structural response can be carried out, in various ways, through different energy balances derived from the equations of motion. For a single degree of freedom system of mass $m$subjected to a seismic input defined by a time-history of acceleration $\ddot{u}_g(t)$, the equation of motion can be written in terms of the displacement $u(t)$ of the structure relative to the ground as

$$\mathrm{m\ddot{u}(t) + f(t) = -m\ddot{u}_g(t)} \tag{1}$$

The restoring force f is typicallygiven by the sum of viscous and hysteretic components.

$$\mathrm{f(t) = c\dot{u}(t) + f_H(t)} \tag{2}$$

Uang and Bertero [27] suggested that a useful power balance can be obtained by multiplying Eq. (1) by the velocity v(t) of the structure relative to the ground:

$$\underbrace{m\ddot{u}(t)v(t)}_{P_K(t)} + \underbrace{f(t)v(t)}_{P_a(t)} = \underbrace{-m\ddot{u}_g(t)v(t)}_{P_I(t)} \tag{3}$$

Equaiton (3) shows that, at every time t, the power $P_I$ of the inertia force (input power) must be balanced by the sum of the power $P_a$ spent by the restoring force (internal power) and of the relative kinetic power $P_K$. Integrating power balance eq, (3) with respect to time the following energy balance is obtained:

$$\underbrace{m\frac{v(t)^2}{2}}_{E_K(t)} + \underbrace{\int_0^t f(t)v(t)dt}_{E_a(t)} = \underbrace{-\int_0^t m\ddot{u}_g(t)v(t)dt}_{E_I(t)} \tag{4}$$

Which similarly shows that the total energy input from the ground motion $E_I(t)$ must be balanced by the sum of the relative kinetic energy $E_K(t)$ and by the work spent by the restoring force $E_a(t)$.

## 3 Distance in the Energy-Power Plane as a Metric of Damage

Internal power $P_a$ and the corresponding expended energy $E_a$ provide useful information about the system response. For example, in a conservative system internal work $E_a$ is spent to change the value of the stored elastic energy, whereas for a dissipative system, a part of the $E_a$ is converted into other forms of energy. In view of the following developments, it is useful to normalize Ea and Pawith respect to their maximum values over their time histories

$$\overline{E}_a(t) = \frac{E_a(t)}{\max(E_a)} \overline{P}_a(t) = \frac{P_a(t)}{\max(P_a)} \tag{5}$$

Moreover, it is helpful to define a new quantity D (Eq. 6) which represents the distance between the state of the system from the origin in the energy-power plane:

$$D = \sqrt{\overline{E}_a^2 + \overline{P}_a^2} \tag{6}$$

To discuss the relevance of such a quantity it is useful to illustrate the response, in the energy-power plane, of two single-degree of freedom systems subjected to the simplest harmonic excitation.

Figure 1 shows the response of apurely elastic system (Fig. 1), after the conclusion of the transient regime. Since energy and power oscillates at constant amplitude (Fig. 1d), the system follows a closed trajectory in the energy-power plane (Fig. 1e) which passes, at every cycle, through the origin and the distance D(Fig. 1f)oscillates with constant amplitudearound a fixed mean value.

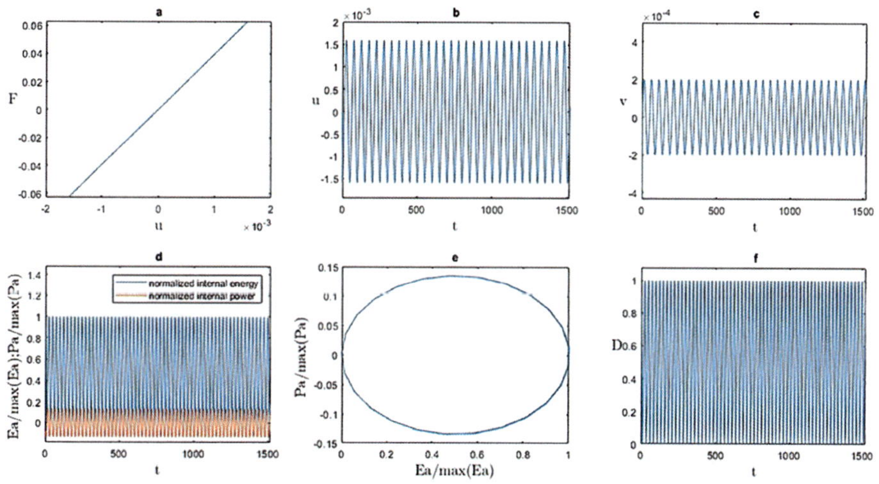


**Fig. 1.** Elastic Response of an SDOF to a constant amplitude sinusoidal signal.

On the other hand, Fig. 2 shows the response of an elasto-plastic system again to a harmonic excitation. In this case whereas power oscillates around a constant value energy Ea shows a clear drift due to the occurrence of a significant hysteretic dissipation (Fig. 2d). This is clearly reflected in the energy-power plan (Fig. 2e) where the systems follows a trajectory which progressively diverges away from the origin. Accordingly, the distance D (Fig. 2f) shows a clear trend towards greater values.

This shows that the distance D is strongly correlated with the dissipation of the system. Similar considerations can be made for the response of an elasto-plastic system subjected to an actual seismic input (Fig. 3).

In this case it is interesting to observe how in the $\overline{E}_a$, $\overline{P}_a$plane (Fig. 3e) the system follows various orbits progressively more distant from each other while approaching the end of the motion. Remarkably, in the last part of the motion ($t > 1500$) the distance D (Fig. 3f) attains an almost constant value while the system is still oscillating. This means that the system finally follows, in the energy-power plane, small orbits far from the origin due to the fact that the signal produced significant hysteretic dissipation. Since the final value of D is less than 1, the system has reached, during the motion energy peaks higher than those on which it stabilized at the end of the motion due to partial elastic recoveries. This can be interpreted as the fact that, despite having accumulated damage, the system has not collapsed. On the other hand if the final value of D would be close to 1 during the last part of the oscillation, this could be interpreted as a sign of collapse.

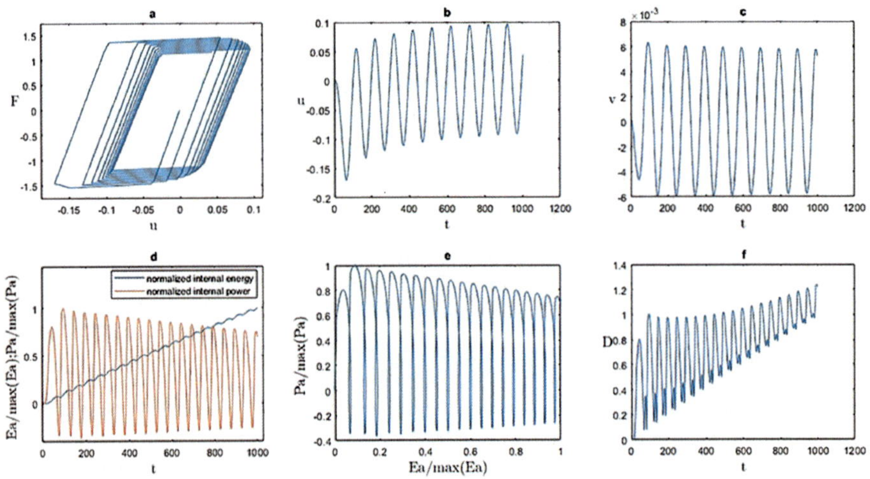

**Fig. 2.** Elastic-plasticresponse of an SDOF to a constant amplitude sinusoidal signal.

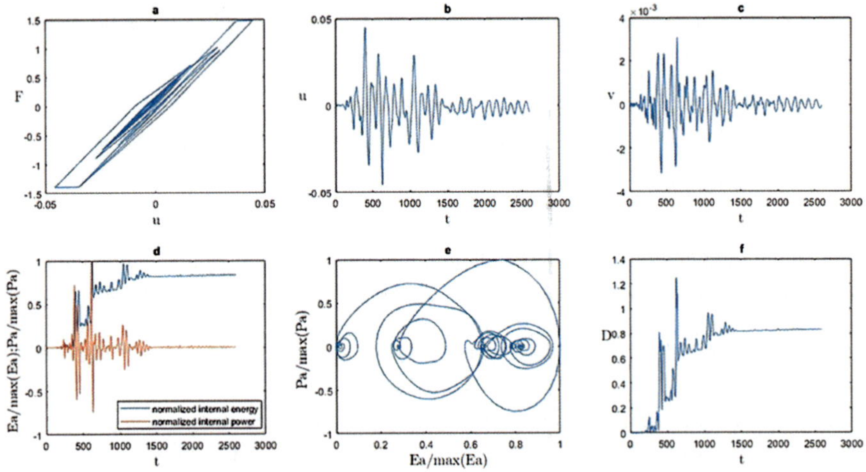

**Fig. 3.** Elastic-plastic responce of a SDOF to a natural signal.

### 3.1 Meaning of the Peaks of D

The two quantities Pa and Ea contain interesting information about on the hysteresis cycles. Figure 4 shows, in a schematic way, the correspondence between the trends of energy, power and the hysteresis cycle.

In the linear branch AB there is a rapid growth in energy and power, followed by a decrease in BC after yielding when the power decreases to zero. In the elastic recovery branch CD the energy decreases due to the partial energy recovery, and the power becomes negative up to thenew peak D. in the region DE a new rapid increase in energy that will reach a velocity peak in E, to then show a slowdown up to F and a new

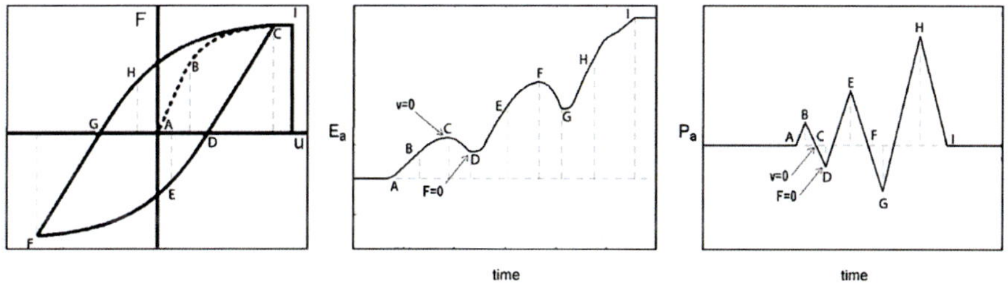


**Fig. 4.** Comparison between the hysteresis cycle, the absorbed energy Ea and its power Pa.

elastic recovery phase up to G. Point I, represents a structural collapse since the force return to zero when Ea is at its maximum value, and Pa is zero, at the end of the motion. These observations can be complex to process analytically and qualitatively, as the energy plateaus are not always so marked, since the systems can also show particularly small areas of cycles. Therefore, the need arises to condense this information into theparameter D which, can provide a qualitative and numerical evaluation of the dynamic problem. In Fig. 5a) we see the function $\overline{P}_a = f(\overline{E}_a)$ obtained by relating $\overline{E}_a$and $\overline{P}_a$with the correspondence of the points with Fig. 4.

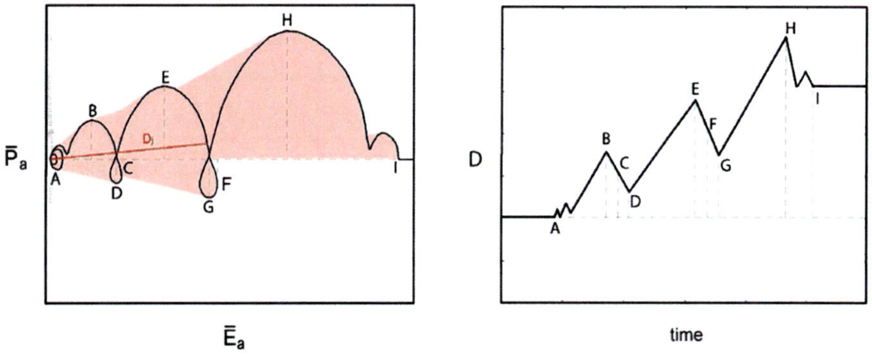


**Fig. 5.** Corrispondence between $\overline{E}_a$, $\overline{P}_a$plane and D values. Pink area is the dominium of D possible values in $\overline{E}_a$, $\overline{P}_a$plane.

The negative part of the cycles indicates a phase with negative velocity, therefore of elastic recovery, while the positive part indicates a phase of positive velocity and therefore an increase in dissipation. It can be observed that also the values of $\overline{E}_a$ decrease in correspondence with the negative values of $\overline{P}_a$ and then return to increase when $\overline{P}_a$ becomes positive again. At the beginning, it can be observed that the $\overline{E}_a\overline{P}_a$ function is concentrated around the starting point describing ever wider cycles and then begin to move away definitively without going back to the origin. This phase correspond to the elastic response. In Fig. 5a) a j-th value of D is shown and in pink the domain of the values assumed by D from the origin to the collapse, while in Fig. 5b) we can see the corresponding trend of D with the corresponding points of Fig. 4. It can be observed that D contains all the accumulated information of Ea and Pa. Through D it will therefore be possible to know the quantity of energy dissipated and the velocity with which it will be dissipated. D micro peaks means anhigher number of micro elastic recovery during he single hysteretic cycle. A fractal tendency has been observed (Fig. 6) in the manifestation of micro peaks which can be used for a classification of materials that takes into account micro-phases of elastic recovery in the non-linear field, particularly evident in materials that show a degradation of stiffness such as reinforced concrete and masonry.

The number of peaks of D corresponds to the number of hysteretic particalcycles could be a relevant information concerning low-cycle fatigue. The amplitude of the single peak is proportional to its contribution to the achievement of damage.

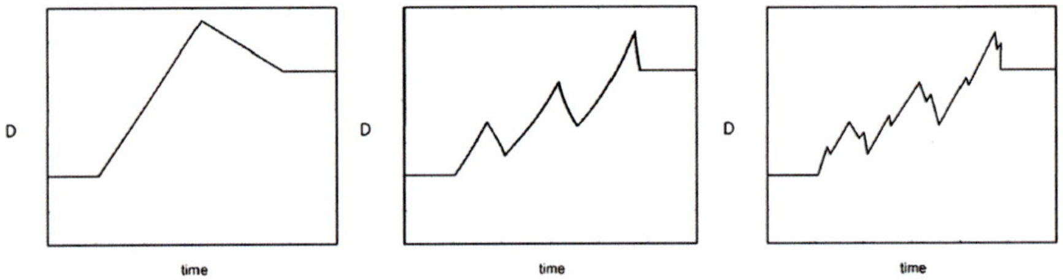


**Fig. 6.** Fractal development of the micro-peaks in D.

A significant interpretation of of D can be given as a measure of the state of damage. To discuss this aspect, Fig. 7 shows the response of an hysteretic SDOF system subjected to two different natural signals. Yielding point is obtained by the model and the correspondent Ea and Pa values are localized. Yielding point for the two systems is localized also on the $\overline{E}_a$, $\overline{P}_a$plane. In the energy-power plot a circle centered in the origin with a radius equal to 0.003 is drawn (Fig. 7). The circle intersect the two orbits in two different points which therefore identify two points that have similar level of damage but different valued of energy and power. This remark could be taken as basis for an heuristical interpretation of the distance D as a measure of damage.

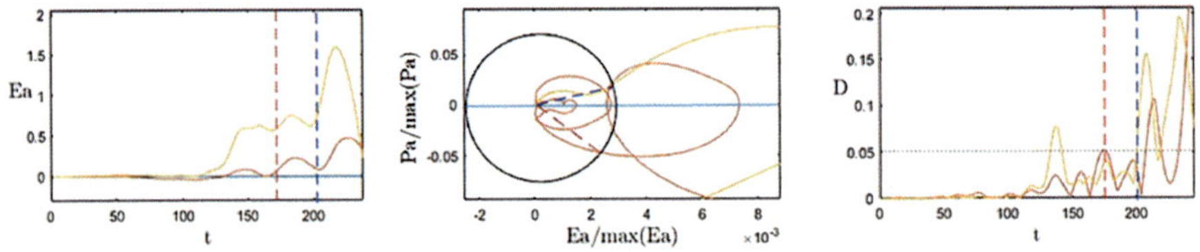


**Fig. 7.** Focus on the elastic phase responce of a SDOFmodelled with Ibarra Medina Krawinkler, subjected to a natural signal.

As it can be seen, this damage limit is reached for different Ea values in the two cases (red and blue line), while D has a very similar value for both cases under examination. In fact, it can be observed in the $\overline{E}_a$, $\overline{P}_a$plane that both the values of the two functions are equidistant from the origin. Therefore, for each system and each energy input, a neighborhood of Ea can be defined, defined by the corresponding Pa peak limits in which there is a similar dissipative condition. In other words, a smaller amount of energy dissipated at a higher velocity leads to the same damage state of a greater amount of energy dissipated at a lower velocity. Thus, a certain damage condition is not identified for a certain value of Ea but for a neighborhood of Ea, so D can be used as a better damage index.

## 4 Experimental Analysis

A sequence of experimental tests on different materials (Fig. 8), regular masonry [34], steel [33] and a set of tests on reinforced concrete pillars [28–32] is shown below. All the samples considered were subjected to increasing cyclic loading until the loss of bearing capacity was reached, close to collapse. In general, this can be viewed from the time history of D which shows a response characterized by a series of peaks with an almost constant frequency, typical of a load history characterized by cycles at a constant frequency. All concrete specimens are shear provided and rectangular. The tabled data (Table 1) of the experimental tests considered are listed below.

Cases 5 and 6, relating to masonry and steel, show significantly lower D peaks than in the experimental tests conducted on reinforced concrete pillars. Cases 1 (Fig. 8) do not reach collapse and do not show a marked degrading trend. In fact, the average trend of D remains linear. It show a progressive incremental value of D.The amplitude of the single D peak shows the contribution of the cycle corresponding to the achievement of a certain state of damage.

Case 2 (Fig. 9) shows a series of jumps in D which correspond to the abrupt changes in the hysteresis loop related to a rapid loss of stiffness. In correspondence with the jumps of D, a wider peak than the others can be observed at the beginning of each new jump; this is because the transition from one specific stiffness condition to another is characterized by a much wider cycle than the others. The average value of D can provide a clue on the type of response shown by the component.

Sample 3 (Fig. 10) shows a series of significantly smaller peaks than in the other experimental tests, with an average linear trend up to the final stage, when it shows a sudden jump. In the hysteresis cycle, it can be seen that after a certain number of cycles the specimen loses the ability to show elastic recovery and goes towards collapse with two final cycles, which can be viewed in D as a final peak greater than the others and a non-linear final average trend. The collapse is reached with a single large final cycle. D in fact shows a series of small constant peaks and a large final jump, thus showing an average linear trend for the entire load history and a final non-linear jump. From these first tests described it can be deduced that D can be used to understand if the system has shown a degrading phase before the collapse. In fact, with no degradation, the hysteresis cycles will show an almost constant growth of their subtended area. In case of degradation, however, the area under them will suddenly grow according to a non-linear rule. This can be observed in D as an average linear trend if no degrading trend is shown and a non-linear average trend in case of degrading trend.

The specimen corresponding to Case 4 (Fig. 11) reachesthe collapse and D shows an average non-linear trend in correspondence of the degrading trend. Collapse is reached in a series of progressive cycles, with olnly a more pronounced cycle near the collapse,

**Table 1.** Experimental tests data.

| Number | 1 | 2 | 3 |
|---|---|---|---|
| **Name** | Ang et al. 1981, No. 3 | Sezen and Moehle No. 1 | Sezen and Moehle No. 4 |
| **Concrete Strength** | 23.6 (MPa) | 21.1 (MPa) | 21.8 (MPa) |
| **Transverse Steel** | Grade: 275;Yield Stress: 320 (MPa) Strength: 434 (MPa) | Grade: 414;Yield Stress: 476 (MPa) Strength: 724 (MPa) | Grade: 414;Yield Stress: 476 (MPa) Strength: 724 (MPa) |
| **Longitudinal Steel** | Grade: 380;Corner Yield Stress: 427 (MPa) Strength: 670 (MPa);Intermediate Yield Stress: 427 (MPa) Strength: 670 (MPa) | Grade: 414;Corner Yield Stress: 434.4 (MPa) Strength: 645 (MPa) Intermediate Yield Stress: 434.4 (MPa) Strength: 645 (MPa) | Grade: 414;Corner Yield Stress: 434.4 (MPa) Strength: 645 (MPa) Intermediate Yield Stress: 434.4 (MPa) Strength: 645 (MPa) |
| **Dimensions** | Width: 400 (mm) Depth: 400 (mm) | Width: 457.2 (mm) Depth: 457.2 (mm) | Width: 457.2 (mm) Depth: 457.2 (mm) |
| **Length** | L-Inflection: 1,600 (mm) L-Measured: 1,600 (mm) | L-Inflection: 1,473.2 (mm) L-Measured: 1,473.2 (mm) | L-Inflection: 1,473.2 (mm) L-Measured: 1,473.2 (mm) |
| **Axial Load** | 1,435 (kN) | 667 (kN) | 667 (kN) |
| **Diameter** | 16 (mm) corner bars \| 16 (mm) intermediate bars | 28.7 (mm) corner bars \| 28.7 (mm) intermediate bars | 28.7 (mm) corner bars \| 28.7 (mm) intermediate bars |
| **Number of Bars** | 12 | 8 | 8 |
| **Perpendicular to Load** | Clear cover: 24.5 (mm) Number of intermediate bars: 2 | Clear cover: 65.1 (mm) Number of intermediate bars: 1 | Clear cover: 65.1 (mm) Number of intermediate bars: 1 |
| **Parallel to Load** | Clear cover: 24.5 (mm) Number of intermediate bars: 2 | Clear cover: 65.1 (mm) Number of intermediate bars: 1 | Clear cover: 65.1 (mm) Number of intermediate bars: 1 |
| **Corner and intermediate** | reinfrocement ratio 0.0151 | reinfrocement ratio 0.0247 | reinfrocement ratio 0.0247 |
| **Region of Close Spacing** | Bar Diameter: 12 (mm) Hoop Sets: 5 Hoop Spacing: 100 (mm) | Bar Diameter: 9.5 (mm) Hoop Sets: 10 Hoop Spacing: 304.8 (mm) | Bar Diameter: 9.5 (mm) Hoop Sets: 10 Hoop Spacing: 304.8 (mm) |
| **Region of Wide Spacing** | | Bar Diameter: 9.5 (mm) Hoop Sets: 0 Hoop Spacing: 0 (mm) | Bar Diameter: 9.5 (mm) Hoop Sets: 0 Hoop Spacing: 0 (mm) |
| **Rectangular and interlocking ties** | reinfrocement ratio 0.028 | reinfrocement ratio 0.002 | reinfrocement ratio 0.002 |
| **Axial load ratio** | 0.38 | 0.151 | 0.146 |
| **Failure Type** | Flexure | Flexure-Shear | Flexure-Shear |
| **Damage Observation** | Concrete Crushing: 10 (mm); Significant Concrete Spalling: 20 (mm); Long Bar Buckling: 50 (mm); Long Bar Fracture: 50 (mm); | Concrete Crushing: 55.9 (mm);Long Bar Buckling: 139.7 (mm) | Long Bar Buckling: 139.7 (mm) |

*(continued)*

**Table 1.** (*continued*)

| Number | 4 | 5 | 6 |
|---|---|---|---|
| Name | Soesianawati et al. 1986, No. 4 | Andrea Penna et all, speciemen CS01 | Popov-EERC-70-3-Specimen1 |
| Concrete Strength | 40 (MPa) | | Material designation: ASTM A36; |
| Transverse Steel | Grade: 275;Yield Stress: 255 (MPa) Strength: 402 (MPa) | | |
| Longitudinal Steel | Grade: 380;Corner Yield Stress: 446 (MPa) Strength: 702 (MPa) Intermediate Yield Stress: 446 (MPa) Strength: 702 (MPa) | | beam size=W18x50; column size=W12X106; beam length to centerline of column (BLE)= 98.56 in; beam effective length=83.12 in; column length to centerline of the beam (CLC)=45 in; |
| Dimensions | Width: 400 (mm) Depth: 400 (mm) | Regular masonry wall; L = 2.50 m; t = 0.175m; h = 2.50 m; | |
| Length | L-Inflection: 1,600 (mm) L-Measured: 1,600 (mm) | | |
| Axial Load | 1,920 (kN) | Experimental conditions: σv = 1.0 Mpa | |
| Diameter | 16 (mm) corner bars \| 16 (mm) intermediate bars | | |
| Number of Bars | 12 | | |
| Perpendicular to Load | Clear cover: 13 (mm) Number of intermediate bars: 2 | | |
| Parallel to Load | Clear cover: 13 (mm) Number of intermediate bars: 2 | | |
| Corner and intermediate | reinfrocement ratio 0.0151 | | |
| Region of Close Spacing | Bar Diameter: 6 (mm) Hoop Sets: 7 Hoop Spacing: 94 (mm) | | |
| Region of Wide Spacing | | | |
| Rectangular and interlocking ties | reinfrocement ratio 0.006 | | |
| Axial load ratio | 0.3 | | |
| Failure Type | Flexure | Rocking-flexure | Fatigue |
| Damage Observation | Concrete Crushing: 16.4 (mm);Significant Concrete Spalling: 24.6 (mm);Long Bar Buckling: 41 (mm) | Cracking conditions: 12mm of top displacement | |

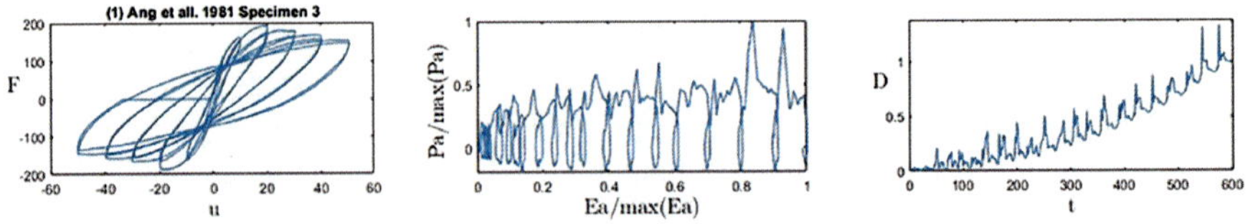


**Fig. 8.** Rectangular concrete column, incomplete flexural failure, low degradation.

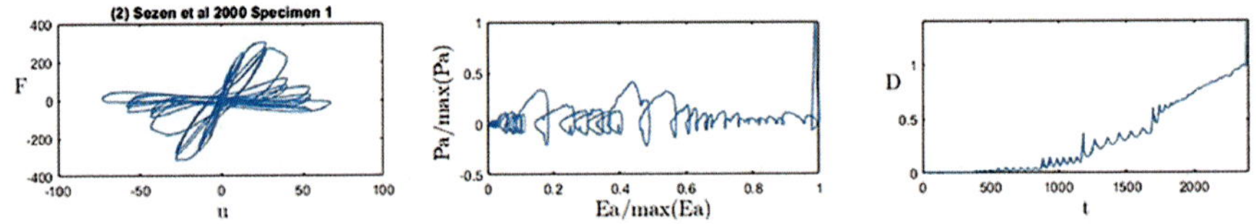


**Fig. 9.** Rectangular concrete column, flexural-shear failure, progressive severe degradation.

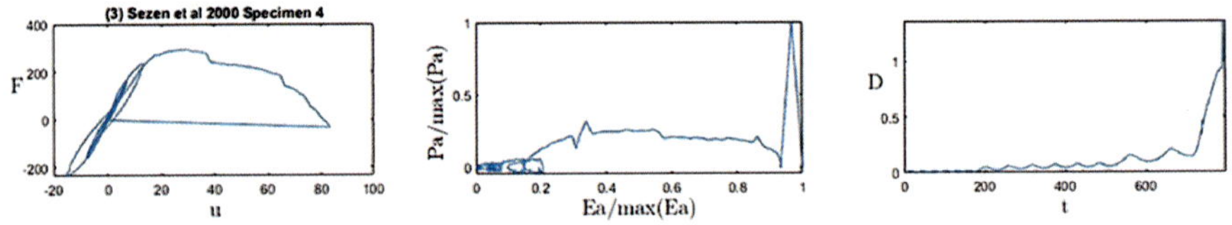


**Fig. 10.** Rectangular concrete column, flexural-shear failure, sudden severe degradation.

and therefore there are no appreciably larger D peaks than in the others towards the end. Major peaks followed by smaller peaks can be observed along the D loading history. These phases correspond to a variation in the structural response. This sequence can be well observed after the 250th time step. From the experimental observations, these phases correspond to the qualitative manifestation of a different damage condition.

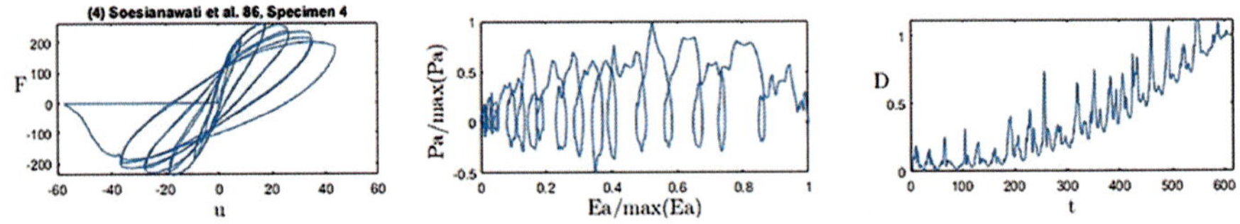


**Fig. 11.** Rectangular concrete column, flexural failure, low progressive degrading response for the first non linear phase, severe degradation near collapse.

Cases 5(Fig. 12) and 6(Fig. 13), relating to masonry and steel, show significantly lower D peaks than in the experimental tests conducted on reinforced concrete pillars.

In the cases of masonry and steel, in fact, the average trend remains linear until the collapse, shown in the case of masonry a brittle failure and in the case of steel a failure due to fatigue. Differently from the other cases described so far, D shows higher peaks, this is because the area under each cycle is greater than in the other cases. It is possible to observe in masonry case a higher peaks amplitude of D in correspondence with the

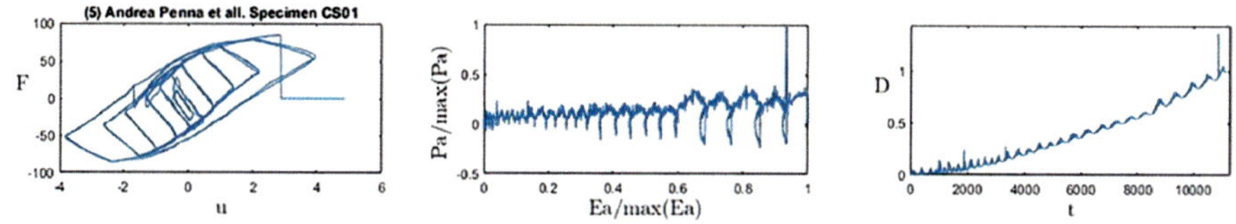


**Fig. 12.** Regular masonry wall, rocking-flexure failure, fragile behaviour.

degrading phase, before the collapse. Building a hysteresis curve could be complex on this kind of hysteresis loop, and approximative, so the information obtained by force-displacement method can be inexat. Collapse is localized for a displacement value lower than the maximum one, and after a final small non degrading phase showed between degranding phase and collapse.

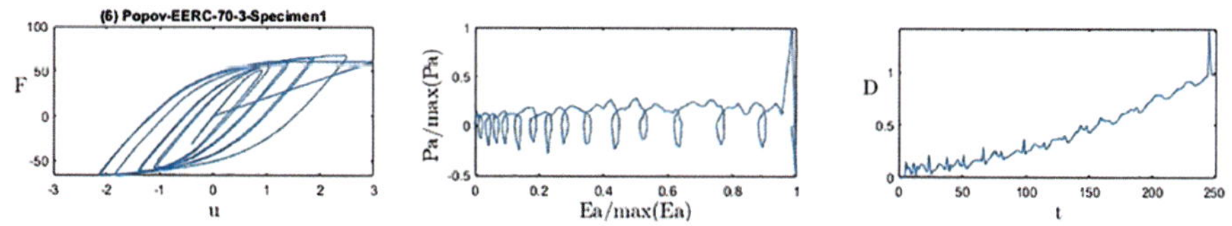


**Fig. 13.** Steel beam-column joint, fatigue failure.

In fatigue failure collapse arrive for displacement values lower than the maximum one. So in we use an hysteresis backbone to characterize the non linear behavior we localize the collapse in correspondence of a wrong value of displacement, because with the backbone method we localize the collapse in correspondence of the maximum displacement. Using D as a damage parameter we don't have this problem, because the last value of D is localized in correspondence with the maximum Ea expressed during the non linear behavior.

All cases except the first one reached a final value of D equal to 1, which means that the system has reached the collapse. In the first case the final value of D reaches a maximum of 0.98, showing that the specimens have reached a high degree of damage and close to collapse, but without collapsing. A behavior confirmed by the qualitative observation of the cycles of hysteresis.

## 5 Conclusions

The proposed methodology is based on the idea that it is possible to represent in a measurable space the dynamic response of a system to a seismic input through two variables (Ea, Pa) of the same energy quantity, the energy absorbed by the system in the form of elastic and plastic work. These two variables are processed to obtain a metrics, D, which condenses the information of Ea and Pa, which characterizes the damage. This metric contains the informations of the hysteretic cycle and give the possibility to characterize the collapse.

The theory is showed with a SDOFsystem and it is being verified on experimental tests on materials different in order to show their pervasiveness. The methodology is based only on energy, so the measures are scalar and are not influenced by load direction, costituive law, dynamic characteristics of the system and characteristics of the seismic signal. This study show that the collapse could not be proportional to the maximux displacement expressed by the system, like in the experimental steel case showed and that the limit states could be influenced by the velocity of immission of energy in the system. The proposed method allows to better characterize the nonlinear behavior of a system with respect to the method of forces and displacements. This is particularly evident in the fatigue failure observable in the experimental cases in the steel column, which is not captured with the backbone curve built on the hysteresis loop.

The plastic work could be calculated both globally and for each structural component. In turn, if the structural component is composed of several elements, as in the case of reinforced concrete, it is possible to evaluate the influence on the capacity of the component of the individual elements. It is therefore possible to carry out an analysis on several levels.

**Acknowledgements.** This work has been supported by the Italian Ministry of Education, University and Research (MIUR), Sapienza University of Rome (Grant n. RG11916B4C241B32) and the Italian Network of University Laboratories of Seismic Engineering (ReLUIS).

## References

1. SEAOC Vision 2000: "Performance based seismic engineering of buildings," vols. I and II: Conceptual framework. Structural Engineers Association of California, Sacramento (CA) (1995)
2. Ghobarah, A.: Performance-based design in earthquake engineering: state of development. Eng. Struct. **23**(8), 878–884 (2001). https://doi.org/10.1016/S0141-0296(01)00036-0
3. Ghobarah, A., Abou-Elfath, H., Biddah, A.: Response-based damage assessment of structures. Earthq. Eng. Struct. Dyn. **28**(1), 79–104 (1999)
4. Cosenza, E., Manfredi, G., Ramasco, R.: The use of damage functionals in earthquake engineering: a comparison between different methods. Earthq. Eng. Struct. Dyn. **22**(10), 855–868 (1993)
5. Bozorgnia, Y., Bertero, V.: Improved shaking and damage parameters for post-earthquake applications. In: SMIP01 Seminar on Utilization of Strong-Motion Data, pp. 1–22 (2001)
6. Cao, V., Ronagh, H.R., Ashraf, M., Baji, H.: A new damage index for reinforced concrete structures. Earthq. Struct. **6**(6), 581–609 (2014)
7. Whittaker, A.S., Deierlein, G., Hooper, J., Merovich, A.: Engineering Demand Parameters for Structural Framing. Redwood City, California (2004)
8. Williams, M.S., Sexsmith, R.G.: Seismic damage indices for concrete structures: a state-of-the-art review. Earthq. Spectra **11**(2), 319–349 (1995)
9. Bracci, J.M., Reinhorn, A.M., Mander, J.B., Kunnath, S.K.: Deterministic model for seismic damage evaluation of reinforced concrete structures. Technical Report NCEER-8-–0033 (1989)
10. ChungY. S., MeyerC., and ShinozukaM., "Seismic damage assessment of reinforced concrete members," Buffalo, NY, 1987

11. DiPasquale, E., Cakmak, A.S.: Seismic damage assessment using linear models. Soil Dyn. Earthquake Eng. **9**(4), 194–215 (1990)
12. Sozen, M.A.: Review of earthquake response of reinforced concrete buildings with a view to drift control. State-of-the-art in Earthquake Eng. 383–418 (1981)
13. Bojórquez, E., Reyes-Salazar, A., Terán-Gilmore, A., Ruiz, S.E.: Energy-based damage index for steel structures. Steel Compos. Struct. **10**(4), 331–348 (2010)
14. Arroyo, D., Ordaz, M.: On the estimation of hysteretic energy demands for SDOF systems. Earthq. Eng. Struct. Dyn. **36**(15), 2365–2382 (2007)
15. Decanini, L.D., Mollaioli, F.: An energy-based methodology for the assessment of seismic demand. Soil Dyn. Earthq. Eng. **21**(2), 113–137 (2001). https://doi.org/10.1016/S0267-7261(00)00102-0
16. Quaranta, G., Mollaioli, F.: Analysis of near-fault pulse-like seismic signals through Variational Mode Decomposition technique. Eng. Struct. **193**(March), 121–135 (2019). https://doi.org/10.1016/j.engstruct.2019.05.003
17. Mollaioli, F., Bruno, S., Decanini, L., Saragoni, R.: Correlations between energy and displacement demands for performance-based seismic engineering. Pure Appli. Geophys. **168**(1–2), 237–259 (2011). https://doi.org/10.1007/s00024-010-0118-9
18. Chou, C., Uang, C.: A procedure for evaluating seismic energy demand of framed structures. Earthq. Eng. Struct. Dyn. **32**(2), 229–244 (2003)
19. Benavent-Climent, A., Zahran, R.: An energy-based procedure for the assessment of seismic capacity of existing frames: Application to RC wide beam systems in Spain. Soil Dyn. Earthq. Eng. **30**(5), 354–367 (2010). https://doi.org/10.1016/j.soildyn.2009.12.008
20. Teran-Gilmore, A., Jirsa, J.O.: Energy demands for seismic design against low-cycle fatigue. Earthq. Eng. Struct. Dyn. **36**(3), 383–404 (2007)
21. Benavent-Climent, A.: An energy-based damage model for seismic response of steel structures. Earthq. Eng. Struct. Dyn. **36**(8), 1049–1064 (2007). https://doi.org/10.1002/eqe.671
22. Deniz, D., Song, J., Hajjar, J.F.: Energy-based seismic collapse criterion for ductile planar structural frames. Eng. Struct. **141**, 1–13 (2017). https://doi.org/10.1016/j.engstruct.2017.02.051
23. Zhou, H., Li, J.: Effective energy criterion for collapse of deteriorating structural systems. J. Eng. Mech. **143**(12), 04017135 (2017)
24. Szyniszewski, S., Krauthammer, T.: Energy flow in progressive collapse of steel framed buildings. Eng. Struct. **42**, 142–153 (2012). https://doi.org/10.1016/j.engstruct.2012.04.014
25. Meigooni, F.S., Mollaioli, F.: Simulation of seismic collapse of simple structures with energy-based procedures. Soil Dyn. Earthq. Eng. **145**, 106733 (2021). https://doi.org/10.1016/j.soildyn.2021.106733
26. Pathak, S., Khennane, A., Al-Deen, S.: Power based seismic collapse criterion for ductile and non-ductile framed structures. Bull. Earthq. Eng. **18**(13), 5983–6014 (2020). https://doi.org/10.1007/s10518-020-00925-w
27. Uang, C., Bertero, V.: Evaluation of seismic energy in structures. Earthq. Eng. Struct. Dyn. **19**(1), 77–90 (1990)
28. Zahn, F.A., Park, R., Priestley, M.J.N.: Design of Reinforced Concrete Bridge Columns for Strength and Ductility, Report 86-7, 380 p. Department of Civil Engineering, University of Canterbury, Christchurch, New Zealand (1986)
29. Sezen, H., Moehle, J.P.; Seismic behavior of shear-critical reinforced concrete building columns. In: Seventh U.S. National Conference on Earthquake Engineering, Boston, Massachusetts, 21–25 July 2002
30. Soesianawati, M.T., Park, R., Priestley, M.J.N.: Limited Ductility Design of Reinforced Concrete Columns. Report 86-10, 208 p. Department of Civil Engineering, University of Canterbury, Christchurch, New Zealand (1986)

31. Ang, B.G., Priestley, M.J.N., Park, R.: Ductility of Reinforced Bridge Piers Under Seismic Loading. Report 81-3, 109 p. Department of Civil Engineering, University of Canterbury, Christchurch, New Zealand (1981)
32. Gill, W.D., Park, R., Priestley, M.J.N.: Ductility of Rectangular Reinforced Concrete Columns With Axial Load. Report 79-1, 136 p. Department of Civil Engineering, University of Canterbury, Christchurch, New Zealand (1979)
33. PopovEgor P. (1983), "seismic moment connections for moment-resisting steel frames". Report to Sponsors the American Iron and Steel Institute National Science Foundation
34. Magenes, G., Morandi, P., Penna, A.: In-plane cyclic tests of calcium silicate masonry walls. Periodic activity report on ESECMaSE (2008)

# Relation Between Input Energy and Equivalent Monotonic Response Curve

C. Leangheng[1(✉)], P. Doung[2], and S. Leelataviwat[1]

[1] King Mongkut's University of Technology Thonburi, Bangkok, Thailand
chea.lean@mail.kmutt.ac.th
[2] Institute of Technology of Cambodia, Phnom Penh, Cambodia

**Abstract.** This study investigates the relation between the input energy and the equivalent monotonic response curve. The relationship is represented by a parameter called the energy factor, which is defined as the ratio of the energy required to push the structure under monotonic loading up to the maximum displacement due to ground motion to the elastic input energy. Single-degree-of-freedom systems were analyzed under selected ground motions. The input energy and the maximum inelastic response were used to calculate the energy factor for each case. The results indicated that the obtained energy factor varied mainly according to the ductility level and were fairly constant except in a short period range. An expression was then derived to predict the energy factor as a function of period, strength, and ductility level. By using the energy factor, the ties between the structural strength, the maximum deformation, and input energy are established. The concept can then be used to design a structure, given the input energy, by calculating the required strength to limit the deformation within the selected target. Finally, the application of the energy factor in an energy-based plastic design of a structure is illustrated.

**Keywords:** Energy factor · energy-based plastic design · monotonic response

## 1 Introduction

It is well known that structures designed by elastic, force-based methods may undergo large inelastic deformation during major earthquakes. To achieve more predictable structural performance, strength, deformation, and energy dissipation must become an integral part of the design process. One energy-based design concept that has been used since the early days of energy-based design development was the energy balance concept pioneered by Housner (1956). In this concept, the energy that contributes to damage (assumed as the elastic vibrational energy $\frac{1}{2}MS_v^2$, where $M$ is the mass and $S_v$ is the pseudo-velocity) is taken equal to the energy under the monotonic response curve. The energy balance concept was later modified by Lee and Goel (2001). In the modified energy balance concept, the energy balance is applied between the fraction of the energy $\frac{1}{2}MS_v^2$ and the work needed to push the structure monotonically up to the target maximum displacement. A new parameter called the energy factor was introduced. This


H. Varum et al. (Eds.): IWEBSE 2023, LNCE 236, pp. 87–97, 2023.
https://doi.org/10.1007/978-3-031-36562-1_7

factor is defined as the ratio of the energy required to push the structure under monotonic loading up to the maximum displacement due to a ground motion to the energy $\frac{1}{2}MS_v^2$. By introducing the energy factor, the ties between the structural strength, the maximum deformation, and input energy are established. The concept can be used to design a structure, given the input energy. Several studies have been conducted on the energy factor and its relationship to other system properties. Predictive models have also been proposed (Lee and Goel 2001, Merter and Uçar 2020, Uçar and Merter 2022). The predictive equations for the energy factor model were also established for the non-EPP hysteretic behavior and were a subject of research in various studies (Ke et al. 2016, Zhai et al. 2017, Ke et al. 2018, Wang et al. 2018, Ji et al. 2020, Zhou et al. 2021).

In recent years, the input energy has been a subject of extensive research. Earthquake input energy has been studied in detail, and several parameters that affect input energy, such as the peak ground acceleration (PGA), the peak ground velocity (PGV), the predominant periods, damping ratio, ductility factor, etc., were identified (Gholami et al. 2022). Equations have also been proposed for estimating the energy input. Hence, it is now able to express the input energy more precisely, given key ground motion parameters. The energy balance concept can now be utilized using the elastic input energy from the ground motion. This study investigates the relation between the elastic input energy and the equivalent monotonic response curve. A parameter similar to the energy factor was developed. The new energy factor is defined as the ratio of the energy under the monotonic response curve to the elastic input energy. Single-degree-of-freedom (SDOF) systems were analyzed under selected ground motions. The input energy and the maximum inelastic response were used to calculate the energy factor for each case. An expression was then derived to predict the energy factor as a function of period, strength, and ductility level. Finally, the application of the energy factor in an energy-based design of example structures is illustrated.

## 2 Relation Between Input Energy and Equivalent Monotonic Response Curve

The well-known energy contents of an SDOF system subjected to seismic input can be obtained by integrating the governing equation of motion

$$\int_0^t m\ddot{u}\dot{u}dt + \int_0^t c\dot{u}\dot{u}dt + \int_0^t ku\dot{u}dt = -\int_0^t m\ddot{u}_g(t)\dot{u}dt \tag{1}$$

for an elastic system, and

$$\int_0^t m\ddot{u}\dot{u}dt + \int_0^t c\dot{u}\dot{u}dt + \int_0^t f_s(u)\dot{u}dt = -\int_0^t m\ddot{u}_g(t)\dot{u}dt \tag{2}$$

for an inelastic system, where $m$, $k$, $c$, and $f_s$(u). Are the mass, stiffness, damping coefficient, and restoring force, $u$, $\dot{u}$, and $\ddot{u}$ are the relative displacement, velocity, and acceleration, and $\ddot{u}_g$(t) is the ground acceleration. The right-hand side of the above equations represents the input energy. In this study, the relation between the elastic input energy and the equivalent monotonic response curve is investigated. The relationship is represented by a parameter called the energy factor, which is defined as the ratio of the energy required to push the structure under monotonic loading up to the maximum displacement due to ground motion to the elastic input energy:

$$\gamma^* = \frac{E_{mono}}{E_I'} \tag{3}$$

where $\gamma^*$ is called the energy factor, $E_I'$ is the input energy of the elastic system, and $E_{\text{mono}}$ is the energy of the equivalent monotonic response. The energy factor is illustrated in Fig. 1. The energy factor provides the ties between the structural strength, the maximum deformation, and the input energy. Once the energy factor has been established, a structure can be designed given the input energy and the required target maximum deformation.

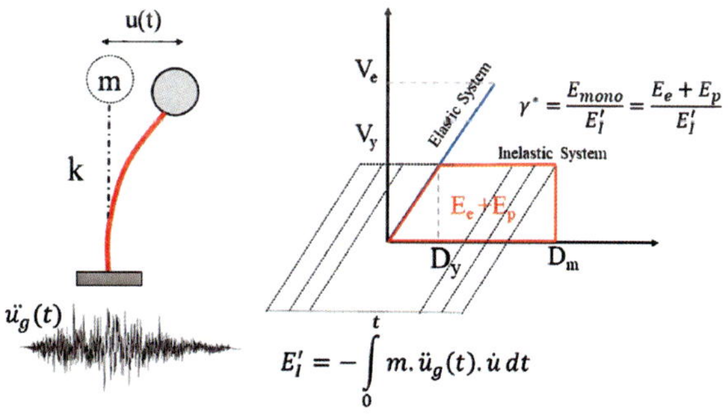


**Fig. 1.** Concept of Energy Factor

## 3 Nonlinear Time History Analysis of SDOFs

SDOF systems were analyzed under selected ground motions. The input energy and the maximum inelastic response were used to calculate the energy factor for each ground motion. Two sets of ground motions (Chopra and Chintanapakdee 2004) representing large-magnitude-small-distance (LMSR) and large-magnitude-large-distance (LMLR) ground motions were used. The LMSR and LMLR sets included ground motions with the magnitudes between 6.6–6.9 and distances of 13–30 km and 30–60 km for the LMSR and LMLR sets, respectively. All records were selected from earthquakes in California, USA, with a site condition classified by $V_{s30}$ as site class D (stiff soil with 180 m/s < $V_{S30} \leq$ 360 m/s) (Chopra and Chintanapakdee 2004). The response spectra of the two sets of ground motions are shown in Fig. 2.

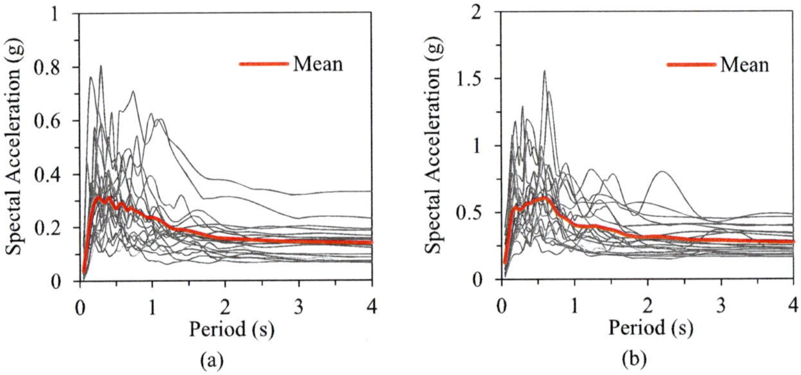


**Fig. 2.** Mean spectral acceleration: (a) LMLR and (b) LMSR ground motions

An elastic-perfectly-plastic (EPP) hysteretic model was specified together with a damping ratio of 5%. Response analyses were carried out for ductility ratios $\mu = 2$, 3 and 4 using BiSpec V2.2 software (Hachem 2012). For each response case, the monotonic response energy and the elastic input energy of the corresponding elastic system were calculated. The mean values for the energy factor are shown in Fig. 3a and 3b. The results indicated that the energy factor varied according to the ductility level and was fairly constant except in a short period range. As the ductility ratio increased, the value of the energy factor tended to decrease. The energy factor was extremely sensitive at a short period range but became fairly constant after the period of 0.2 s. Hence, within the practical period range for most building structures, the energy factor could be considered as almost constant.

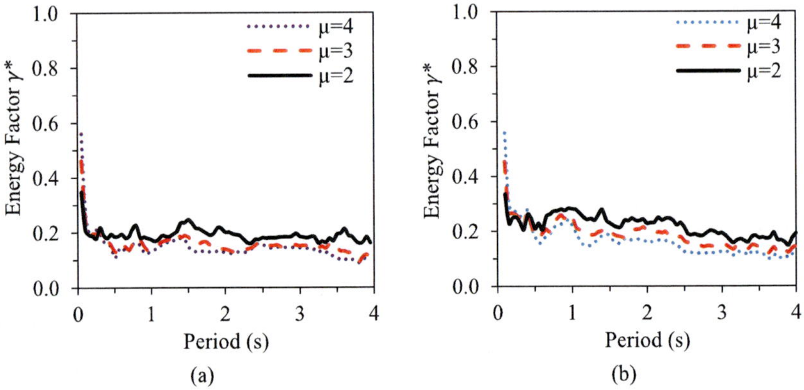


**Fig. 3.** Energy factor from nonlinear time history analysis with different ductility levels: (a) LMLR and (b) LMSR ground motions

## 4 Estimation of the Energy Factor

The energy factor can be estimated from the input energy and from the monotonic response curve given the ductility and the structural strength (Lee and Goel 2001). The input energy demand on a linear elastic SDOF can be computed as:

$$E_I' = \frac{1}{2} M V_{eq}^2 \tag{4}$$

where $V_{eq}$ is the equivalent velocity. A recent study by Alici and Sucuoğlu (2017) indicated that the equivalent velocity is correlated with the pseudo velocity ($S_v$). The ratio $V_{eq}/S_v$ is represented in this study by $\eta$:

$$\frac{V_{eq}}{S_v} = \eta \tag{5}$$

where $\eta$ is a period- and damping-dependent function. The function $\eta$ as proposed by Alici and Sucuoğlu (2017) can be written as:

$$\eta = a.e^{-bT} + c \tag{6}$$

where $T$ is the period of the structure and *a, b, and c* are numerical constant factors depending on the period and damping ratio. Using Eq. (5), the input energy can be expressed as a function of $S_v$

$$E_I' = \frac{1}{2} \eta^2 M S_v^2 \tag{7}$$

Substituting Eq. (7) into the equation expressing the energy factor, the following equation can be obtained:

$$\gamma^* = \frac{E_e + E_p}{E_I'} = \frac{\frac{1}{2} V_y D_y + V_y (D_m - D_y)}{\eta^2 \frac{1}{2} M S_v^2} = \frac{\frac{1}{2} V_y D_y + V_y (D_m - D_y)}{\eta^2 \frac{1}{2} V_e D_e} \tag{8}$$

where $M$ is the mass of the system, $V_y$ is the yield strength, $D_y$ is the yield displacement, $D_m$ is the maximum inelastic displacement, $V_e$ is the required strength of the corresponding elastic system, and $D_e$ is the maximum displacement of the corresponding elastic system. Expressing the above equation in terms of the well-known ductility ratio ($\mu = D_m/D_y$) and the yield strength reduction factor ($R_\mu = V_e/V_y$.) gives (Lee and Goel 2001)

$$\gamma^* = \frac{2\mu - 1}{\eta^2 R_\mu^2}. \tag{9}$$

Given the period and the required ductility, the corresponding yield strength reduction factor ($R_\mu$) can be obtained from any $R_\mu$-$\mu$-$T$ relationship. In this study, the well-known $R_\mu$-$\mu$-$T$ relationship proposed by Newmark and Hall (1982) was utilized to compute the energy factor. The values of the energy factor computed using Eq. (9) are compared with the values from the time history analysis results in Figs. 4 and 5 for the LMLR and LMSR ground motions.

Overall, Eq. (9) predicts the values of the energy factor reasonably well. The energy factor from Eq. (9) remains fairly constant except in the short period range. The small peaks around the period of 0.5 s. were due primarily to the $R_\mu$-$\mu$-$T$ relationship used in calculating t energy factor. It should be noted that other smooth $R_\mu$-$\mu$-$T$ relationship functions could also be used. In addition, the soil type could also affect the energy factor. $R_\mu$-$\mu$-$T$ relationship functions developed for other soil types could also be applied. The values of the energy factor from Eq. (9) based on the $R_\mu$-$\mu$-$T$ relationship proposed by Newmark and Hall (1982) slightly underestimates the mean energy factor values for LMSR ground motion case, especially for the ductility ratio of 4. However, in actual design, the target ductility ratio is normally kept less than 4. Therefore, the predictive equation is sufficiently accurate for practical purposes.

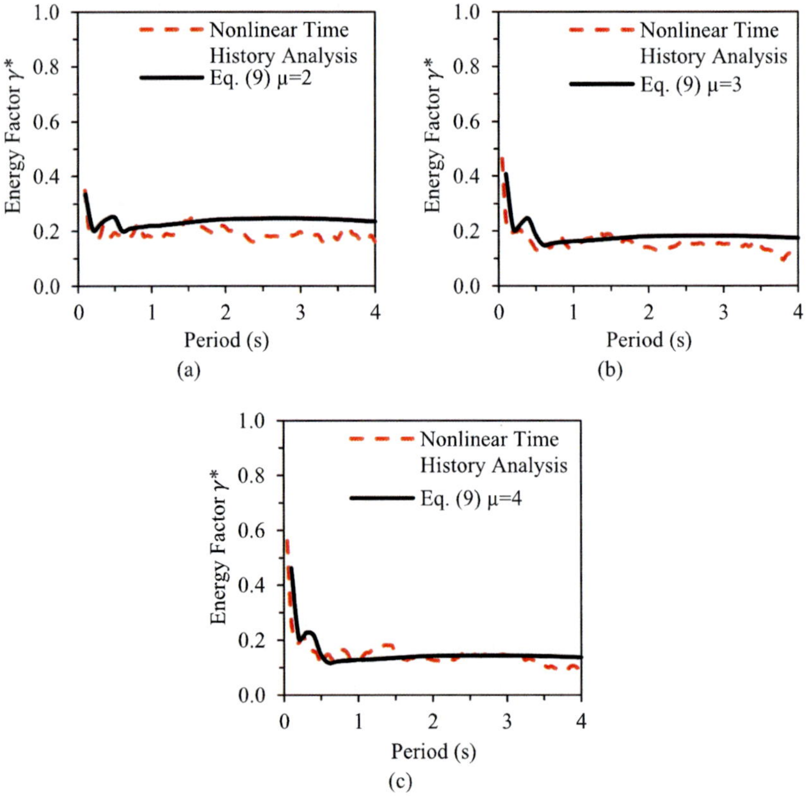


**Fig. 4.** Computed and predicted energy factor spectra for LMLR ground motions: (a) ductility ratio $\mu = 2$; (b) ductility ratio $\mu = 3$; and (c) ductility ratio $\mu = 4$

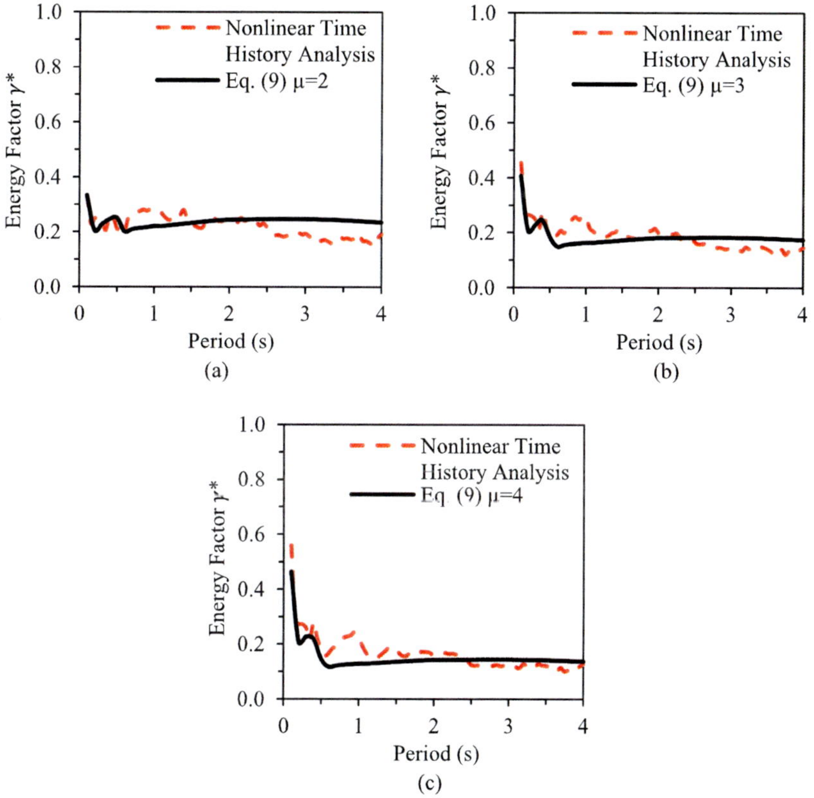


**Fig. 5.** Computed and predicted energy factor spectra for LMSR ground motions; (a) ductility ratio $\mu = 2$; (b) ductility ratio $\mu = 3$; and (c) ductility ratio $\mu = 4$

## 5 Application

The above energy balance concept, with some simplification, can be directly applied in the seismic design of building structures. The energy balance concept presented above is strictly applied only for the SDOF system. The energy contents and energy balance concept for multi-degree-of-freedom (MDOF) systems are still a subject of current research. A previous study has indicated that the concept can be applied to individual modes of the MDOF system (Leelataviwat et al. 2009). However, combining the individual modal energy results becomes complex for inelastic MDOF systems and warrants further research. Since it is normally agreed that the first mode dominates the response, especially for low-rise structures, the above energy concept can be applied to the entire structure for low-rise buildings. The higher mode contribution is compensated by assuming that the entire mass of the structures is used in the energy balance equation:

$$\gamma^* \frac{1}{2} M V_{eq}^2 = \frac{1}{2} V_y u_y^* + V_y (u_m^* - u_y^*) \tag{10}$$

where $M$ is the mass of the structural system and $u_y^*$ and $u_m^*$ are the displacements at the center of the restoring force at the yield and maximum states, respectively. The energy balance concept for an MDOF steel frame system with a sway mechanism is illustrated in Fig. 6. For design purposes, it is customary to express the deformation in terms of the story drift. The corresponding energy balance equation can be rewritten as

$$\gamma^* \frac{1}{2} M V_{\text{eq}}^2 = \frac{1}{2} V_y h_e^* \theta_y + V_y h_p^* (\theta_m - \theta_y) \tag{11}$$

where $h_e^*$ and $h_p^*$ are the heights at the center of the restoring force during the elastic and plastic response states, respectively, $\theta_m$ is the maximum drift, and $\theta_y$ is the yield drift. Past research suggests that the lateral force distribution during elastic and inelastic responses can be different (Lee and Goel 2001). Therefore, $h_e^*$ and $h_p^*$ may be different as well.

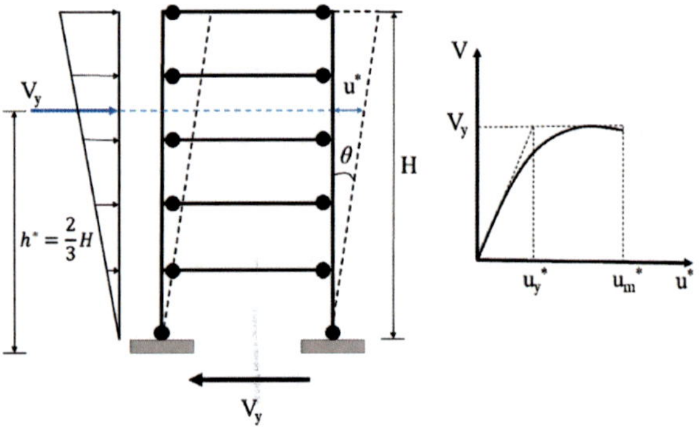


**Fig. 6.** Plastic mechanism of a moment frame system and idealized elastic–plastic force–displacement relationship

It should be noted that for an elastic system, the concept of yield displacement is no longer valid. However, by expressing the equivalent velocity as $V_{\text{eq}} = \eta S_v$ and using Eq. (9) together with the assumption that the elastic displacement is equal to $V_y/M\,(\frac{2\pi}{T})^2$, the above equation reduces to $V_y = MS_a$, where $S_a$ is the spectral acceleration, the same as the code-based elastic design base shear. As an example, the above energy balance concept was applied to compute the required design base shear for a set of example steel moment frames used in a previous study by Lee and Goel (2001). The example frames consisted of one-bay regular frame structures (2- to 10-story frames). The typical bay width was 7.62 m (25 feet), and the typical story height was 4.27 m (14 feet). The story mass was uniform with a typical story weight of 845.16 kN (190 kips). The seismic hazard was specified by Uniform Building Code (1994) design spectrum for seismic zone 4, standard occupancy and soil type S3. The base shear values were calculated for plastic rotation ($\theta_p$=$\theta_m - \theta_y$) of 0% (elastic), 0.5%, 1%, 1.5%, and 2%. The yield rotation was assumed to be 1% for all frames. The fundamental period $T$ was calculated using ASCE/SEI 7-22 (2022). As a first approximation, the height of the center of the restoring force was taken as 2/3 of the building height for both the elastic and inelastic states ($h_e^* = h_p^*$ = 2/3H). The equivalent velocity was calculated by $V_{eq} = \eta S_v$ with the pseudo-velocity calculated from the specified spectral acceleration spectrum. Thus, the base shear of the system is as follows:

$$\frac{V_y}{W} = \frac{\gamma^*\left(\frac{1}{2}\frac{\eta^2}{g}\right)\left(\frac{S_a}{2\pi}T\right)^2}{\frac{2}{3}H(\theta_p + \frac{1}{2}\theta_y)} \tag{12}$$

where $H$ is the building height, and other parameters are as defined earlier. The results of the base shear from Eq. (12) are compared with the results calculated by the Performance-Based Plastic Design (PBPD) method by Goel and Chao (2008) in Fig. 7. The plot indicates that the base shear from Eq. (12) is slightly less than the design base shear from the PBPD method.

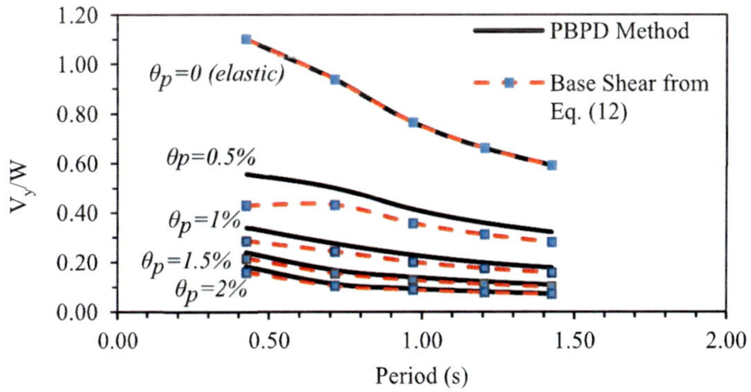


**Fig. 7.** Base shear results from Eq. (12) compared the results from the PBPD method

It should be noted that the above design base shear is the base shear at yield. Hence, the above design base shear is intended for plastic design where the yield strength and yield mechanism can be ensured. The PBPD method is also based on an energy concept and plastic design, hence, it can be compared directly with the proposed method. It should be noted that, in using Eqs. 11 and 12 to calculate the base shear strength, the energy demand can be specified directly by the equivalent velocity, or it can be obtained from the conventional code-base spectral acceleration spectrum. Hence, the proposed method can be used in both energy-based and conventional code-based design frameworks.

## 6 Conclusions

In this study, the relation between the input energy and the equivalent monotonic response curve as expressed by the energy factor was investigated for SDOF systems with EPP hysteretic behavior under two different sets of ground motions. The values of the energy factor for different ground motions were calculated from nonlinear time history analyses.

An equation was proposed to compute the energy factor as a function of period, ductility, and strength reduction factor. The main findings are the following.

- The energy factor varied according to the ductility factor and was fairly constant except in the short period range. As the ductility ratio increased, the value of the energy factor tended to decrease.
- The proposed equation to predict the energy factor is reasonably accurate for practical purposes.
- By using the energy factor, the connections between the required structural strength, the deformation, and the input energy are established. The proposed energy-based design method could be utilized for computing the required base shear of a structural system, given the target ductility. The method can be applied in both energy-based and conventional code-based design frameworks.

**Acknowledgments.** This research is financially supported by the Petchra Pra Jom Klao Scholarship from the King Mongkut's University of Technology Thonburi (KMUTT) to the first author and the Thailand Research and Innovation Fundamental Fund (Advance Construction Toward Thailand 4.0 Project) to the Construction Innovations and Future Infrastructures Research Center at KMUTT.

## References

Alici, F.S., Sucuoğlu, H.: Prediction of input energy spectrum: attenuation models and velocity spectrum scaling. In: 16th World Conference on Earthquake (16 WCEE), Chile (2017)

American Society of Civil Engineers (ASCE), ASCE Standard: Minimum Design Loads and Associated Criteria for Buildings and Other Structures: ASCE/SEI 7-22 (2022)

Chopra, A.K., Chintanapakdee, C.: Inelastic deformation ratios for design and evaluation of structures: single-degree-of-freedom bilinear systems. J. Struct Eng. **130**(9), 1309–1319 (2004)

Gholami, N., Garivani, S., Askariani, S.S.: State-of-the-art review of energy-based seismic design methods. Arch. Computat. Meth. Eng. **29**, 1965–1996 (2022). https://doi.org/10.1007/s11831-021-09645-z

Goel, S.C., Chao S.H.: Performance-Based Plastic Design Earthquake-Resistant Steel Structures. International Code Council (ICC), Country Club Hills (2008)

Hachem M.M.: BiSpec Professional (Version 2.2), www.eqsols.com (2012)

Housner, G.W.: Limit design of structures to resist earthquake. In: Proceedings of the 1st World Conference on Earthquake Engineering, USA (1956)

Ji, D., Wen, W., Zhai, C.: Constant-ductility energy factors of sdof systems subjected to mainshock-aftershock sequences. Earthq. Spectra **37**(2), 1078–1107 (2020)

Ke, K., Chuan, Q., Ke, S.: Seismic energy factor of self-centering systems subjected to near-fault earthquake ground motions. Soil Dyn. Earthq. Eng. **84**, 169–173 (2016)

Ke, K., Zhao, Q., Yam, M.C., Ke, S.: Energy factors of trilinear SDOF systems representing damage-control buildings with energy dissipation fuses subjected to near-fault earthquakes. Soil Dyn. Earthq. Eng. **107**, 20–34 (2018)

Lee, S.S., Goel, S.C.: Performance-Base Design of Steel Moment Frame Using Target Drift and Yield Mechanism. Research Report UMCEE 01-17. University of Michigan (2001)

Leelataviwat, S., Saewon, W., Goel, S.C.: Application of energy balance concept in seismic evaluation of structures. J. Struct Eng. **135**(2), 113–121 (2009)

Merter, O., Ucar, T.: Energy modification factor of single-degree-of-freedom systems based on real ground motion records. In: Proceedings of 2nd International Symposium of Engineering Applications on Civil Engineering and Earth Sciences (IEACES), Turkey (2020)
Newmark, N.M., Hall, W.J.: Earthquake Spectra and Design. Earthquake Engineering Research Institute, California (1982)
Uçar, T., Merter, O.: Predictive model for constant-ductility energy factor spectra of near- and far-fault ground motions based on gauss-newton algorithm. J. Earthq. Eng. **26**(15), 7689–7714 (2022)
Uniform Building Code: International Conference of Building Officials, Whittier, California (1994)
Wang, F., Ke, K., Zhang, H., Yam, M.C.: Constant-ductility-based energy factor demands of oscillators with modified clough hysteretic model. Soil Dyn. Earthq. Eng. **115**, 36–40 (2018)
Zhai, C., Ji, D., Wen, W., Lei, W., Xie, L.: Constant ductility energy factors for the near-fault pulse-like ground motions. J. Earthq. Eng. **21**(2), 343–358 (2017)
Zhou, X., Zhang, H., Ke, K., Guo, L., Yam, M.C.: Damage-Control Steel Frames Equipped with SMA Connections and Ductile Links Subjected to Near-Field Earthquake Motions: A Spectral Energy Factor Model. Engineering Structures 239 (2021)

# Comparison of Energy-Based, Force-Based and Displacement-Based Seismic Design Approaches for RC Frames in the Context of 2G of Eurocode 8

Amadeo Benavent Climent[1](✉), Jesús Donaire-Ávila[2], and Fabrizio Mollaioli[3]

[1] Technical University of Madrid, 28006 Madrid, Spain
amadeo.benavent@upm.es
[2] University of Jaén, 23700 Linares, Jaén, Spain
[3] Sapienza University of Rome, 00197 Rome, Italy

**Abstract.** The second generation of Eurocode 8 (2G EN1998) considers two main approaches to verify conventional RC bare frames: the force-based approach and the displacement-based approach. In addition, for structures with energy dissipation devices a method based on the energy balance is considered. Although the formulation of the energy-based approach in 2G EN1998 is intended to be used only for structures with energy dissipation systems, it can be easily extended to conventional structures by removing the contribution of the energy dissipation devices and adding few details concerning the calculation of some terms involved in the equations. This paper presents a case study of a RC bare frame designed/verified applying the energy-based methodology and compares the resulting design with that obtained with the force-based and the displacement-based approaches. The response of the structure of the structure dimensioned with the energy-based approach under several ground motions is obtained through non-linear time history analyses and compared with the predictions.

**Keywords:** Force-Based Design · Displacement-Based Design · Energy-Based design

## 1 Introduction

The second generation of Eurocode 8 (2G EN1998) [1, 2] considers two main approaches to verify that a new conventional reinforced concrete (RC) bare frame possesses the adequate combination of strength, deformation capacity and cumulative energy dissipation capacity under prescribed levels of seismic loading: the Force-Based (FB) approach and the Displacement-Based (DB) approach. In addition, for structures with energy dissipation systems, and Energy-Based (EB) approach is considered. Although the latter is intended to be used for structures with energy-dissipations system, it could be applied to a conventional structure by removing the contribution of the energy dissipation devices (EDDs) and adding few details concerning the calculation of some terms involved in the


H. Varum et al. (Eds.): IWEBSE 2023, LNCE 236, pp. 98–111, 2023.
https://doi.org/10.1007/978-3-031-36562-1_8

equations that establish the energy balance. These details consider that, for economic reasons, it is not reasonable to require that a conventional bare RC frame exhibits the same levels of performance in terms of damage control than a structure with EDDs. The main role of the EDDs is to reduce significantly or even avoid the damage under rare (i.e. design level of seismic hazard) ground motions. A fortiori, in case of frequent earthquakes the main structure of a building with EDDs should remain in the elastic range and fully operational (OP). Clause 6.8.2.2.(9) prEN1998-1-1 clarifies that when required for specific structural members of a structure with antiseismic devices that should remain in the "elastic range or quasi-elastic range", it should be verified that these members comply with the condition of the damage-limitation (DL) limit state. The condition of the DL limit state for a structural member is that the decrease in stiffness must be limited. The condition of the OP limit state raises this requirement stating that the structural member shall be only slightly damaged—clause 4.3. (1) prEN1998-1-1. It is not quantitatively established in prEN1998-1-1 the stiffness reduction that is considered as "limited" or"slight damage", but in case of structures with EDDs this reduction should zero or very small – i.e. far below the one half of the corresponding stiffness of the uncracked members indicated in clause 5.1.3(3) of prEN1998-1-2). Moreover, if the frame interacts with the infills, the inter story drift at the DL limit state, $d_{r,DL}$, must satisfy $d_{r,DL} \leq \lambda_{ns} h_s$ where $\lambda_{ns}$ is a coefficient that accounts for the sensitivity of the ancillary element to interstory drift and $h_s$ is the storey height.

This paper presents a case study of a RC bare frame designed/verified applying the FB, the EB and the DB methodology. The formulation used for the EB analysis is basically the same included in the last draft of prEN1998-1-2 for structures with EDDs but specialized for the case without EDDs. The response of the structure dimensioned with the EB approach is also compared with the results of time-history analyses. In all verifications the last drafts of 2G of Eurocode 8 (prEN1998-1-1 and prEN1998-1-2) have been used.

## 2 Background and Formulation of the Energy-Based Approach for Structures with EDDs in prEN1998-1-2

The energy approach is based on the fundamental energy balance given by Eq. (1):

$$E_e + E_\xi + E_H = E_I \tag{1}$$

where $E_e$ is the elastic vibrational energy (i.e. kinetic energy plus elastic strain energy), $E_H$ is the plastic strain energy, $E_\xi$ is the energy dissipated by damping, and $E_I$ the total energy input. $E_D = E_I - E_\xi$ is what Housner [3] called the energy that contributes to damage. Equation (1) can be established at the instant of maximum lateral displacement $t_m$ and at the end of the ground motion $t_o$. $E_D(t_o)$ can be estimated from the spectral velocity $S_V$ (=$S_e T/2\pi$) by $E_D = m S_v{}^2/2$ where $m$ is the total mass of the building, $T$ the period and $S_e$ the elastic acceleration response spectrum [3, 4]. In highly nonlinear systems $E_D(t_m) \leq E_D(t_o)$ and for design purposes it can be assumed that $E_D(t_m) = E_D(t_o)$. The EB approach adopted in prEN1998-1-2:2022 for the verification of structures with displacement-dependent EDDs is based on establishing the energy balance given by

Eq. (1) at $t = t_m$ with $E_D(t_m) = E_D(t_o)$, and uses the formulation developed by Akiyama [4, 5]. It is basically the same formulation adopted by the current Building Law of Japan [6]. The approach assumes that the structure responds in the nonlinear range only in the first mode. The contribution of the higher modes is obtained by applying conventional modal response analysis.

A building with energy dissipation systems (Fig. 1), is composed of two systems in parallel: (a) the main structural system composed of the primary and secondary structural elements; and (b) the EDD with the auxiliary elements that connect the EDDs with the main structure.

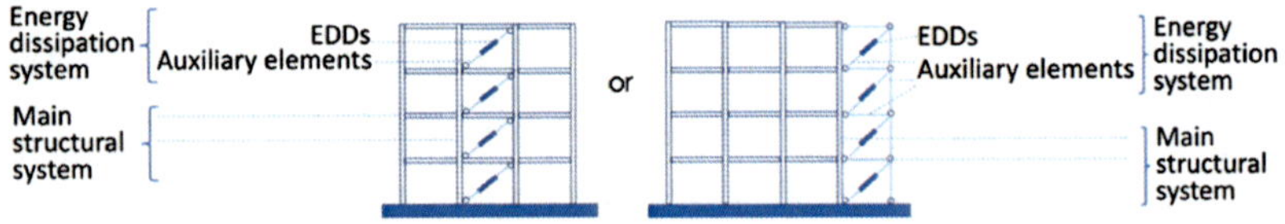


**Fig. 1.** Building with energy dissipation systems

The primary role of the main structural system is to sustain the gravity loading when it is subjected to lateral displacements. The primary role of the energy dissipation system is to absorb most of the energy input by the earthquake. Under frequent earthquakes, the main structure must remain elastic (undamaged). Under rare ground motions (design earthquake), most of the energy must be absorbed by the EDDs, although the primary elements of the main structure can also contribute to dissipate energy through (limited) plastic deformations. The main reason to use the technology of EDDs is to control the damage on the main structure. It would make little sense to design a building with EDDs that does not remain fully operational after a frequent earthquake, or that endures severe damage on the main structure under a rare earthquake. Accordingly, a building with energy dissipation systems must satisfy the following main requirements.

*Verification for Operational (OP) and Damage Limitation (DL) Limit States (LS)*

The seismic action associated to the OP and DL LSs is a "frequent" earthquake characterized by an elastic 5% damped response spectrum which ordinates are 0,5 times those of the reference response spectrum defined in clause 5.2 of prEN1998-1-1:2021 (the "design" earthquake). It must be verified that the amount of energy that the overall structure can absorb while the primary seismic members of the main structure remain elastic is equal or larger than the energy input by a frequent earthquake as given by Eq. (2):

$$\sum\nolimits_{k=1}^{N}\left[E_{pes,k} + E_{des,k} + E_{dHk}\right] \geq \left\{\frac{1}{2}m\left[\frac{T_1}{2\pi}0,5S_e\eta\right]^2\right\} \tag{2}$$

$E_{pes,k}$ and $E_{des,k}$ are the elastic strain energy stored by the primary seismic members of the main structure and by the EDDs, respectively, at the $k$-th storey while the main structure remains in the elastic range. In structures with EDDs $E_{pes,k}$ can be estimated by $E_{pes,k} = V_{pD,k}\, d_{D,k}/2$ and $E_{des,k} = V_{dy,k}\, d_{dy,k}\, /2$. $E_{dH,k}$ is the energy dissipated in form of plastic strain by the EDDs at the $k$-th storey while the primary seismic members of the main structural system at the $k$-th story remain in the elastic range. $E_{dH,k}$ can be estimated by $E_{dH,k} = 4V_{dy,k}\,(d_{D,k} - d_{dy,k})$ [6]. In above expressions, $T_1$ is the fundamental period of the whole structure in elastic conditions, $S_e$ is the spectral acceleration value of the 5% damped elastic response spectrum that characterizes the design earthquake, and $\eta$ the damping correction factor ($\eta = 1$ for 5% damping). The meaning of the rest of variables is shown in Fig. 2 that represents the storey shear vs inter storey drift curve of the $k$-th story. This curve is obtained with a pushover analysis for a pattern of lateral loads proportional to the first mode, using mean values of material properties and the deformation with the strength models of prEN1998-1-1:2021 and including second order effects. It is worth recalling that to prevent damage on the main structure under frequent earthquakes, $V_{pD,k}$ must be taken as the maximum storey shear for which the primary seismic members of the main structural system at the $k$-th storey remain in the elastic domain. In case of RC structures, $V_{pD,k}$ should be determined with the uncracked properties of members. That is, $V_{pD,k}$ should be taken as the storey shear associated with the inception of new flexural cracks caused by lateral loads (i.e. excluding previous cracks owing to gravity loads combined with stresses caused by restrained shrinkage, thermal strains or other imposed deformations) in the structural members of the $k$-th story. In case of steel structures, $V_{pD,k}$ should be determined with the elastic modulus of the sections. This will result in small values for $V_{pD,k}$ and $d_{D,k}$ but for a structure with EDDs this is not a problem because the left member of Eq. (2) can be easily increased by an appropriate selection of the lateral strength $V_{dy,k}$ and yield displacement $d_{dy,k}$ of the EDDs.

*Verification for Significant Damage (SD) LS*

The seismic action associated to the SD LS is the "design" earthquake characterized by the elastic 5% damped reference response spectrum defined in clause 5.2 of prEN1998-1-1:2021. This is the "reference seismic action" corresponding to a reference return period for consequence class CC2 named $T_{ref}$ (=475 years as recommended value). The amount of energy that the primary seismic members of the main structure at the $k$-th storey can dissipate through plastic deformations before one of them reaches the significant damage (SD) limit state, $E_{pH,k,SD}$, must be larger than the maximum energy dissipation demand at the $k$-th story under the design seismic action, $E_{pH,k,max}$. A rational design of a building with EDDs should keep $E_{pH,k,max}$ far below $E_{pH,k,SD}$ in order to limit the damage in the main structure. $E_{pH,k,max}$ is estimated with Eq. (3).

$$E_{pH,k,max} = E_{H,k}\frac{V_{py,k}\left(d_{r,k,1} - d_{py,k}\right)}{V_{py,k}\left(d_{r,k,1} - d_{py,k}\right) + V_{dy,k}\left(d_{r,k,1} - d_{dy,k}\right)} \tag{3}$$

Here, $E_{H,k}$ is the total plastic strain energy to be dissipated by the building at the $k$-th storey estimated from $E_H$. To estimate $E_H$, the shear force vs interstory drift of the structure at the $k$-th story is idealized with an elastic perfectly plastic model (dashed line in Fig. 2) and $E_e$ is estimated by summing up the areas of the triangles o-f-g in Fig. 2. Recalling that $E_H = E_D - E_e$ and $E_D = mS_v{}^2/2$, Eq. (4) is obtained.

$$E_H = \frac{1}{2}m[\frac{T_{s1}}{2\pi}S_e(T_{s1}, 5\%)\eta(T_{s1}, \xi_I)]^2 - \sum_{l=1}^{l=N} \alpha_l \overline{m}_l mgd_{py,l}/2 \tag{4}$$

Here, $T_{s1}$ is the period that gives the largest $E_D$ in the range between $T_1$ and $1.4T_1$, $\overline{m}_k$ is the normalized mass above the $k$-th storey calculated from the mass of each level $m_l$, and $\alpha_k$ is the normalized total shear strength of the $k$-th storey, $V_{py,k} + V_{dy,k}$, as defined by Eq. (5).

$$\overline{m}_k = \sum_{l=k}^{N} m_l/m; \quad \alpha_k = \left(V_{py,k} + V_{dy,k}\right)/(\overline{m}_k mg) \tag{5}$$

$E_H$ is distributed among the storey as follows [4–6]:

$$E_{H,k} = \frac{s_k(p_k p_{t,k})^{-n}}{\sum_{l=1}^{N} s_l(p_l p_{t,l})^{-n}} E_H \tag{6}$$

Here $s_k$ is given by Eq. (7) and represents the distribution of plastic strain energy among the stories if the normalized lateral strength of the $k$-th storey, $\alpha_k$, is equal to the "optimum" value $\alpha_{opt,k}$ that would make the damage to distribute uniformly along the heigh of the building.

$$s_k = \left\{\overline{m}_k + \frac{2T_1}{1+3T_1}\left[\sqrt{\overline{m}_k} - \overline{m}_k^2\right]\right\}^2 \frac{d_{py,k}(V_{py,1} + V_{dy,1})}{d_{py,1}(V_{py,k} + V_{dy,k})} \tag{7}$$

In developing Eq. (7), the distribution of $\alpha_{k,opt}$, i.e. $\overline{\alpha}_{k,opt} = \alpha_{k,opt}/\alpha_1$ of the Building Law of Japan [6] given by Eq. (8) is adopted.

$$\overline{\alpha}_{k,opt} = 1 + \left[\frac{1}{\sqrt{\overline{m}_k}} - \overline{m}_k\right]\frac{2T_1}{1+3T_1} \tag{8}$$

Parameters $p_k$ and $p_{t,k}$ in Eq. (6) take into account that the distribution of $E_H$ among the storeys is governed by the distribution of lateral strength and by torsional effects [4, 5]. More precisely, $p_k$ (=$\alpha_k/\alpha_{k,opt}$) expresses the deviation of the normalized lateral strength of the storey from the "optimum" value $\alpha_{k,opt}$. Parameter $n$ in Eq. (6) captures the proneness of the main structure to concentrate the damage. If the main structure satisfies the global and local ductility conditions established by prEN1998-1-2:2022 then $n = 4$ otherwise $n = 8$ must be adopted [6]. Using the expressions for $\alpha_k$ and $\alpha_{k,opt}$ defined above $p_k$ can be rewritten as follows:

$$p_k = \frac{(V_{py,k} + V_{dy,k})}{(V_{py,1} + V_{dy,1})}\left\{\overline{m}_k + \frac{2T_1}{1+3T_1}\left[\sqrt{\overline{m}_k} - \overline{m}_k^2\right]\right\}^{-1} \tag{9}$$

The scattering in material properties and other reasons cause unavoidable variations on the actual value of the normalized lateral strength $\alpha_k$ of each story. These deviations significantly affect the distribution of the plastic strain energy among storeys and must be taken into account. A simple way to do that is to consider different possible scenarios in which the lateral strength of the storeys are affected by a factor $p_{d,k}$. The following set of scenarios can be considered. In the first scenario (CASE 1) it is assumed that the lateral strength of the 1$^{st}$ storey is $p_d$ times the expected value, while in the rest of storeys it remains unchanged (i.e. $p_{d,k} = p_d$ for $k = 1$ and $p_{d,k} = 1$ for $k \neq 1$). In the second scenario (CASE 2) the lateral strength of the 2$^{st}$ storey is assumed $p_d$ times the expected value, while in the rest of storeys it remains unchanged (i.e. $p_{d,k} = p_d$ for $k = 2$ and $p_{d,k} = 1$ for $k \neq 2$), and so on. Torsional effects have also a relevant influence on the distribution of total plastic strain energy $E_H$ among the stories. The proneness of a storey to torsional effects can be characterized by the ratio $e_{ox}/r_x$, where $e_{ox}$ is the distance between the center of stiffness and the center of mass, measured along the direction perpendicular to the direction of analysis considered, and $r_x$ is the torsional radius. $p_{t,k}$ has been calibrated in past studies [4–6] and can be estimated with Eq. (10).

$$p_{t,k} = \left\{ \begin{array}{c} 1 \text{ for } (e_{ox}/r_x) \leq 0,15 \\ 1,15 - (e_{ox}/r_x) \text{ for } 0,15 < (e_{ox}/r_x) < 0,30 \\ 0,85 \text{ for } (e_{ox}/r_x) \geq 0,30 \end{array} \right\} \tag{10}$$

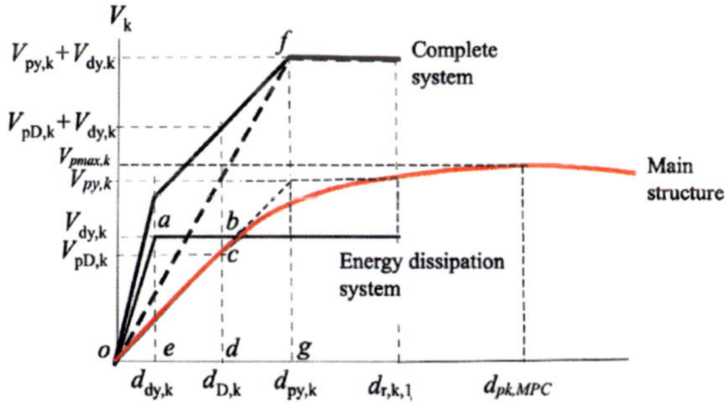


**Fig. 2.** Lateral shear vs inter storey drift of a given story $k$

In addition to above EB main verifications, it must be verified that at the SD LS the design values of the action effects in the seismic design situation calculated with the design seismic forces at the $i$-th level given by Eq. (11) for the first mode and by Eq. (12) for the higher $m$-th mode, (action effects combined with SRSS or CQC rules), do not exceed the corresponding design resistance of structural members.

$$F_{i1} = \left(V_{\mathrm{pD},1} + V_{\mathrm{dy},1}\right)\left\{\left[\overline{m}_k + \frac{2T_1}{1+3T_1}\left(\sqrt{\overline{m}_k} - \overline{m}_k^2\right)\right] - \left[\overline{m}_{k+1} + \frac{2T_1}{1+3T_1}\left(\sqrt{\overline{m}_{k+1}} - \overline{m}_{k+1}^2\right)\right]\right\} \tag{11}$$

$$F_{\mathrm{im}} = m_i \phi_{\mathrm{im}} \Gamma_{\mathrm{m}} S_{\mathrm{e}}(T_{\mathrm{m}}, 5\%)\eta(T_{\mathrm{m}}, \xi_{\mathrm{I}})/q_S \tag{12}$$

The forces in the higher modes are estimated with conventional modal response analysis since it is assumed that in these modes the structure remains elastic. $\phi_{im}$ is the component of the $m$-th mode vector $\boldsymbol{\phi}_m$ at the i-th level; $\boldsymbol{\phi}_m$ must be normalized so that $\phi_{im}$ is equal to 1 at the roof level (this normalization applies to all higher modes). $T_m$ and $\Gamma_m$ are the period and participation factor of the $m$-th mode, $\xi_I$ is the inherent damping of the structure and $q_s$ the behavior factor component of the main structure owing to overstrength that is taken as $q_s$ = 1,5. Finally, the maximum lateral displacements at each level $i$ is estimated by combining (with the SRSS or CQC rules) the maximum displacement in the first mode $d_{i1}$ calculated from the inter storey drifts $d_{r,k,1}$ as given by Eq. (13), with the maximum inter storey drifts in the higher modes $d_{im}$ given be Eq. (14).

$$d_{\mathrm{r},k,1} = d_{\mathrm{py},k} + \left\{\frac{E_{\mathrm{H},k}}{\left[4\left(V_{\mathrm{py},k} + V_{\mathrm{dy},k}\right)\right]}\right\}; \quad d_{i1} = \sum\nolimits_{k=1}^{k=i} d_{\mathrm{r},k,1} \tag{13}$$

$$d_{im} = \Gamma_m S_{De}(T_m, 5\%)\eta(T_m, \xi_1)\phi_{im} \tag{14}$$

Note that in Eq. (13), $d_{r,k,1}$ is estimated assuming that the hysteretic energy $E_{H,k}$ is dissipated in a single cycle of displacement [6]. In Eq. (14), $S_{De}$ is the ordinate of the 5% damped elastic design displacement spectrum at period $T_m$.

## 3 Design of the RC Frame with the Force-Based Approach

The RC bare frame used in this case study is an interior frame with three spans and six stories of a domestic building structure constituted of parallel RC frames spaced 5 m (Fig. 3). The loads considered are 3.5 kN/m$^2$ (dead), 3 kN/m$^2$ (life) and 26 kN/m$^3$ (self-weigh of RC elements). The seismic hazard was characterized by the 5% damping elastic response acceleration spectrum shown with bold red lines in Fig. 4 (left). The mass considered for the seismic design situation is 100% of the permanent loads plus 15% of the variable loads. The characteristic compressive concrete strength, $f_{ck}$, and characteristic yield stress of the reinforcing steel $f_{yk}$ are $f_{ck} = 25$ MPa and $f_{yk} = 500$ MPa. The partial factors considered for concrete $\gamma_c$ and steel $\gamma_s$ are $\gamma_c = 1.5$ and $\gamma_s = 1.15$. The frame was designed using linear elastic analysis for the combination of actions of the non-seismic, $1.35G + 1.5Q$, and seismic, $1.0Q + 0.15Q \pm A_{Ed,ULS}$, design situations. Here, $G$ denotes the permanent actions, $Q$ the variable actions and $A_{Ed,ULS}$ the design value of the seismic actions. In the RC frame designed with the FB approach $A_{Ed,ULS}$ were obtained as established in prEN1998:1–1:2021, assuming ductility class DC2 which corresponding behavior factor is $q = q_R \times q_S \times q_D = 1{,}3 \times 1{,}5 \times 1{,}3 = 2{,}5$. In calculating $A_{Ed,ULS}$, a cracked flexural stiffness of members equal to 50% of the uncracked stiffness was considered —clause 5.1.3(3) of prEN1998-1-2:2022— which gave the periods $T_1 = 1{,}53$s and $T_2 = 0.65$s in the first and second modes. These modes mobilized > 90% of the total mass. prEN1998-1-2022 requires that the formation of a soft-storey plastic mechanism should be prevented. This may be considered satisfied if the condition $\sum M_{Rc} \geq 1{,}3 \sum M_{Rb}$ is fulfilled in all stories. Here, $\sum M_{Rc}$ is the sum of the moment resistance of columns and $\sum M_{Rc}$ that of the moment resistance of beams framing the joint. It is worth noting that above condition is not compulsory in DC2; it is just a way to satisfy the requirement. In this case study, the condition $\sum M_{Rc} \geq 1{,}3 \sum M_{Rb}$ was applied for designing the RC frame with the FB approach. The resulting size and reinforcement of members is shown in Fig. 3 (left).

## 4 Design of the RC Frame with the Energy-Based Approach

A RC frame with the same geometry and size of columns and beams used for the FB approach was employed in the EB analysis. The difference between the RC frames designed with the two approaches is the amount of reinforcement. The EB analysis uses the vibration periods and modes of the elastic structure (i.e. those calculated with the uncracked stiffness of members) giving $T_1 = 1.1$s and $T_2 = 0.36$s in the first and second mode, and a larger fundamental period $T_{s1}$ (between $T_1$ and $1.4T_1$) that takes into account the enlargement due to plastic deformations. The two modes mobilized > 90% of the total mass.

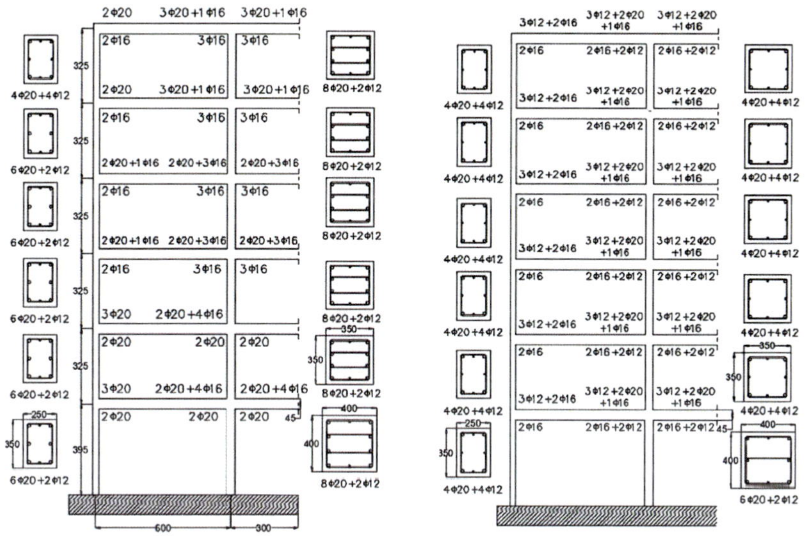


**Fig. 3.** Details of the RC frame used in the forced-based approach (left), and in the EB and DB approach (right)

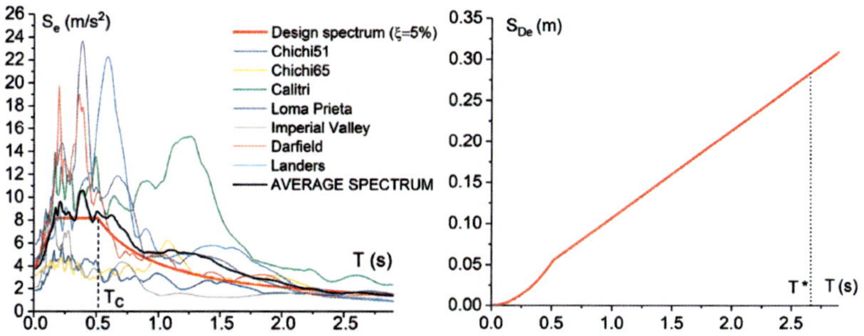


**Fig. 4.** Acceleration spectrum (left) and displacement spectrum (right) for SD LS

First, the size and reinforcement of members was tentatively determined using linear elastic analysis for the actions effects due to the non-seismic design situation, $1.35G + 1.5Q$, and the seismic design situation $1.0Q + 0.15Q \pm A_{Ed,ULS}$. In this initial dimensioning, $A_{Ed,ULS}$ includes only the lateral forces of the second mode ($m = 2$) obtained with Eq. (12), given that $V_{pD1}$ is not yet determined. To determine $V_{pD1}$, a nonlinear static analysis was conducted by applying only the gravity loads, followed (i.e. continuing from the state at the end of this nonlinear analysis) by a first-mode pushover analysis. In the analyses, mean values of material and the moment rotation relation, $M - \theta$, of the plastic hinges shown with solid lines in Fig. 5 were used.

The yield moment $M_y$, the yield rotation $\theta_y$ and the ultimate rotation $\theta_u$ were obtained with the empirical strength models of prEN1998-1-1:2021. The mean values for concrete and steel are $f_{cm} = f_{ck} + 8 = 33$ MPa, $f_{ym} = 1.15 f_{yk} = 575$ MPa. The cracking moment $M_{cr}$ was calculated with $M_{cr} = (f_{ctm} + N/A_t) I_t / y_t$, where $f_{ctm}$ is the mean tensile strength

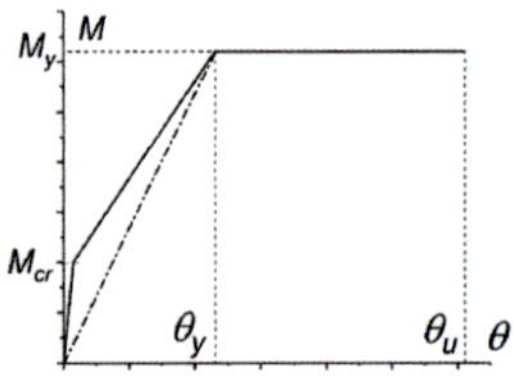


**Fig. 5.** Idealized $M - \theta$ curve

of concrete $f_{ctm} = 0.3 f_{ck}^{2/3} = 2.56$ MPa, $N$ the axial force (positive in compression), $A_t$ and $I_t$ are the cross-sectional area and inertia of the uncracked transformed section and $y_t$ the distance of the extreme tension fibres from the centroid. Here, uncracked transformed section means the concrete section in which any reinforcing bar of area $A_{s,i}$ has been replaced by an equivalent concrete area, $\alpha A_{s,i}$, where $\alpha = E_s/E_c$ is the ratio of moduli of the two materials. The pushover analysis provided an initial tentative value for $V_{pD,1}$ that was substituted in Eq. (11) (with $V_{dy1} = 0$) to calculate the design lateral forces in the first mode $F_{i1}$ and their corresponding action effects. In this case study $V_{pD,1}$ was found to be the base shear that triggered the inception of flexural cracks in the columns of the first story, since beams were already cracked owing to gravity loads. The RC frame was checked again for the design value of the action effects corresponding to the seismic design situation $1.0Q + 0.15Q \pm A_{Ed,ULS}$. Now, the seismic actions $A_{Ed,ULS}$ considered are the lateral forces $F_{i1}$ (first mode) and $F_{i2}$ (second mode) given by Eqs. (11), (12), and their effects were combined with SRSS rule. This check required increasing the amount of reinforcement in some members (the size was kept unchanged). The pushover analysis was then repeated with the new reinforcement, and $V_{pD,1}$ recalculated and replaced in Eq. (11) to obtain new values for $F_{i1}$. The RC frame was re-checked, and the verification was found satisfactory. The condition $\sum M_{Rc} \geq 1{,}3 \sum M_{Rb}$ was not imposed although it was not satisfied in most interior joints. The resulting longitudinal reinforcement is shown in Fig. 3 (right). Figure 6 shows the shear force $V_{p,k}$ versus inter storey drift $d_{r,k}$ of each storey $k$ obtained with the pushover analysis.

Second, the main verification given by Eq. (2) was conducted with the RC frame dimensioned as shown in Fig. 3 (right). $E_{des,k}$ and $E_{dH,k}$ were taken as 0 since there are not EDDs. As discussed in the introduction, it is unreasonable to expect that a conventional structure without EDDs exhibits the same performance levels than a structure with EDDs because this is economically unviable. That is, it is not reasonable to design the RC frame so that it remains perfectly elastic (i.e. with interstory drifts below $d_{pD,k}$ in Fig. 2) under a frequent earthquake. In fact, the verification of the DL limit state under frequent earthquakes in conventional RC frames applying the FB approach allows inter storey drifts up to $\lambda_{ns} h_s$ and admits certain reduction of member's stiffness. In this case study, it has been assumed that the building has ductile ancilliary elements attached to the structure and thus $\lambda_{ns} = 0.0075$. The interstory drift corresponding to $0.0075 h_s$ is plot in the storey capacity curves of Fig. 6 and it can be seen that the corresponding reduction of the initial stiffness $K_e$ ranges between $0{,}38 K_e$ and $0{,}46 K_e$. Thus, in conventional RC frames without EDDs and for the purpose of verifying Eq. (2), it is reasonable to take for $E_{pes,k}$ the area below the $k$-th storey capacity curve up to the maximum allowed inter storey drift for the DL LS ($0.0075 h_s$ here) instead of $E_{pes,k} = V_{pD,k}\, d_{D,k}/2$. The value

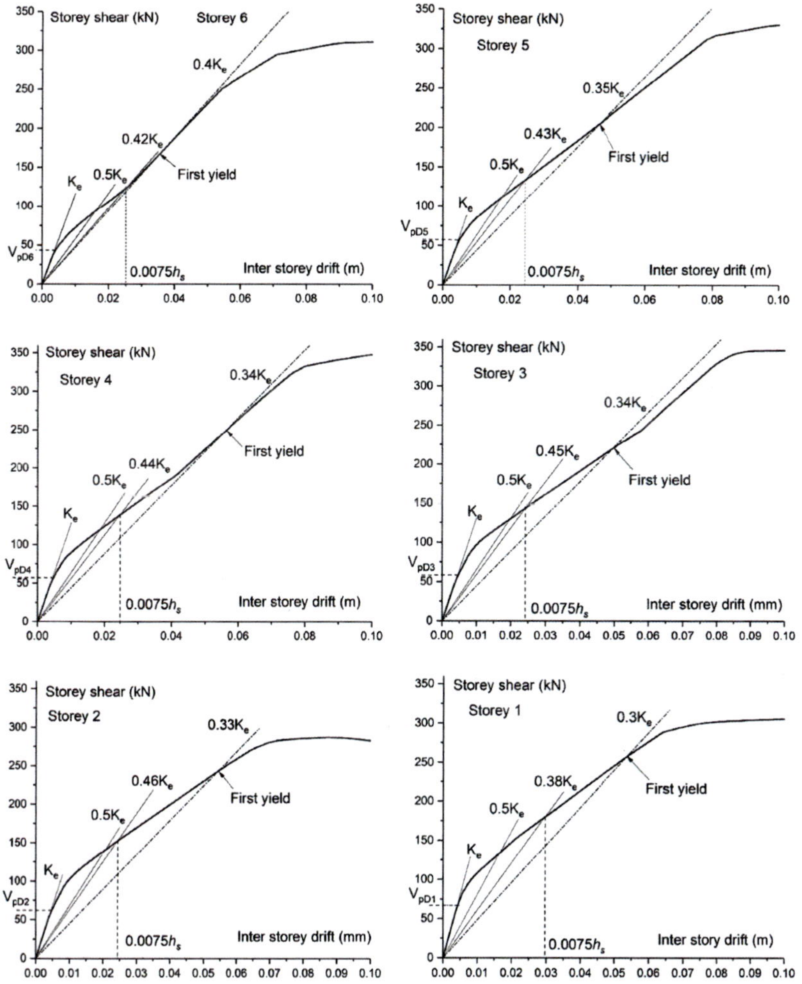

**Fig. 6.** Capacity curve of each storey used in the EB approach

$\Sigma E_{pes,k}$ (=14.8 kNm) calculated in this way satisfies Eq. (2), yet very closely to the energy demand given by the second member of Eq. (2) (14.7 kNm).

Third, the main verification $E_{pH,k,SD} \geq E_{pH,k,max}$ was conducted with the RC frame pre-dimensioned as shown in Fig. 3 (right). The plastic strain energy dissipation demand $E_{pH,k,max}$ was estimated with Eqs. (3)–(10) adopting $p_{t,k} = 1$ for all storey since a 2D frame is studied. Figure 7(left) shows the plastic strain energy dissipation demands $E_{pH,k,max}$ determined for different scenarios with $p_d = 0.85$. Case 0 corresponds to the scenario where $p_{d,k} = 1$ in all stories. As for the plastic strain energy dissipation capacity $E_{pH,k,SD}$, a lower bound value of $E_{pH,k,SD}$ can be obtained using the deformation criteria and strength models established in chapter 7 of prEN1998-1-1:2021, together with the verification criteria of clause 6.7.2(1) of prEN1998-1-1:2021. The former gives $M_y$, $\theta_y$ and the ultimate plastic rotation $\theta_u^{pl}$ (=$\theta_u - \theta_y$) (see Fig. 5); the later establishes the allowed plastic rotation at the SD LS, $\theta_{SD}^{pl}$ as $\theta_{SD}^{pl} = \alpha_{SD,\theta}\theta_u^{pl}/\gamma_{Rd,SD,\theta}$, where $\alpha_{SD,\theta}$ (=0,5) is the portion of the plastic part of the ultimate deformation that corresponds to the attainment of SD, and $\gamma_{Rd,SD,\theta}$ (=1,35 for rectangular sections) is a partial factor on resistance at SD given in clause 10.4.5(3) prEN1998-1-2:2022. To estimate the lower bound of $E_{pH,k,SD}$ the following procedure can be applied. Step one, the $M_y$ and $\theta_{SD}^{pl}$ of each potential plastic hinge $h$ is calculated, $M_{h,y}$ and $\theta_{h,SD}^{pl}$ herein. Step two, the frame is decomposed into "story frames" [4, 5]. In this decomposition, the structural properties $S_o$ of a beam located between storeys $k$ and $k+1$ are decomposed in two parts by the ratio $d_k$ so that $S_o d_k$ is assigned to the $k$-th storey and $S_o(1-d_k)$ to the $(k+1)$-th storey. $d_k$ is given by Eq. (15) [4, 5] where $h_k$ is the heigh of the $k$-th storey.

$$d_k = (h_k\overline{m}_k\overline{\alpha}_{k,opt})/(h_k\overline{m}_k\overline{\alpha}_{k,opt} + h_{k+1}\overline{m}_{k+1}\overline{\alpha}_{k+1,opt}) \tag{15}$$

Step three, the plastic mechanism of each story frame under lateral forces is analyzed to determine the location of the plastic hinges and the relative rotation of each hinge, $\theta_h$, when the story frame is subjected to a given inter storey drift $d_{r,k,1}$. The plastic hinge $h$ that first attains its corresponding $\theta_{h,SD}^{pl}$ determines the inter story drift $d_{r,k,1,\mathrm{SD}}$ for which the story reaches the SD LS. Step four, the rotation of each plastic hinge corresponding to $d_{r,k,1,\mathrm{SD}}$ is determined and the lower bound value of $E_{pH,k,SD}$ is simply obtained by summing up the plastic strain energy dissipated by each plastic hinge under monotonic loading up to $d_{r,k,1,\mathrm{SD}}$. It is worth emphasizing that the plastic strain energy dissipation capacity of the storey obtained in this way is a lower bound because the ultimate energy dissipation capacity of structural members under cyclic loading is in general larger than under monotonic loading, the increment being influenced by the characteristics of the history of loading [7]. The discussion on the relation between energy dissipation capacity under monotonic and under cyclic loading is beyond the scope of this paper, but larger values of this lower bound can be used in practical design based on experimental studies and tests. It can be seen in Fig. 7(left) that the predicted plastic strain energy dissipation demand is safely below the capacity, being about one half in the first story. Figure 7 (right) shows the maximum interstory drifts estimated with Eq. (13) for CASE 0 and the envelope for all cases. It can be seen that the maximum interstory drift is less sensible to variations of the lateral strength than the plastic strain energy. Moreover, the predicted inter storey drifts are below the limit of 2% of story heigh required by prEN1998-1-2:2022 and thus the RC frame would pass the verification without need of increasing the size/reinforcement of members. The maximum displacement at roof relative to the ground in the first mode $d_{roof1}$ obtained with Eq. (13) is $d_{roof1} = 0.3562$ m for CASE 0. It remains basically unchanged (0.3564) after combining with that of the second mode $d_{roof,2}$ calculated with Eq. (13) with the SRSS rule.

## 5 Design of the RC Frame with the Displacement Approach

The same RC frame used for the EB approach (CASE 0) is used to conduct the validation applying the DB method. The non-linear static analysis described in prEN1998-1-1:2021 is applied to obtain the response at the SD LS. A first-mode pushover analysis is conducted with the $\mathrm{M}-\theta$ relationships idealized now with two segments as shown

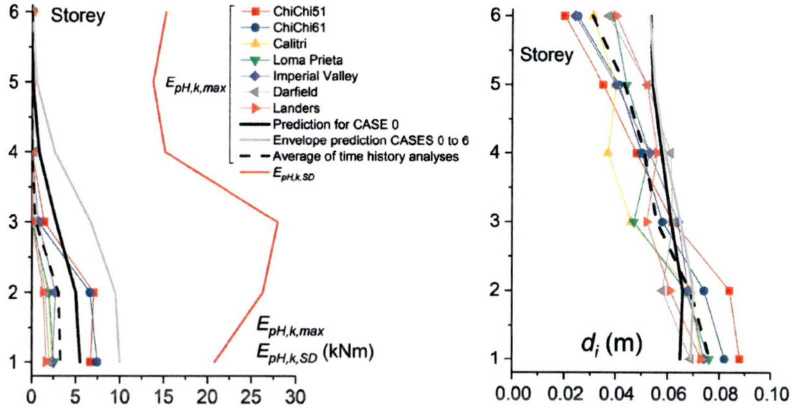


**Fig. 7.** EB approach and results of time history analyses: energy dissipation demand vs capacity (left); inter storey drifts (right)

with dashed lines in Fig. 5. The resulting curve that relates the base shear $V_b$ with the horizontal displacement at roof level $d_{roof}$ was obtained until any primary member attains its ultimate local deformation $\theta_u$ (Fig. 5), that corresponds to a roof displacement $d_{roof,u}$. The peak shear force and the corresponding roof displacement are denoted by $V_{b,m}$ and $d_{roof,m}$. The curve was transformed through the first-mode participation factor $\Gamma$ to obtain the force-displacement $F^*$-$d^*$ curve of the equivalent SDOF system depicted in Fig. 8, where $F^* = V_b/\Gamma$, $F^*_m = V_{b,m}/\Gamma$, $d^* = d_{roof}/\Gamma$, $d^*_m = d_{roof,m}/\Gamma$, $d^*_y = (2E^* - F^*_m d^*_m)/(k^* d^*_m F^*_m)$. Here, $k^*$ is the secant stiffness to the point of the capacity curve where first yielding of the primary structure occurs – clause 6.5.3(3) – and $E^*$ is the area under the transformed capacity curve $F^*$-$d^*$ up to the point ($d^*_m$, $F^*_m$). In this case study, the period of the equivalent SDOF system $T^*$(=2.65s $= 2\pi\sqrt{m^*/k^*}$, where $m^* = \Sigma m_i \phi_i$) is larger than $T_c$ (see Fig. 4) and thus the target displacement $d^*_t$ of the equivalent SDOF system is $d_t{}^* = S_{De}(T^*) = 0{,}282$m, and the corresponding roof displacement $d_{roof,max} = d_t{}^*\Gamma = 0.355$ m. It is worth noting that this displacement is very close to that obtained with the energy based approach (=0.356 m). At the target displacement $d_{roof,max} = d_t{}^*\Gamma = 0.355$ m, the pushover analysis gave plastic rotations at the hinges located at the bottom end of the columns of the first (ground) storey and at the hinges located at the ends of the beams of the first and second levels. The maximum total rotations in these hinges at $d_{roof,max} = d_t{}^*\Gamma = 0.355$ m ranged between 37% and 78% of the corresponding maximum value $\theta_{SD} = (\theta_y + \alpha_{SD,\theta}\theta_u^{pl})/\gamma_{Rd,SD,\theta}$ allowed by pr1998-1-1:2021. The predicted maximum inter storey drifts are plot with black solid bold lines in Fig. 8 (right); in the first and second storeys their values are 34% and 42% larger than the 2% of story heigh required by prEN1998-1-2:2022. Thus, the RC verified with the DB approach should be re-dimensioned.

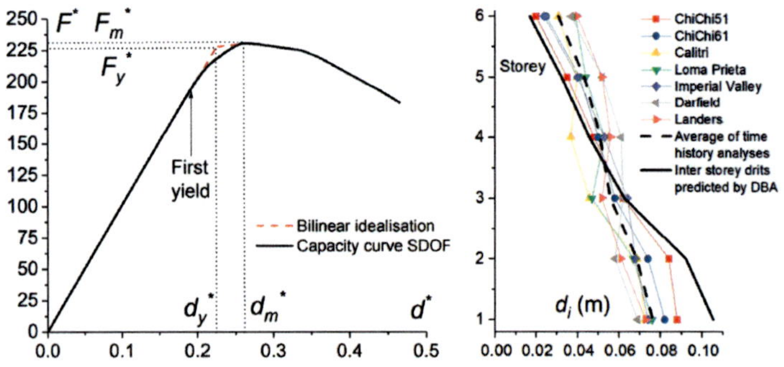


**Fig. 8:** DB approach: capacity curve (left) and predicted maximum drifts (right)

# 6 Time History Analyses

The response of the RC frame verified with the EB (CASE 0) approach, and with the DB approach (Fig. 3 right) was subjected to seven accelerograms which 5% damped elastic acceleration spectra are depicted in Fig. 4 (left). The records were scaled by factor λ in order to satisfy the scaling criteria of prEN1998-1-1:2021 and the additional condition that the energy $E_D$ at instant of maximum lateral deformation $E_D$ ($t_m$) was equal or larger than the value adopted in the EB approach (59 kNm). The name of the records used and the values of λ are: CHICHI-TCU051-E (λ = 1.0), CHICHI-TCU065-E (λ = 0.4), NORTHR-0637-270 (λ = 0.7), LOMAP-G02000 (λ = 1.63), IMPVALL-H-EDA270 (λ = 0.6), RSN6897-DARFIELD-DSLCN27W (λ = 1.75), LANDERS-MVH000 (λ = 2.0), CALITRI-NS (λ = 3.0).

Figure 7 shows with lines and symbols the plastic strain energy dissipation demand on each story, $E_{pH,k,max}$ (left) and the maximum interstory drifts (right). In calculating $E_{pH,k,max}$ the energy dissipated by each beam was split in two parts with the ratio $d_k$ of Eq. (15); the portion $d_k$ was assigned to the storey below the beam and the portion (1-$d_k$) to the story above the beam. The black bold dashed lines indicate the averaged value for the seven records and the black bold solid lines the prediction of the EB method. Regarding the plastic energy dissipation demand (Fig. 7 left), it can be seen that the prediction is larger than the average value obtained in the time history analyses in all storeys. For the first and second storey that dissipated most of the energy, the prediction is 65% larger than the averaged demand, and 29% smaller than the maximum demand. Therefore, the EB method provides a safe-side estimation of the average energy dissipation demand. As for the maximum inter storey drifts (Fig. 7 right), the ratio between averaged response and prediction ranges between 0.58 (top storey) and 1.17 (first storey), the coefficient of variation (COV) being 0.206. The EB approach predicts reasonably well the maximum drift.

Figure 8 (right) compares the averaged response obtained with the time history analyses (bold black dashed line) with the prediction provided by the DB approach (bold black solid line) in terms of inter storey drifts. The ratio between averaged response and prediction ranges between 1.79 (top storey) and 0.72 (first storey), the coefficient of variation (COV) being 0.835.

## 7 Conclusions

A RC frame was designed and verified applying the FB approach, the EB approach and the DB approach, as established in the last drafts of the 2G of Eurocode 8. The RC frame designed with the last two approaches was also verified through time history analyses. The following conclusions can be put forth.

The volume of reinforcing steel required by the RC frame designed with the FB approach is about 23% larger than that required by the counterpart frame verified with the EB approach. This increment occurs mostly in columns and is due to imposing the condition $\sum M_{Rc} \geq 1{,}3\sum M_{Rb}$. In the RC frame verified with the EB approach the condition $\sum M_{Rc} \geq 1{,}3\sum M_{Rb}$ is not applied, but in turn it is directly and quantitatively verified that the plastic strain energy dissipation demand is smaller than the capacity. In our opinion, the energy verification is a much more clear and straightforward way to avoid the negative consequences of the formation of a soft-storey plastic mechanism than imposing the condition $\sum M_{Rc} \geq 1{,}3\sum M_{Rb}$. The RC frame successfully verified with the EB approach did not pass the verification with the DB approach because the later method overestimated the maximum lateral drifts of the first and second stories by about 38%. The plastic strain energy dissipation demand at each storey predicted with the EB approach is reasonably close and safe-side above the average values obtained with time history analyses. Comparing the average inter storey drifts obtained with time history analyses with those predicted by the EB and by the DB approaches, the former gives better results with smaller COV of the error.

**Acknowledgements.** This work has been supported by the Spanish Government under Project MEC_PID2020-120135RB-I00 and received funds from the European Union (Fonds Européen de Dévelopment Régional).

## References

1. Eurocode 8 Design of structures for earthquake resistance Part 1–1: General rules and seismic actions; prEN1998-1998-1-1:2021. European Committee for Standardization; Brussels, Belgium (2021)
2. Eurocode 8 Design of structures for earthquake resistance Part 1–2: Buildings; prEN1998-1998-1-2:2022. European Committee for Standardization; Brussels, Belgium (2021)
3. Housner, G.W.: Limit design of structures to resist earthquakes. In: Proceedings of the First World Conference on Earthquake Engineering. International Association for Earthquake Engineering (IAEE), Berkeley, CA (1956)
4. Akiyama, H.: Earthquake-Resistant Limit-State Design for Buildings. University of Tokyo Press, Tokyo, Japan (1985)
5. Akiyama, H.: Earthquake-Resistant Design Method for Buildings based on energy Balance. Ed. Gihodo Shuppan Ltd. (1999)
6. Building Standard Law: The Building Standard Law of Japan, Building Research Institute; Tokyo, Japan (2009)
7. Benavent-Climent, A.: An energy-based damage model for seismic response of steel structures. Earthq. Eng. Struct. Dyn. **36**, 1049–1064 (2007)

# Physics-Based Approach to Define Energy-Based Seismic Input: Application to Selected Sites in Central Italy

Fabio Romanelli[1] (✉), Franco Vaccari[1], Cristina Cantagallo[2], Guido Camata[2], and Giuliano F. Panza[3,4]

[1] Department of Mathematics and Geosciences, University of Trieste, Trieste, Italy
romanel@units.it
[2] University "G. D'Annunzio", Chieti-Pescara, Italy
[3] Accademia Dei Lincei, Rome, Italy
[4] Associate to the National Institute of Oceanography and Applied Geophysics - OGS, Sgonico, Italy

**Abstract.** For relevant engineering purposes a viable reliable alternative to standard estimates of seismic hazard is represented by the use of scenario earthquakes, characterized not only in terms of magnitude, distance and faulting style, but also taking into account the complexity of the kinematic source rupturing process. Multi-scenario-based NDSHA (Neo-Deterministic Seismic Hazard Assessment) effectively accounts for the tensor nature of earthquake ground motions, formally described as the tensor product of the earthquake source functions and the Green's functions of the transmitting anelastic medium, naturally supplying realistic time series, readily applicable to engineering analysis. Furthermore, it does not rely on scalar empirical ground motion attenuation models (GMPEs), as these are often both weakly constrained by available observations and fundamentally unable to account for the tensor nature of earthquake ground motions. This methodology has been successfully applied to many urban areas worldwide for the purpose of seismic microzoning, to strategic buildings, lifelines and cultural heritage sites. Some examples for selected sites in Central Italy, affected by the seismic sequence started in 2016, are here discussed, analyzing the energy content of the synthetic strong motion time histories, including the ones that predicted the spectral characteristics of the ground shaking due to the 24 August 2016 event. Specifically, the energy content obtained from (1) linearly scaled real records, (2) spectral matched records, (3) artificial accelerograms and (4) physics-based NDSHA ground motions, selected for a bridge hypothetically located at Amatrice, is compared. The comparison shows that the generation of artificial accelerograms (3) and partially also the spectral matching process (2) are characterized by an increase of the ground motion energy parameters with respect to the physics-based synthetic signals (4) and the linearly scaled records (1).

**Keywords:** Seismic Hazard · Ground Motion · Seismic Input · NDSHA · Central Italy


H. Varum et al. (Eds.): IWEBSE 2023, LNCE 236, pp. 112–128, 2023.
https://doi.org/10.1007/978-3-031-36562-1_9

## 1 Introduction and Methodology

The modeling of seismic input described in this work is carried out using the methodology known as NDSHA (Neo-Deterministic Seismic Hazard Assessment), based on the calculation of realistic synthetic seismograms, already used, since more than two decades, for seismic hazard estimation at regional and local scales both in Italy and abroad (e.g.: Panza et al., 2001; Panza et al., 2004; Panza et al., 2012; Panza and Peresan, 2016; Panza et al., 2022 and bibliography cited therein). It is a method based on the physics of seismic wave generation and propagation that requires the definition of the characteristics of the source and anelastic medium in which the waves propagate; the method is in line with paragraph 3.2.3.6 of Norme Tecniche per le Costruzioni 2018 (NTC 2018). It enables quick parametric tests to examine how each model component affects the shaking recorded at the sites under consideration.

For seismic hazard mapping on a national scale, a "Regional Scale Analysis" is performed using many possible sources and simplified realistic structural models, which are representative of rock substrate (bedrock) conditions. When a detailed analysis is needed, a "Site Specific Analysis" can be carried out that allows for the consideration of structural and topographic heterogeneities as well as the impact of the fault rupture process on the features of the generated wavefield.

The software, developed at the University of Trieste in the framework of a large international cooperation (for a recent review see Panza et al. 2022), allows both first-order seismic zonation of a large territory and detailed modeling of the seismic input to a specific site taking into account in detail the kinematic characteristics of the rupture process on the fault (Gusev and Pavlov, 2006; Gusev, 2011) and the site effects caused by local stratification. The multimodal summation technique (Panza, 1985; Florsch et al., 1991) is used to generate the signals when the site of interest is located in the far field relative to the epicenter zone, whereas the discrete wave number method (Pavlov, 2009) permits the generation of seismic input even for sites located above the fault (i.e., near field). The impact of lateral heterogeneities in the media on seismic wave propagation is effectively approximated along profiles using a hybrid methodology based on multimodal summation and the finite difference technique (Fäh et al., 1993). For recent applications of this methodology in Italy, see Panza et al. (2015); Fasan et al. (2016); De Natale et al. (2019); Nunziata and Costanzo (2022); Romanelli et al. (2022).

The recent evolution of the package of programs for the calculation of realistic synthetic seismograms and their use in the engineering environment is based on the experience gained in the field of seismic verification and is the result of close collaboration between seismologists and engineers. The development was aimed primarily at making it easy to use for issues related to seismic input estimation, for seismic verifications of pre-existing structures or for the design of new structures. To this end, a web application was also created (Vaccari, 2016; Vaccari and Magrin, 2022), which hides the complexity of the computational engine behind a friendly graphical user interface.

## 2 Seismic Input Definition at Amatrice (Central Italy)

Panza et al. (2012) published (continuing their 2001 work) several national-scale seismic hazard maps for the Italian territory, based on the NDSHA methodology. Synthetic seismograms are calculated at the knots of a 0.2° × 0.2° pitch grid covering the national territory. At each site, synthetic seismograms generated from all sources that fall within seismogenic zones (e.g., Meletti et al., 2004) and seismogenic nodes (e.g., Gorshkov et al., 2002) within a reasonable limit of epicentral distance are calculated. The largest shaking value in terms of displacement, velocity and acceleration is then represented in each map, with intentionally coarse clustering classes consistent with the real resolving power of the available data, signifying the limited precision that can be associated with a zoning performed over a relatively large territory.

The kinematic characteristics of the source rupture process and of the local layering are considered in this Section, targeted at a site in Amatrice, where a hypothetical bridge is considered, for which the model knowledge can be adequately deepened. The spatial variability of ground motion (asynchronous ground motion) may generate a larger structural demand than the synchronous one. In some cases, the 2016 earthquake caused structural and non-structural damage due to differential displacements and large superstructure movements, which the asynchronous ground motion could have worsened (e.g. Romanelli et al., 2004; Di Sarno et al., 2019).

### 2.1 Definition of Structural Models

Along the lines of what was done by Fasan et al. (2016) for Norcia, modeling has been carried out for the Amatrice site using both the lithospheric structure adopted for calculations at the national scale by Panza et al. (2012, Model 6) and models in which the surface part of those structures (representing bedrock conditions) is replaced by layers with comparatively slower wave propagation velocities, which provide a simplified but realistic representation of the soil characteristics at the site of interest. Figure 1a shows the lithospheric structure used for the Amatrice area in the calculations for Models 5 and 6 in Panza et al. (2012) and Fig. 1b depicts the uppermost kilometer, where the stratification includes two low-velocity layers, typical of the site conditions at Amatrice, defined starting from the information available in the microzonation study of Chiaretti and Nibbi (2017).

### 2.2 Definition of Seismogenic Sources

In the modeling of seismic hazard at the national scale, for the characterization of seismic sources the NDSHA methodology allows to consider both the definition of seismogenic zones, such as the ZS9 defined by Meletti et al. (2004), and the seismogenic nodes identified applying pattern recognition to a morphostructural analysis of the considered territory (Gorshkov et al., 2002; 2004; 2009; Gorshkov and Soloviev, 2022). Sources are distributed within seismogenic zones and nodes on a 0.2° x 0.2° grid offset by 0.1° from the grid on which the sites where synthetic seismograms are to be computed are placed. The magnitude associated with each source is derived by processing the seismicity image provided by a reliable earthquake catalog. The seismicity defined by

a) 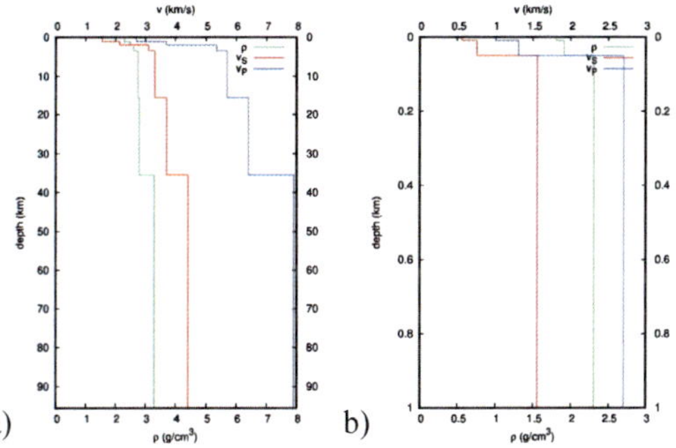
b)

**Fig. 1.** a) Bedrock structure for the Amatrice area extracted from the model used to define seismic hazard at the national scale by Panza et al. (2012). b) Upper kilometer of the model in Fig. 1a, in which the uppermost layering has been modified introducing two low-velocity layers to realistically represent the average local geological situation in Amatrice.

the events reported in the catalog is first discretized on the source grid, assigning to each point the maximum magnitude of the events that belong to the $0.2^\circ \times 0.2^\circ$ cell centered on that point. A "smoothing" operation is then performed by assigning the magnitude of the central cell to grid points that fall within a window having a radius equal to three cells, if this leads to an increase in magnitude for the points belonging to the window; the operation is repeated for all grid points (as described in Panza et al., 2001). Figure 2a shows the initial image of the seismicity for the area of interest, taken from the CPTI04 catalog (Gruppo di lavoro CPTI, 2004), while in Fig. 2b we see the result of the discretization and "smoothing" operation within the ZS9 seismogenic zones. The 2016 seismic sequence is not included in the CPTI04 catalog because its most recent event is from October 31, 2002. Figure 3a shows the geometry, size (circles of 25 km radius) and magnitude associated to the seismogenic nodes identified in the area (Gorshkov et al., 2002; 2004; 2009; Gorshkov and Soloviev, 2022); Fig. 3b shows the result of the seismicity discretization and "smoothing" operation jointly applied to the seismogenic nodes and zones. In Fig. 4 are shown the focal mechanisms associated with seismogenic nodes (a) and ZS9 seismogenic zones (b). All sources that fall within the same node or seismogenic zone share the focal mechanism. Where nodes overlap seismogenetic zones, two sources are considered on the same grid point, each with its own focal mechanism.

## 2.3 Multi-scenario Calculation

Figure 5 illustrates an example of applying the NDSHA approach to the seismic hazard estimate at the Amatrice site (equivalent to Model 5 of Panza et al. (2012)). The sources are located on the sites depicted in Fig. 3b, and synthetic seismograms are generated at the four grid points closest to Amatrice (Fig. 5a). In this modeling, realistic synthetic seismograms (cutoff frequency = 10 Hz) were generated by the modal summation technique for epicentral distances greater than 20 km, and by the discrete wave number method for shorter paths, thus optimizing the use of computational resources. At all four locations, the maximum acceleration value, obtained for the source closest to each

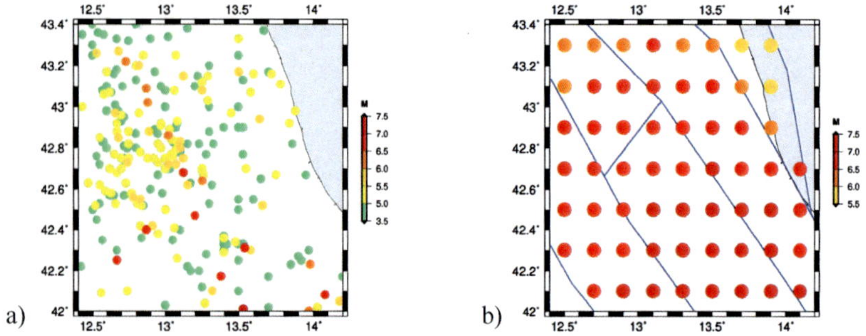


**Fig. 2.** Application of the seismicity discretization method to the area centered on Amatrice. CPTI04 catalog (CPTI04 Working Group, 2004); ZS9 seismogenic zones (Meletti et al., 2004).

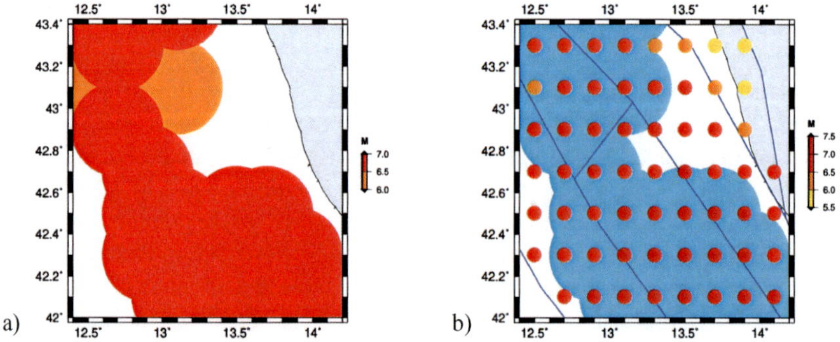


**Fig. 3.** Application to the area centered on Amatrice of the seismicity discretization procedure, applied jointly to seismogenic zones and nodes (circles of 25 km radius). CPTI04 catalog (CPTI04 Working Group, 2004); seismogenic nodes from Gorshkov et al. (2002; 2004; 2009) and Gorshkov and Soloviev (2022); ZS9 seismogenic zones from Meletti et al. (2004).

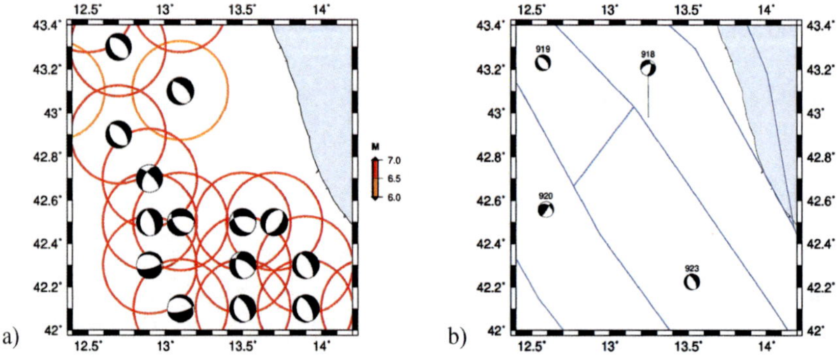


**Fig. 4.** Focal mechanisms associated with seismogenic nodes (a) and seismogenic zones (b). Data used by Panza et al. (2012) for their Models 4–7.

site (epicentral distance ~ 14 km), falls in the 0.3–0.6 g class (Fig. 5b), whose range represents the quality (real resolving power) of the available data.

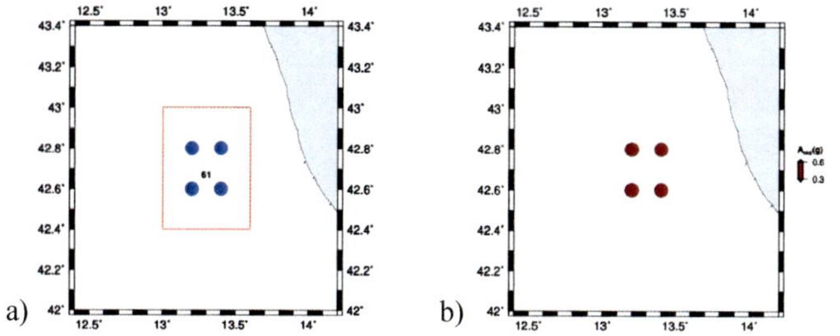


**Fig. 5.** (a) Location of the four points of the 0.2° × 0.2° cell that encompasses Amatrice. The red rectangle defines the structural model considered: it is a portion of polygon 61 model used by Panza et al. (2012); (b) peak values of the resultant horizontal acceleration, obtained for a single realization of the rupture process at the earthquake source.

The slip distribution on the fault, the location of the rupture nucleation point, and the rupture propagation velocity on the fault plane are all defined by the source time functions used to describe the energy release at the hypocenter. These functions also have a random component. The result shown in Fig. 5 comes from a single random realization of the rupture process. Moreover, different realizations, consistently with the resolving power of the available data, may cause a jump of one acceleration class (either up or down), particularly for sources that are near the observation point. To properly explore the model parameter space, multiple scenarios are generated, considering different realizations of the rupture process on the fault, and different structural models. To deal with the uncertainty on the rupture model of the fault, which is very important especially for short epicentral distances, multiple temporal source functions are generated with the NDSHA approach for many random variations in the slip distribution on the fault plane, the location of the initial rupture point, and the rupture propagation velocity. To include the effects of the fault rupture on the time series, these functions are then convolved with the accelerograms produced by a point and instantaneous earthquake source model (Size and Time Scaled Point Source (STSPS) approximation used by Panza et al. (2012) for Models 2–7). Calculations are also repeated under the assumption of bilateral or unilateral rupture, placing the initial point of rupture near the center of the fault or near one edge of the fault, respectively.

To summarize the results, statistics are then performed on the response spectra obtained from the synthetic accelerograms, according to the MCSI scheme (Maximum Credible Seismic Input, as described in Fasan et al. (2016) and Rugarli et al. (2019)). In the specific case of Amatrice, 100 realizations of unilateral rupture and 100 of bilateral rupture were taken into account, for a total of 200 realizations for each of the sources considered. In this simulation, we produced realistic synthetic accelerograms at the grid point that was the closest to Amatrice, the one at coordinates 42.6N, 13.2E. The modeling was repeated also taking into consideration the two structural models shown in Fig. 1, namely the CELLE bedrock models used by Panza et al. (2012) and its variant including the layering that can be reasonably assumed in Amatrice. The results are shown in

Fig. 6, where it is evident the increase in spectral acceleration values for the structural model in which the local stratigraphy, characterized at the surface by Vs velocities of 550 m/s, is considered. The increase in values is close to a factor of 2, consistent with a one-degree change in macroseismic intensity (MCS) (Cancani, 1904; Lliboutry, 2000); in other words, in the present case, site effects may be responsible for a one-degree (or more) increase in MCS intensity.

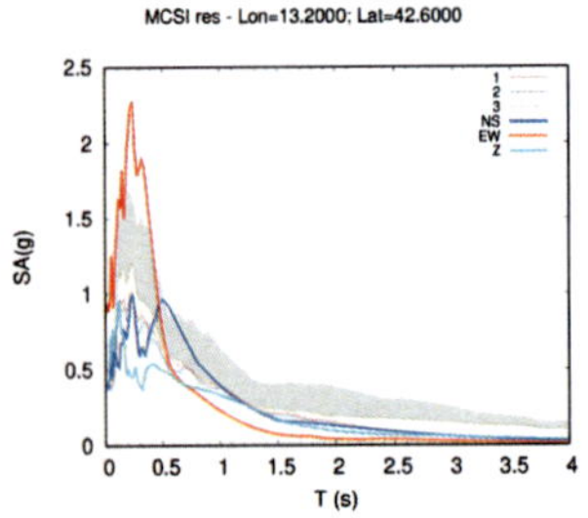


| | source | $M_w$ | edi (km) | depth (km) | strike (°) | dip (°) | rake (°) | sre (°) | slon (°E) | slat (°N) |
|---|---|---|---|---|---|---|---|---|---|---|
| 1 | nn 3045 un | 7.0 | 13.8 | 15.0 | 175 | 50 | 320 | 138 | 13.1 | 42.5 |
| 2 | nn 3046 un | 7.0 | 13.8 | 15.0 | 207 | 40 | 263 | 243 | 13.3 | 42.5 |
| 3 | nn 3045 bn | 7.0 | 13.8 | 15.0 | 175 | 50 | 320 | 138 | 13.1 | 42.5 |

a)

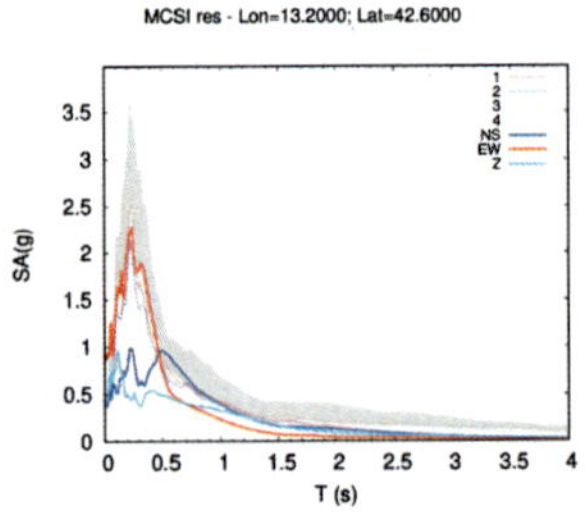


| | source | $M_w$ | edi (km) | depth (km) | strike (°) | dip (°) | rake (°) | sre (°) | slon (°E) | slat (°N) |
|---|---|---|---|---|---|---|---|---|---|---|
| 1 | nn 3045 un | 7.0 | 13.8 | 15.0 | 175 | 50 | 320 | 138 | 13.1 | 42.5 |
| 2 | nn 3046 un | 7.0 | 13.8 | 15.0 | 207 | 40 | 263 | 243 | 13.3 | 42.5 |
| 3 | nn 3045 bn | 7.0 | 13.8 | 15.0 | 175 | 50 | 320 | 138 | 13.1 | 42.5 |
| 4 | nn 3046 bn | 7.0 | 13.8 | 15.0 | 207 | 40 | 263 | 243 | 13.3 | 42.5 |

b)

**Fig. 6.** MCSI spectra obtained using the bedrock structural model (a) and the model fitted to the local stratigraphy of Amatrice (b). The numbered curves refer to the medians of the spectral acceleration associated with the sources whose median dominates over the others at least at one period of the response spectrum. The gray area of the graphs ranges from the $50^{th}$ to the $95^{th}$ percentile of the spectral acceleration distribution for each period of the response spectrum. NS, EW and Z response spectra are obtained from the accelerograms recorded at the AMT station of the National Accelerometer Network for the August 24, 2016 event.

Figure 6 shows also the comparison of the MCSI spectra with the NS, EW and Z response spectra obtained from the accelerograms recorded for the August 24, 2016 event ($M_w$ 6.0, as reported by INGV; $M_w$ 6.2 as reported by USGS, i.e. equal within global standard error 0.2–0.3 (Båth, 1973; Bormann et al., 2007)) at the AMT station of the National Accelerometric Network. For the observed response spectrum to have the EW component fall within the $95^{th}$ percentile of the MCSI spectrum, local stratification must be taken into account (upper limit of the gray area). A selection, made around the $90^{th}$ percentile at the periods of T = 0.2 s, T = 0.6 s and T = 1.0 s, of eleven pairs (horizontal components) acceleration time-histories extracted from the MCSI spectra is shown in Fig. 7a, 7b and 7c, respectively, which are analyzed in Sect. 3.

The modeling presented in this section does not specifically target the event under consideration, but rather, it is an extraction made at the grid point closest to Amatrice of calculations performed earlier (e.g. Panza et al., 2012) for a first-order zonation of the entire national territory considering the set of sources defined within the seismogenic zones and seismogenic nodes.

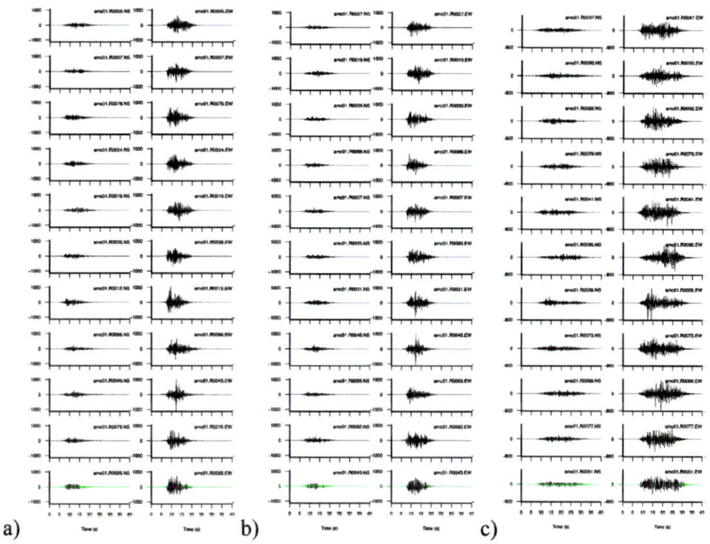

**Fig. 7.** Accelerograms extracted around the 90th percentile of the MCSI shown in Fig. 6b at periods a) T = 0.2 s, b) T = 0.6 s and c) T = 1 s.

## 3 Comparison Between Simulated, Scaled, Matched and Artificial Records

Seismic assessment of structures and more specifically of bridges via non-linear dynamic analysis requires proper seismic input selection. Seismic codes suggest different procedures to select ground motion signals, most of those assuming spectral compatibility to the elastic design spectrum as the main criterion (Iervolino et al., 2008, 2009). As indicated in Lin (2018), the main issues related to the use of spectrum-compatible accelerograms are: (i) types of accelerograms to be used in the analysis (i.e. real or artificial), and (ii) method for the selection and scaling of the spectrum-compatible accelerograms. In literature, the following four types of time histories are used as the seismic input, namely: (1) Linearly Scaled Records (LS-R); (2) Spectral Matched Records (SM-R); (3) Artificial Accelerograms (AA); and (4) Physics-Based NDSHA ground motions. LS-R (1) and SM-R (2) are real earthquake records modified to represent the characteristics of ground motions at the considered location. Spectral matching can be carried by using scaled or unscaled records, and the corresponding modified records are defined as Spectral Matched Scaled Records SM-SR (2a) and Spectral Matched Unscaled Records SM-UR (2b), respectively. Artificial accelerograms (3) are signals obtained using the theory of random vibration. AASHTO (2014) 4.7.4.3.4b appears to encourage the use of real records (eventually linearly scaled) with respect to spectral matched or artificial accelerograms, as it notes that "*Analytical techniques used for spectrum matching shall be demonstrated to be capable of achieving seismologically realistic time series that are similar to the time series of the initial time histories selected for spectrum matching*". Similarly, Eurocode 8 (EC8) Part 2 Sect. 3.2.3 (2005) suggests using real records as the primary choice for the seismic input. More specifically, this code states that "*When the required number of pairs of appropriate recorded ground motions is not available, appropriate modified recordings or simulated accelerograms may replace the missing recorded motions.*". Similarly, the US Standard ASCE/SEI 7–10 (ASCE, 2010) states clearly that real (i.e. recorded) accelerograms are preferred for the time-history analysis while artificial accelerograms can also be used if the real ones are not available. All the considered regulatory codes for bridges contemplate therefore as primary choice the use of real or modified real records to perform time history analysis. As described in Sect. 2, NDSHA synthetic accelerograms (4) are generated on the base of a physics-based approach that simulates ground motions founded on realistic source and ray-path propagation models (as prescribed by NTC 2018 (§3.2.3.6)). For this reason, the NDSHA synthetic signals are compared with LS-R (1), SM-SR (2a), SM-UR (2b) and AA (3). To have the same number of accelerograms obtained from the simulated signals, for all considered selection sets, eleven pairs of horizontal accelerograms are considered. The comparison is done in terms of energy-based Ground Motion Parameters (GMPs), which are potential indicators of how much energy it will be released from the earthquake.

### 3.1 Record Selection Based on Spectrum-Compatibility

The definition of the ground motion records is performed according to the EC8 - Part 2 provisions (§3.2.3 *Time-history representation*). To this purpose, the spectrum corresponding to the ground motion recorded during the Amatrice earthquake of 24 August 2016 is assumed as target spectrum (Magnitude $M_w = 6.0$; Epicentral Distance $R =$ 8.5 km; station code AMT; $V_{s30} = 670$ m/s; Soil class EC8 = B). Ground motion records are selected using the accelerograms recorded on soil class B (EC8 – Part 1) stored in the Engineering Strong Motion Database v.2.0 (Luzi et al., 2020) with the partial support of the software SeismoSelect v.2022.

After having performed a preliminary selection of 23 ground motion records based on seismological, known characteristics of the considered area, for each selected record, two 5%-damped response spectra of the corresponding horizontal components are considered. Subsequently, as indicated in the EC8 - Part 2, the Square Root of the Sum of Squares (SRSS) of the horizontal spectral components is calculated.

$$S_a(T_i) = \sqrt{S_{aX}(T_i)^2 + S_{aY}(T_i)^2} \tag{1}$$

Then, all the SRSS spectra are scaled to the Target Spectral acceleration amplified by 1.3 times to have a minimum deviation, $\delta$, from the target spectrum and a minimum scaling factor. A single scale factor is applied to both horizontal components of each record.

The deviation of the single SRSS records from the target spectrum amplified by 1.3 times is defined in Iervolino et al. (2008), and modified to accommodate the seismic code requirement for bridges (EC8 - Part 2):

$$\delta = \sqrt{\frac{1}{N}\sum_{i=1}^{N}\left(\frac{S_a(T_i) - 1.3 \cdot S_{a,TS}(T_i)}{1.3 \cdot S_{a,TS}(T_i)}\right)^2} \tag{2}$$

$S_a(T_i)$ is the pseudo-acceleration ordinate of the SRSS spectra corresponding to the period $T_i$, $S_{a,TS}(T_i)$ is the value of the spectral ordinate of the target spectrum at the same period, and $N$ is the number of points sampling the spectra within the considered range of periods.

Eleven scaled SRSS spectra are selected so that, according to EC8 - Part 2, in the range of periods between 0.12 s and 0.2 s, their average spectrum is not lower than 1.3 times the 5%-target spectrum. In addition to this requirement, an upper bound limit is added to the spectrum-compatibility criteria, verifying that the average spectrum is lower than the 130% of the target spectrum amplified by 1.3 times. Table 1 shows the list of the selected ground motion records with the corresponding relevant seismological parameters and scaling factors. In this work, the 1.3 target spectrum is compared with the average value of the eleven SRSS ground motion spectra, according with EC8 - Part 2, also if other efficient approaches are used in literature (Cantagallo et al., 2019).

The eleven pairs of records so selected are then matched to the target spectral shape by wavelet transform. The wavelet transform basically consists of using modulating functions, selectively located in time to modify the spectrum of the signal, where and when it is needed, to match the target spectrum (Hancock et al., 2006). The matching of the time histories is performed with the software SeismoMatch v.2022 using the wavelets algorithm proposed by Al Atik and Abrahamson (2010) and a spectrum matching range [0.1–3.0] s. To observe the variation of the energy-based GMPs, the spectral matching is carried out first on the eleven pairs of linearly scaled records (SM-SR) and then on the corresponding unscaled records (SM-UR).

Finally, eleven pairs of Artificial Accelerograms (AA) are generated using the code SIMQKE based on random vibration theory (Gasparini and Vanmarke, 1976). The generation is carried out using a duration of the stationary part equal to 10 s, a total duration of the accelerograms equal to 20 s and one cycle to smoothen the target spectrum.

Figure 8 shows the comparison between the application of the spectrum-compatibility criteria applied on LS-R (1), SM-SR (2a), SM-UR (2b) and AA (3). Grey curves correspond to the combination SRSS of the horizontal spectra corresponding to each record, the red line is the average value of the eleven SRSS spectra, the black line is the considered target spectrum amplified by 1.3 times according to EC8 - Part 3, the black dashed line is the upper bound, and the black dotted lines correspond to the spectrum-compatibility range.

To verify if the different types of selected accelerograms have comparable intensity values and spectral shapes, the synthetic sets of signals SS (4) corresponding to T = 0.2 s (SS0.2 - 4a), T = 0.6 s (SS0.6 - 4b) and T = 1.0 s (SS1.0 - 4c) at the 90$^{th}$ percentile are compared with the Amatrice target spectrum according to the spectrum-compatible criteria provided in EC8 - Part 2 (Fig. 9). The selected periods fall within the spectrum-compatibility range of a hypothetical bridge located at Amatrice and characterized by a first period of vibration $T_1 = 0.6$ s. Figure 9 shows that the Physics-Based NDSHA ground motions SS (4), are comparable with the LS-R (1) and SM-R (2) because in all considered cases (4a, 4b and 4c) their average spectra are comparable with the corresponding target spectra. Since in (4b) this condition is satisfied over the whole spectrum-compatibility range, SS0.6 are assumed as reference SS for the comparison between the energy-based GMPs.

**Table 1.** Ground motion records selected according to the EC8 - Part 2 spectrum-compatibility criteria

| n | Event ID | Station ID | Nation | Event date/time | Event Latitude/Longitude | Magnitude | Vs30 (m/sec) | Epicentral Distance | Scaling Factor |
|---|---|---|---|---|---|---|---|---|---|
| 2 | EMSC-20160824_0000006 | NRC | IT | 24/08/2016 01:36 | 42.6983/13.2335 | 6.0/$M_w$ | 498 | 15.3 | 1.98 |
| 4 | EMSC-20161030_0000029 | AMT | IT | 30/10/2016 06:40 | 42.8322/13.1107 | 6.5/$M_w$ | 670 | 26.4 | 1.62 |
| 6 | GR-1993–0011 | PYR1 | GR | 26/03/1993 11:58 | 37.6843/21.4367 | 5.4/$M_w$ | 473 | 1.3 | 3.02 |
| 7 | GR-1993–0027 | PAT2 | GR | 14/07/1993 12:31 | 38.2024/21.7701 | 5.6/$M_w$ | 381 | 4.9 | 4.65 |
| 8 | GR-1995–0047 | AIGA | GR | 15/06/1995 00:15 | 38.4043/22.2719 | 6.5/$M_w$ | 524 | 23.6 | 1.30 |
| 10 | IR-1990–0004 | A6211 | IR | 20/06/1990 21:00 | 36.9893/49.3462 | 7.4/$M_w$ | 621 | 34.4 | 1.22 |
| 11 | IR-1994–0020 | ST3297 | IR | 20/06/1994 09:09 | 29.0142/52.6413 | 5.9/$M_w$ | 672 | 6.5 | 1.26 |
| 15 | IT-1997–0006 | NCR | IT | 26/09/1997 09:40 | 43.0310/12.8620 | 6.0/$M_w$ | 555 | 10.9 | 1.37 |
| 19 | IT-2009–0009 | AQV | IT | 06/04/2009 01:32 | 42.3420/13.3800 | 6.1/$M_w$ | 474 | 4.9 | 1.22 |
| 20 | IT-2009–0009 | AQG | IT | 06/04/2009 01:32 | 42.3420/13.3800 | 6.1/$M_w$ | 696 | 5.0 | 1.59 |
| 22 | TK-1977–0003 | 3506 | TR | 16/12/1977 07:37 | 38.4140/27.1882 | 5.3/$M_L$ | 771 | 9.5 | 4.52 |

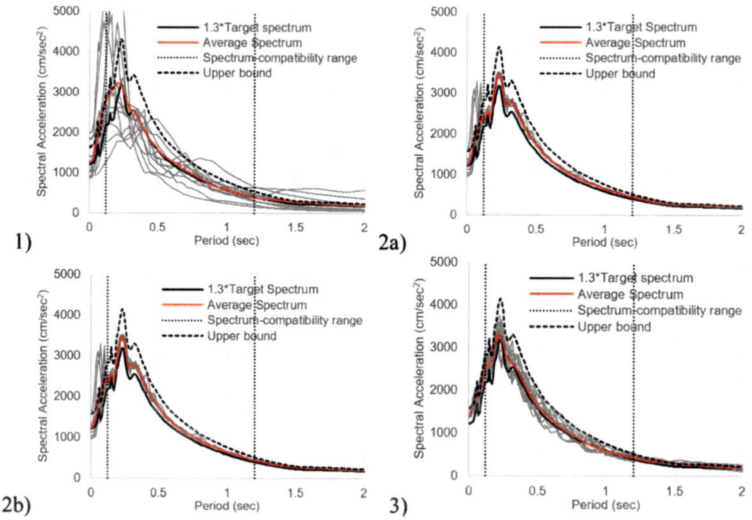


**Fig. 8.** Selection of accelerograms according to the spectrum compatibility criteria using LS-R (1), SM-SR (2a), SM-UR (2b) and AA (3).

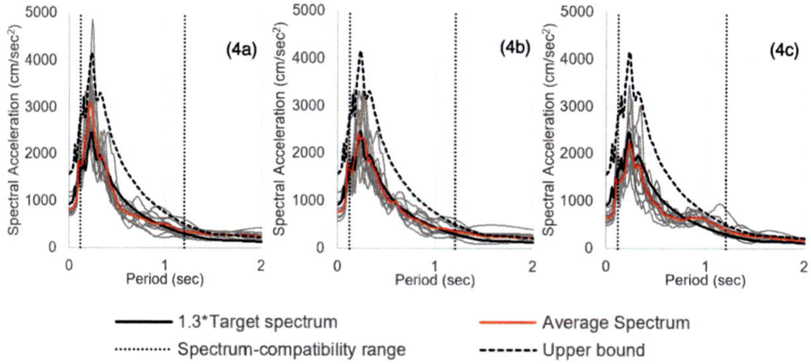


**Fig. 9.** Comparison between the Amatrice target spectrum (black continuous line) and the average spectrum (red line) obtained from the SRSS spectra (in grey) of the eleven pairs of simulated signals calculated for T = 0.2 s (4a), T = 0.6 s (4b) and T = 1.0 s (4c).

## 3.2 Comparison Between Energy-Based Ground Motion Parameters

The Energy-Based GMPs of the synthetic records obtained in Sect. 2.3 are finally compared with the corresponding GMPs of the real, modified and artificial accelerograms. The considered GMPs are the Arias Intensity (Arias, 1970), the Specific Energy Density, the Acceleration Spectrum Intensity (Von Thun et al., 1988) and the Housner Intensity (Housner, 1952).

Arias Intensity (AI) is computed based on the time integral of the squared acceleration according to Eq. (3), where $a(t)$ is the ground motion acceleration at time $t$, $t_{\max}$ is the total duration of the ground motion, and $g$ is the acceleration of gravity.

$$AI = \frac{\pi}{2g}\int_0^{t_{max}} [a(t)]^2 dt \tag{3}$$

Specific Energy Density (SED) is the integral of the square of the velocity time history $v(t)$ for the total duration of the ground motion $t_{max}$.

$$SED = \int_0^{t_{max}} [v(t)]^2 dt \tag{4}$$

Acceleration Spectrum Intensity (ASI) is defined as the integral of the pseudo-spectral acceleration $S_a$ of a ground motion from 0.1 to 0.5 s.

$$ASI = \int_{0.1}^{0.5} S_a(\xi = 5\%, T)dT \tag{5}$$

Housner Intensity (HI) is the integral of the pseudo-velocity spectrum $S_v$ over the period range from 0.1 to 2.5 s.

$$HI = \int_{0.1}^{2.5} S_V(\xi = 5\%, T)dT \tag{6}$$

The energy-based GMPs are obtained for each selected accelerogram. For signals derived from real records, the comparison between the GMPs is carried out on the single ground motion components and then the average values obtained for each selected set are compared, as indicated in Fig. 10. The results indicate that the spectral matching process (2) and the generation of artificial accelerograms (3) generate a significant increase of SED and AI, respectively. In all cases, the SS have the lower energy-based GMPs.

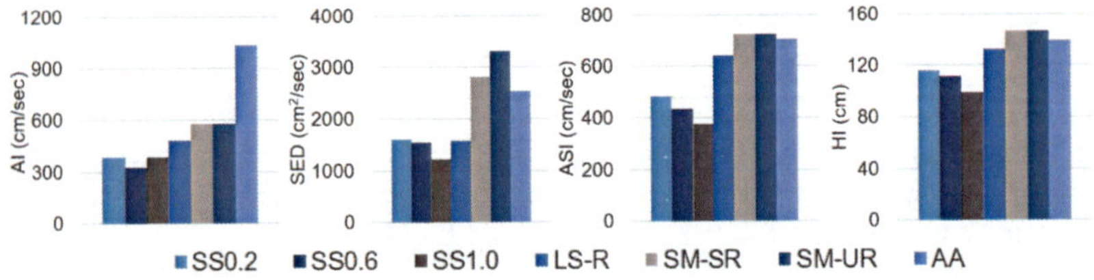


**Fig. 10.** Comparison between the average values of the energy based GMPs (Arias Intensity AI, Specific Energy Density SED, Acceleration Spectrum Intensity ASI and Housner Intensity HI) obtained for the considered set of accelerograms: 1) LS-R: linearly scaled records; 2a) SM-SR: spectral matched scaled records; 2b) SM-UR: spectral matched unscaled records; 3) AA: artificial accelerograms; 4a) SS0.2: synthetic signals for T = 0.2 s; 4b) SS0.6: synthetic signals for T = 0.6 s; 4c) SS1.0: synthetic signals for T = 1.0 s.

Finally, the GMPs obtained from LS-R, SM-SR, SM-UR and AA are compared with those obtained from SS0.6, assumed as reference signals (see Fig. 11). This comparison shows that AA have a very high AI with respect to SS0.6. Conversely, LS-R show the lowest energy content for all considered GMPs.

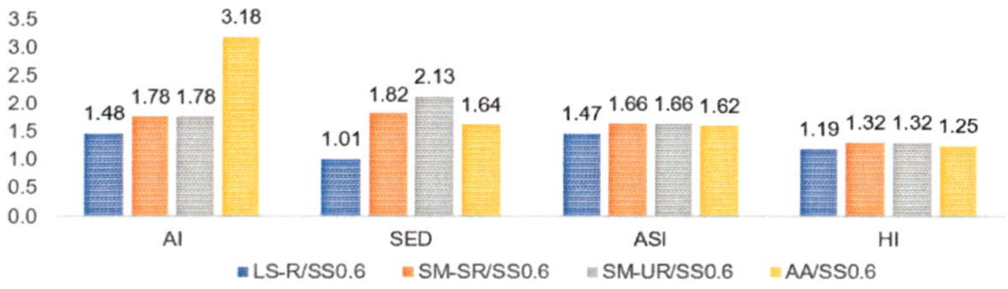


**Fig. 11.** Ratios between the average values of energy-based GMPs (Arias Intensity AI, Specific Energy Density SED, Acceleration Spectrum Intensity ASI and Housner Intensity HI) obtained from Linearly Scaled Records LS-R (1), Spectral Matched Scaled Records SM-SR (2a), Spectral Matched Unscaled Records SM-UR (2b), Artificial Accelerograms AA (3) and the corresponding parameters obtained from the physics-based NDSHA synthetic signals for T = 0.6 s (SS0.6).

# 4 Conclusions

In this work the Energy-Based Ground Motion Parameters (GMPs) obtained from the physics-based NDSHA ground motions (4) are compared with the corresponding GMPs obtained from Linearly Scaled Records LS-R (1), Spectral Matched Records SM-R (2) and Artificial Accelerograms AA (3). To this purpose a set of eleven pairs of horizontal ground motion components is selected and linearly scaled according to the spectrum-compatibility criteria defined in EC8 - Part 2, obtaining the target spectrum from the ground motion recorded during the Amatrice earthquake of 24 August 2016 (reported with Magnitude $M_w = 6.0$; Epicentral Distance $R = 8.5$ km; station code AMT; $V_{s30} = 670$ m/s; Soil class EC8 = B). LS-R are subsequently matched to the target spectrum by using wavelet transforms and obtaining the so-called Spectral Matched Scaled Records SM-SR (2a). The same matching process is carried out on the corresponding unscaled records, obtaining in this case the Spectral Matched Unscaled Records SM-UR (2b). Finally, a set of artificial accelerograms AA (3) is generated by using the random vibration theory. The comparison between the different sets of accelerograms is carried out comparing Arias Intensity (AI), Specific Energy Density (SED), Acceleration Spectrum Intensity (ASI) and Housner Intensity (HI) obtained from LS-R (1), SM-SR (2a), SM-UR (2b), AA (3) and SS (4). The comparison shows that:

- the generation of AA (3) produces high values of AI with respect to the other groups of records;
- the spectral matching process (2) and the generation of artificial signals (3) produce signals with a significative energy content in terms of SED and ASI;
- the ratio between the average values of the GMPs obtained from the different sets of records and from the synthetic signals obtained for T = 0.6 s (4b - SS0.6) shows that LS-R (1) produce the lowest increase of the energy content with respect to SS0.6. Conversely, SM-SR (2), SM-UR (3) and AA (3) generate a considerable increase of the records' energy, especially in terms of AI and SED.

The results applied to the considered groups of records indicate therefore that the application of SS and LS-R is preferable to spectral matched or artificial accelerograms as their unrealistic high energy content could unproperly modify the structural response.

Future developments of this work should provide the application of the considered sets of records to real case studies.

## References


AASHTO L.R.F.D.: Bridge design specifications. American Association of State Highway and Transportation Officials (2014)

Al-Atik, L., Abrahamson, N.A.: An improved method for nonstationary spectral matching. Earthq. Spectra **26**(6), 601–617 (2010)

Arias, A.: A measure of earthquake intensity. In: Seismic design for nuclear power plants. Hansen, R.J., (ed.) pp. 438–483. MIT Press, Cambridge, Massachusetts (1970)

ASCE/SEI 7–10, Minimum design loads for buildings and other structures. American Society of Civil Engineers (2010)

Båth, M.: Introduction to Seismology. Birkhäuser Verlag, Basel (1973)

Bormann, P., Liu, R.F., Ren, X., Gutdeutsch, R., Kaiser, D., Castellaro, S.: Chinese national network magnitudes, their relation to NEIC magnitudes, and recommendations for new IASPEI magnitude standards. Bull. Seismol. Soc. Am. **97**, 114–127 (2007)

Cancani, A.: Sur l'emploi d'une double échelle sismique des intensités. G. Beitr. Ergänzungsband **2**, 281–283 (1904)

Cantagallo, C., Camata, G., Spacone, E.: A probability-based approach for the definition of the expected seismic damage evaluated with non-linear time-history analyses. J. Earthq. Eng. **23**(2), 261–283 (2019)

Chiaretti, F., Nibbi, L.: Microzonazione Sismica di Livello 3 del Comune di Amatrice ai sensi dell'Ordinanza del Commissario Straordinario n. 24 registrata il 15 maggio 2017 al n. 1065. Comune di Amatrice, Provincia di Rieti, Regione Lazio (2017)

De Natale, G., et al.: Seismic risk mitigation at Ischia island (Naples, Southern Italy): an innovative approach to mitigate catastrophic scenarios. Eng. Geology **261**, 105285 (2019)

Di Sarno, L., da Porto, F., Guerrini, G., Calvi, P.M., Camata, G., Prota, A.: Seismic performance of bridges during the 2016 central Italy earthquakes. Bull. Earthq. Eng. **17**(10), 5729–5761 (2018)

Eurocode 8 (EC8): design provisions for earthquake resistance of structures, Part 2-bridges, EN 1998–2:2005 (2005)

Fäh, D., Iodice, C., Suhadolc, P., Panza, G.F.: A new method for the realistic estimation of seismic ground motion in megacities: the case of Rome. Earthq. Spectra **9**, 643–667 (1993)

Fasan, M., Magrin, A., Amadio, C., Romanelli, F., Vaccari, F., Panza, G.F.: A seismological and engineering perspective on the 2016 Central Italy earthquakes. Int. J. Earthq. Impact Eng. **1**(4), 395–420 (2016)

Florsch, N., Fäh, D., Suhadolc, P., Panza, G.F.: Complete synthetic seismograms for high-frequency multimode SH-waves. Pure Appl. Geophys. **136**, 529–560 (1991)

Gasparini, D.A., Vanmarcke, E.H.: Simulated earthquake motions compatible with prescribed response spectra. Evaluation of Seismic Safety of Buildings Report No. 2, Department of Civil Engineering, Massachusetts Institute of Technology, USA (1976)

Gorshkov, A.I., Panza, G.F., Soloviev, A.A., Aoudia, A.: Morphostructural zonation and preliminary recognition of seismogenic nodes around the Adria margin in peninsular Italy and Sicily. J. Seismol. Earthq. Eng. **4**, 1–24 (2002)

Gorshkov, A.I., Panza, G.F., Soloviev, A.A., Aoudia, A.: Identification of seismogenic nodes in the Alps and Dinarides. Boll. Soc. Geol. It. **123**, 3–18 (2004)

Gorshkov, A.I., Panza, G.F., Soloviev, A.A., Aoudia, A., Peresan, A.: Delineation of the geometry of the nodes in the Alps-Dinarides hinge zone and recognition of seismogenic nodes ($M \geq 6.0$). Terra. Nova. (2009). https://doi.org/10.1111/j.1365-3121.2009.00879.x

Gorshkov, A.I., Soloviev, A.A.: Morphostructural zoning for identifying earthquake-prone areas. In: Panza, G.F., Kossobokov, V.G., Laor, E., De Vivo, B., (eds.) Earthquakes and sustainable infrastructure (1st Edition): Neodeterministic (NDSHA) approach guarantees prevention rather than cure, pp. 135–149. Elsevier (2022)
Gruppo di lavoro CPTI: Catalogo Parametrico dei Terremoti Italiani (CPTI04). Istituto Nazionale di Geofisica e Vulcanologia (INGV), Bologna (2004)
Gusev, A.A., Pavlov, V.: Wideband simulation of earthquake ground motion by a spectrum-matching, multiple-pulse technique. In: First European Conference on Earthquake Engineering and Seismology (a joint event of the 13th ECEE & 30th General Assembly of the ESC). Geneva, Switzerland, 3–8 September 2006. Paper Number: 408 (2006)
Gusev, A.A.: Broadband kinematic stochastic simulation of an earthquake source: a refined procedure for application in seismic hazard studies. Pure Appl. Geophys. **168**, 155–200 (2011)
Hancock, J., et al.: An improved method of matching response spectra of recorded earthquake ground motion using wavelets. J. Earthquake Eng. **10**, 67–89 (2006)
Housner, G.V.: Spectrum intensities of strong motion earthquakes. In: Proceedings of the Symposiumon Earthquake and Blast Effects on Structures, EERI, Oakland California, 20–36 (1952)
Iervolino, I., Maddaloni, G., Cosenza, E.: Eurocode 8 compliant real record sets for seismic analysis of structures. J. Earthquake Eng. **12**(1), 54–90 (2008)
Iervolino, I., Maddaloni, G., Cosenza, E.: A note on selection of time-histories for seismic analysis of bridges in Eurocode 8. J. Earthquake Eng. **13**(8), 1125–1152 (2009)
INGV 2016/08/24 event dedicated page: http://cnt.rm.ingv.it/en/event/7073641. Last accessed 14 Mar 2023
Lin, L.: Selection of spectrum-compatible accelerograms for seismic analysis of bridges. Struct. Infrastruct. Eng. **14**(5), 523–537 (2018)
Lliboutry, L.: Quantitative geophysics and geology. Springer, Berlin (2000)
Luzi, L., et al.: ORFEUS Working Group 5 Engineering Strong Motion Database (ESM)(Version 2.0). Istituto Nazionale di Geofisica e Vulcanologia (2020)
Meletti, C., et al.: Zonazione sismogenetica ZS9. Istituto Nazionale di Geofisica e Vulcanologia (INGV) (2004)
NTC 2018. Ministry of Infrastructure and Transport, Updating Technical Standards for Construction; Official Gazette: Rome, Italy (In Italian) (2018)
Nunziata, C., Costanzo, M.R.: Application of NDSHA to historical urban areas. In: Panza, G.F., Kossobokov, V.G., Laor, E., De Vivo, B. (eds.) Earthquakes and sustainable infrastructure (1st Edition): Neodeterministic (NDSHA) approach guarantees prevention rather than cure, pp. 195–213. Elsevier (2022)
Panza, G.F.: Synthetic seismograms: the Rayleigh waves modal summation. J. Geophys. **58**, 125–145 (1985)
Panza, G.F., Romanelli, F., Vaccari, F.: Seismic wave propagation in laterally heterogeneous anelastic media: theory and applications to seismic zonation. Adv. Geophys. **43**, 1–95 (2001)
Panza, G.F., Paskaleva, I., Nunziata, C. (eds.): Seismic ground motion in large urban areas. PAGEOPH Topical volume, Birkhauser Verlag (Basel, Boston, Berlin), p. 368 (2004)
Panza, G.F., La Mura, C., Peresan, A., Romanelli, F., Vaccari, F.: Seismic hazard scenarios as preventive tools for a disaster resilient society. Adv. Geophys. **53**, 93–165 (2012)
Panza, G.F., Romanelli, F, Vaccari, F., Altin, G., Stolfo P.: Vademecum for the seismic verification of existing buildings: application to some relevant buildings of the Trieste province. Province of Trieste (2015). https://www.xeris.it/resources/Vademecum/ProvinciaTS_Vademecum_verifica_sismica_luglio2015_ENG.pdf
Panza, G.F., Peresan, A.: Difendersi dal terremoto si può. EPC Editore (2016). https://www.epc.it/Prodotto/Editoria/Libri/Difendersi-dal-terremoto-si-puo%27/3342

Panza, G.F., Kossobokov, V.G., Laor, E., De Vivo, B. (eds.) Earthquakes and Sustainable Infrastructure (1st Edition): Neodeterministic (NDSHA) Approach Guarantees Prevention Rather than Cure, p. 672. Elsevier (2022)

Pavlov, V.M.: Matrix impedance in the problem of the calculation of synthetic seismograms for a layered-homogeneous isotropic elastic medium. Izv., Phys. Solid Earth **45**, 850–860 (2009)

Romanelli, F., Panza, G.F., Vaccari, F.: Realistic modelling of the effects of asynchronous motion at the base of bridge piers. J. Seismolog. Earthq. Eng. **6**(2), 19–28 (2004)

Romanelli, F., Altin, G., Indirli, M.: Spreading NDSHA application from Italy to other areas. In: Panza, G.F., Kossobokov, V.G., Laor, E., De Vivo, B. (eds.). Earthquakes and sustainable infrastructure (1st Edition): neodeterministic (NDSHA) approach guarantees prevention rather than cure. pp. 175–194. Elsevier (2022)

Rugarli, P., et al.: Neo-deterministic scenario-earthquake accelerograms and spectra: a NDSHA approach to seismic analysis. Engineering Dynamics and Vibrations, pp. 187–241 (2019)

Seismosoft: SeismoSelect v.2022 – A computer program for the selection and scaling of ground motion records (2022). https://seismosoft.com/

Seismosoft. SeismoMatch v.2022—A Computer Program for Spectrum Matching of Earthquake Records (2022). https://seismosoft.com/

USGS 2016/08/24 event page: https://earthquake.usgs.gov/earthquakes/eventpage/us10006g7d/executive#origin. Last accessed 14 Mar 2023

Vaccari, F.: A web application prototype for the multi scale modelling of seismic input. In: D'Amico, S. (ed.) Earthquakes and Their Impact on Society. pp. 563–584. Springer International Publishing (2016)

Vaccari, F., Magrin, A.: A user-friendly approach to NDSHA computation. In: Panza, G.F., Kossobokov, V.G., Laor, E., De Vivo, B. (eds.) Earthquakes and Sustainable Infrastructure (1st Edition): Neodeterministic (NDSHA) Approach Guarantees Prevention Rather than Cure, pp. 215–237. Elsevier (2022)

Von Thun, J.L., Rochim, L.H., Scott, G.A., Wilson, J.A.: Earthquake ground motions for design and analysis of dams. Earthquake Engineering and Soil Dynamics II - Recent Advances in Ground-Motion Evaluation, Geotechnical Special Publication, Vol. **20**, pp. 463-481 (1988)

# Experimental Assessment of the Energy that Contributes to Damage

David Escolano-Margarit[1](✉), Leandro Morillas-Romero[2], and Amadeo Benavent-Climent[1]

[1] Universidad Politécnica de Madrid, Madrid, Spain
d.escolano@upm.es
[2] Universidad de Granada, Granada, Spain

**Abstract.** In the energy-based approach the effect of the earthquake on the structures is characterized in terms of an energy input and a its proper estimation is of key importance. This energy input, $E_I$, flows into the structure and is partially dissipated through the inherent mechanism of damping of the structure, $W_\xi$, and the rest ($E_I - W_\xi$) is what Housner defined as the energy that contributes to damage $E_D$. This inherent damping of the structure results from internal friction in materials and, in case of RC structures, to light cracking in regions that remain essentially elastic, but does not accounts for the energy dissipated through plastic deformations. $E_D$ can be estimated from $E_I$ by using empirical expressions proposed in the literature, or from the relative velocity response spectrum of the ground motion. This paper compares the results of several shake table tests conducted by the authors on different types of structures (i.e. RC frames and RC waffle flat plates without and with energy dissipation devices) with the prediction provided by the two approaches above in terms of the ratio $E_D/E_I$. The evolution of $E_D/E_I$ as a the dissipated plastic strain energy increases is also examined.

**Keywords:** Energy-Based Design · Energy-Input · Damping Energy

## 1 Introduction

How the seismic energy input, $E_I$, flows in a structural system has been a critical question [1] since the very basis of Energy Based Design (EBD) were postulated in the middle of last Century. In an elastic system, when the movement of the system fades away, all the energy input $E_I$ has been dissipated through the inherent mechanism of damping of the structure. In inelastic systems, the energy input dissipates partially through the inherent damping mechanism $W_\xi$ and the rest through cumulative inelastic deformations $E_H$ among structural members. The difference ($E_I - W_\xi$) is what Housner [1] defined as the energy that contributes to damage $E_D \approx E_I - W_\xi$.

It is worth noting that when the movement of the system fades away the elastic vibrational energy is ≈0 and $E_D \approx E_I - W_\xi$. $E_D$ is an engineering demand parameter of paramount importance that must be properly accounted for in a design process. This paper evaluates $E_D/E_I$ by means of the results of 15 years of shake table experimentation at the University of Granada. The experimental results are compared with the prediction provided by empirical formulations available in the literature.


H. Varum et al. (Eds.): IWEBSE 2023, LNCE 236, pp. 129–137, 2023.
https://doi.org/10.1007/978-3-031-36562-1_10

### 1.1 State-of-the-art

In 1956, Housner [1] proposed the simple yet revealing concept of correlating the energy that contributes to damage to the spectral velocity Sv of the ground motion as follows:

$$E_D = \frac{1}{2} M S_v^2 \tag{1}$$

where $M$ is the total mass of the system.

A few years later, based on the results of numerical analysis of inelastic systems Akiyama [2] proposed Eq. 2 that expresses the energy that contributes to damage VD as a fraction of the total energy input VE both in the form of an equivalent velocity:

$$\frac{V_D}{V_E} = \frac{1}{1 + 3\xi + 1.2\sqrt{\xi}} \tag{2}$$

where, $V_D = (2E_D/M)^{0.5}$ $V_E = (2E_I/M)^{0.5}$ and $\xi$ is the damping ratio.

Akiyama also concluded that $V_D$ provides a good approximation to $S_v$, proving Housner's hypothesis. He also showed that for inelastic systems with short natural periods$V_D$ is greater than $S_v$ due to the elongation of the fundamental period caused by the plastic deformations. Based on numerical analyses on nonlinear SDOF systems Benavent-Climent et al. [3] refined Akiyama's expression considers the effect of the level of cumulative plastic deformations and proposed Eq. 3:

$$\frac{V_D}{V_E} = \frac{1.15\eta}{(0.75 + \eta)(1 + 3\xi + 1.2\sqrt{\xi})} \tag{3}$$

where $\eta = E_H/Q_y\delta_y$ is the cumulative plastic deformation ratio, $Q_y$ is the yield strength and $\delta_y$ the yield displacement of the system.

Fajfar and Vidic [4] analised the hysteretic to energy input ratio $E_D/E_I$ through parametric studies with SDOF systems built with different hysteretic models and subjected to several levels of ductility demand. From their parametric studies the following equation was proposed:

$$\frac{E_H}{E_I} = c_E \frac{(\mu - 1)^{c_H}}{\mu} \tag{4}$$

where $c_E$ and $c_H$ are constants that depend on the hysteretic model, the level od damping and the displacement ductility $\mu$ (i.e. the maximum displacement divided by the yield displacement). For a 5% damping ratio and for stiffnessdegrading models, $c_E$ and $c_H$ have been calibrated to be 1.13 and 0.82 respectively.

Moreover, Fajfar and Vidic concluded that the most influential parameter on the $E_H/E_I$ ratio is damping, increasing as the $E_H/E_I$ ratio decreases over the whole period range. They also pointed out the moderate influence of ductility and the hysteretic behaviour and a small influence of the natural period and the ground motion characteristics.

Decanini and Mollaioli [5] proposed a design spectra reproduced in Fig. 1 for the $E_H/E_I$ ratio of the absolute input energy [6] which is comparable to the relative input energy considered in this study. Among their findings it is worth notice the depedency of the $E_H/E_I$ ratio with ductility. For moderate levels of ductility (i.e. $\mu < 3$) $E_H/E_I$ increases with $\mu$. Whereas for large values of ductility (i.e. $\mu > 3$) this growth trend diminishes, reducing the dependency of $E_H/E_I$ ratio to ductility.

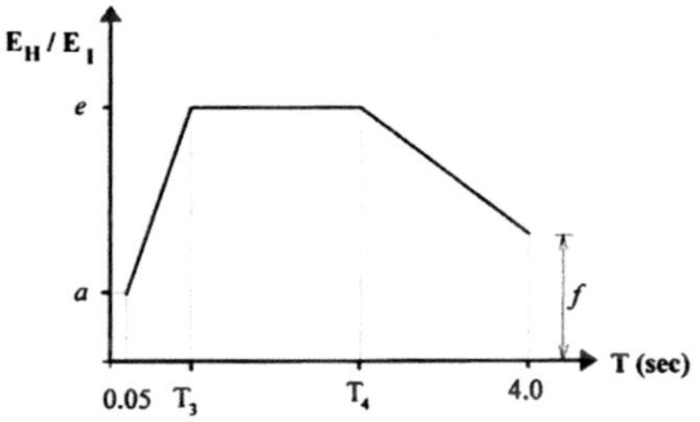


Table 11
Parameters characterizing the design $E_H/E_I$ spectra. EPP model. 5% viscous damping

| Soil class | $\mu$ | a | e | f | $T_3$ (s) | $T_4$ (s) |
|---|---|---|---|---|---|---|
| S1 | 2 | 0.13 | 0.45 | 0.38 | 0.15 | 1.6 |
| | 4 | 0.25 | 0.62 | 0.47 | 0.15 | 1.0 |
| | 6 | 0.30 | 0.65 | 0.48 | 0.15 | 1.0 |
| S2 | 2 | 0.10 | 0.48 | 0.40 | 0.30 | 2.00 |
| | 4 | 0.12 | 0.62 | 0.50 | 0.25 | 1.40 |
| | 6 | 0.15 | 0.66 | 0.52 | 0.20 | 1.20 |
| S3 | 2 | 0.01 | 0.50 | 0.47 | 0.60 | 2.4 |
| | 4 | 0.03 | 0.65 | 0.60 | 0.50 | 2.2 |
| | 6 | 0.05 | 0.70 | 0.62 | 0.45 | 1.8 |

**Fig. 1.** Design spectral shape and parameters characterizing the EH/EI ratio reproduced from [5]

## 2 Experimental Database

The results of almost 15 years of shake table experimentation at the University of Granada and the Polytechnic University of Madrid were collected and curated in the database presented in Table 1. For each test, a small description and view of the specimen and experimental set-up can be found below Table 1. In each test, the specimen was subjected to several seismic simulations of increasing amplitude up to collapse. The name of the test specimen is denoted with capital letters. The number after this name indicates the scaling factor (in %) applied to the reference accelerogram reproduced by the shake table in each seismic simulation. The peak base acceleration of the reference accelerogram was 0.16 g. The results in Table 1 show the energies cumulated over the set of increasing amplitude seismic simulations.

**Table 1.** Experimental database

| TEST | $V_E$ | $V_D$ | $E_I$ | $E_\xi$ | $E_H$ | *Mass* | $\xi$ | $E_H/E_I$ | $V_D/V_E$ | $\mu$ |
|---|---|---|---|---|---|---|---|---|---|---|
| | (m/s) | (m/s) | (kNm) | (kNm) | (kNm) | (kNs$^2$/m) | (%) | | | |
| FD-50 | 0,18 | 0,00 | 0,20 | 0,20 | 0,00 | 12,45 | 2,60 | 0,00 | 0,00 | 0,27 |
| FD-100 | 0,40 | 0,00 | 1,00 | 1,00 | 0,00 | 12,45 | 2,60 | 0,00 | 0,00 | 0,57 |
| FD-200 | 2,03 | 0,61 | 25,65 | 23,34 | 2,32 | 12,45 | 2,60 | 0,09 | 0,30 | 1,14 |
| FD-300 | 2,22 | 1,76 | 30,68 | 11,40 | 19,28 | 12,45 | 2,60 | 0,63 | 0,79 | 1,95 |
| FD-350 | 3,40 | 2,85 | 71,96 | 21,40 | 50,56 | 12,45 | 2,60 | 0,70 | 0,84 | 2,48 |
| WFP-100 | 0,61 | 0,46 | 2,05 | 0,88 | 1,16 | 11,00 | 1,78 | 0,57 | 0,75 | 0,47 |
| WFP-200 | 1,53 | 1,24 | 12,87 | 4,42 | 8,46 | 11,00 | 2,47 | 0,66 | 0,81 | 0,89 |
| WFP-300 | 2,70 | 2,02 | 40,10 | 17,65 | 22,44 | 11,00 | 6,29 | 0,56 | 0,75 | 1,52 |
| WFP-350 | 3,95 | 2,98 | 85,81 | 36,97 | 48,84 | 11,00 | 7,82 | 0,57 | 0,75 | 2,70 |
| WFPD-100 | 0,38 | 0,00 | 0,80 | 0,80 | 0,00 | 11,13 | 1,80 | 0,00 | 0,00 | 0,41 |
| WFPD-200 | 1,03 | 0,37 | 5,90 | 5,14 | 0,76 | 11,13 | 1,80 | 0,13 | 0,36 | 0,80 |
| WFPD-300 | 2,06 | 1,42 | 23,62 | 12,39 | 11,22 | 11,13 | 1,80 | 0,48 | 0,69 | 1,15 |
| WFPD-350 | 3,12 | 2,42 | 54,17 | 21,58 | 32,59 | 11,13 | 1,80 | 0,60 | 0,78 | 1,60 |
| WFPD-400 | 4,22 | 3,46 | 99,10 | 32,48 | 66,62 | 11,13 | 1,80 | 0,67 | 0,82 | 1,92 |
| WFPD-450 | 5,38 | 4,57 | 161,08 | 44,85 | 116,22 | 11,13 | 1,80 | 0,72 | 0,85 | 2,20 |
| WFPD-500 | 6,59 | 5,73 | 241,68 | 58,96 | 182,72 | 11,13 | 1,80 | 0,76 | 0,87 | 2,27 |
| WFPD-550 | 7,84 | 6,93 | 342,06 | 74,80 | 267,26 | 11,13 | 1,80 | 0,78 | 0,88 | 2,40 |
| WFPD-600 | 9,26 | 8,32 | 477,19 | 91,96 | 385,22 | 11,13 | 3,50 | 0,81 | 0,90 | 3,79 |
| WFP-2C-35 | 0,32 | 0,13 | 0,63 | 0,52 | 0,10 | 12,23 | 2,40 | 0,17 | 0,41 | 0,4 |
| WFP-2C-50 | 0,55 | 0,26 | 1,85 | 1,44 | 0,41 | 12,23 | 2,40 | 0,22 | 0,47 | 0,51 |
| WFP-2C-100 | 1,06 | 0,61 | 6,87 | 4,60 | 2,28 | 12,23 | 4,10 | 0,33 | 0,58 | 0,78 |
| WFP-2C-200i | 1,65 | 0,93 | 16,65 | 11,36 | 5,29 | 12,23 | 4,50 | 0,32 | 0,56 | 1,57 |
| WFP-2C-200 | 2,73 | 1,82 | 45,57 | 25,32 | 20,26 | 12,23 | 5,70 | 0,44 | 0,67 | 2,56 |
| WFP-2C-300 | 3,82 | 2,82 | 89,23 | 40,60 | 48,63 | 12,23 | 9,20 | 0,54 | 0,74 | 5,15 |
| WFP-2C-NITI-15 | 0,09 | 0,00 | 0,05 | 0,05 | 0,00 | 12,95 | 3,60 | 0,00 | 0,00 | 0,18 |
| WFP-2C-NITI-25 | 0,26 | 0,12 | 0,44 | 0,34 | 0,09 | 12,95 | 3,60 | 0,21 | 0,46 | 0,33 |
| WFP-2C-NITI-50 | 0,99 | 0,63 | 6,35 | 3,78 | 2,57 | 12,95 | 3,60 | 0,40 | 0,64 | 0,84 |
| WFP-2C-NITI-100 | 1,74 | 1,11 | 19,60 | 11,63 | 7,98 | 12,95 | 3,60 | 0,41 | 0,64 | 2,02 |
| WFP-2C-NITI-130i | 2,44 | 1,59 | 38,55 | 22,18 | 16,37 | 12,95 | 3,70 | 0,42 | 0,65 | 2,53 |
| WFP-2C-NITI-130 | 2,94 | 1,92 | 55,97 | 32,10 | 23,87 | 12,95 | 3,70 | 0,43 | 0,65 | 2,90 |
| WFP-2C-NITI-160 | 3,56 | 2,38 | 82,06 | 45,38 | 36,68 | 12,95 | 3,90 | 0,45 | 0,67 | 3,01 |
| WFP-2C-NITI-190 | 4,26 | 2,93 | 117,51 | 61,92 | 55,59 | 12,95 | 4,00 | 0,47 | 0,69 | 3,04 |
| WFP-2C-HD-35 | 0,52 | 0,34 | 1,65 | 0,95 | 0,71 | 12,23 | 12,00 | 0,43 | 0,65 | |
| WFP-2C-HD-500 | 0,97 | 0,75 | 5,75 | 2,31 | 3,44 | 12,23 | 14,00 | 0,60 | 0,77 | |
| WFP-2C-HD-100 | 1,79 | 1,13 | 19,59 | 11,78 | 7,81 | 12,23 | 16,00 | 0,40 | 0,63 | |
| WFP-2C-HD-200 | 2,90 | 1,82 | 51,43 | 31,17 | 20,26 | 12,23 | 19,00 | 0,39 | 0,63 | |
| WFP-2C-HD-300 | 4,21 | 3,21 | 108,38 | 45,37 | 63,01 | 12,23 | 16,00 | 0,58 | 0,76 | |

**Experiment FD** [7]. This investigation presented a shake table test to provide empirical data on the performance of reinforced concrete (RC) frames with hysteretic dampers [8]. A 2/5-scale RC frame with hysteretic dampers was tested following sequence of uniaxial seismic simulations that represented different levels of seismic hazard. The frame was designed only for gravity loads whereas the dampers provided the necessary lateral strength and stiffness to resist the seismic action.

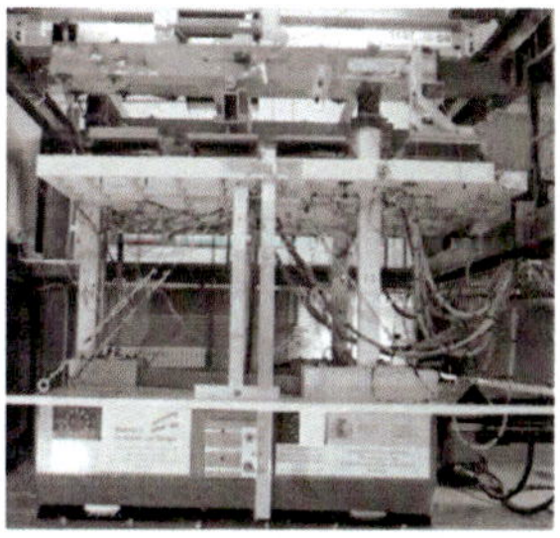

**Experiment WFP**. [9]. This investigation presented the results of several shaking table tests on a scaled reinforced concrete waffle–flat plate structure. It was representative of a typical construction designed in the Mediterranean area. The structure was tested against a sequence of four uniaxial seismic simulations of increasing magnitude which were associated to different seismic hazard levels characterized by the mean return period from 95 to 2475 years.

**Experiment WFPD** [10]. This investigation studied the seismic performance of waffle flat plate structures with hysteretic dampers [8] through shaking table tests on a 2/5-scale structure. The structure was designed only for gravity loads, whereas the lateral strength and stiffness was provided by the dampers at each story. The structure + dampers system was tested against a sequence of uniaxial seismic simulations which were representative of frequent to very rare earthquakes.

**Experiment WFPD-2C** [11]. This investigation studied the response of a reinforced concrete waffle flat plate (WFP) structure under bi-axail seismic loading through shake table tests. The structure was designed using current building codes typically prescribed in regions of moderate seismicity such as the Mediterranean area. It was subjected to as series of far-field ground motions of increasing acceleration which represented very frequent, frequent, design, and very rare earthquakes at the site.

**Experiment WFP-2C-NITI** [12]. This investigation evaluated a new hysteretic energy dissipation device [13] combined with a self centering system, installed on a RC structure. The structure was waffle-flat plate designed only to resist gravitational loads. It was tested on a shake tables against two horizontal components of a ground motion, which were scaled in amplitude to represent different levels of seismic hazard at the building site.

**Experiment WFP-2C-HD** [14]. This investigation evaluated the seismic performance of a new hybrid energy dissipation device [15], that combines a low-cost viscoelastic part and a metallic yielding part. Its performance was assed on a reinforced concrete waffle-flat plate structure, designed for gravitational loads only. The structure + damper system was subjected to bidirectional shake table tests that represented frequent, design and maximum credible earthquakes.

## 3 Discussion of the Results

As discussed in the state-of-the-art section, the reported parameters that mainly affect the relation between the energy that contributes to damage or the hysteretic energy with the total input energy are the damping ratio and the ductility demand on the structural system. The experimental validation of such conclusions is discussed next.

***The $V_D$/$V_E$ relation***. Figure 2a shows the discrete values of the energy that contributes to damage $V_D$ against the energy input $V_E$, expressed in terms of equivalent velocities, toguether with a linear regression of the data (solid line) hat can be approximated with 5:

$$V_D = 0.80 V_E \tag{5}$$

As can be observed in the figure, Eq. 5 fits closely the data for lower values of $V_D$ and $V_E$ (i.e. for low levels of plastic deformations), whereas for larger values of $V_D$ and $V_E$, the points tend to deviate from the regression line. More precisely, as the level of input energy increases (i.e. for increasing levels of plastic deformations) the increment of $V_D$ tends to be larger than the increment in $V_E$.

Since $V_D = (2(E_I - W_\xi)/M)^{0.5}$ and $V_E = (2(E_I)/M)^{0.5}$ this means that the energy dissipated through damping $W_\xi$ compared to the energy that contributes to damage $V_D$ decreases with increasing levels of plastic deformations. This fact was already pointed out by Akiyama [2]: "*...when large inelastic deformations occur, the energy absorption due to damping decreases...*".

***The role of the damping ratio.*** Figure 2b shows the $V_D/V_E$ ratio against the damping ratio $\xi$. As can be observed, the $V_D/V_E$ ratio has a fairly stable value within the 0.6 to 0.8 for $\xi > 5\%$, whereas the dispersion becomes very large for $\xi > 5\%$. Taking into account that in real RC structures $\xi$ is about 5% or larger, Fig. 2b suggests that $\xi$ has a small influence on the $V_D/V_E$ ratio.

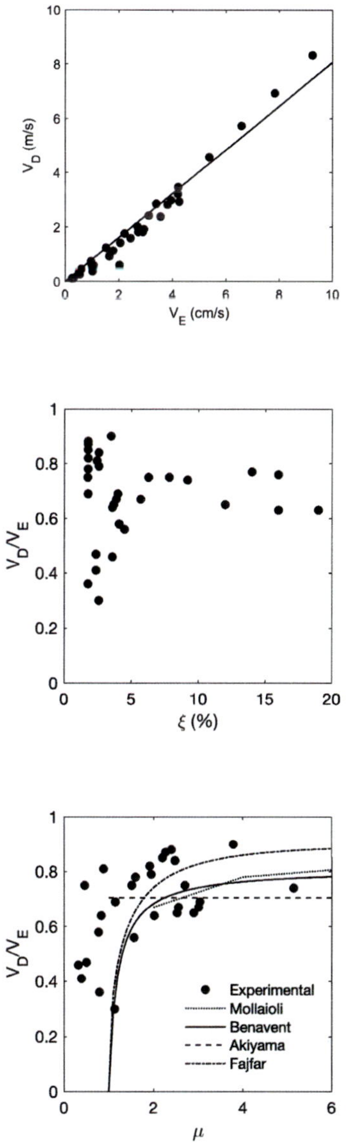


**Fig. 2.** Main parameter relationship $V_D/V_E$ (top), $V_D/V_E$ and $\xi$ (middle) $V_D/V_E$ and $\mu$ (bottom).

***The role of ductility***. Figure 2c shows the $V_D/V_E$ ratio against the ductility demand. As can be observed there is a logarithmic trend on the $V_D/V_E$ ratio with increasing values of ductility, suggesting that for lower values ($\mu < 2$) the $V_D/V_E$ ratio depends greatly on the ductility level, whereas for larger values ($\mu > 2$) its influence diminishes being almost imperceptible.

The effectiveness in predicting the $V_D/V_E$ ratio using the models described in section 1.1 was assessed by comparing their results with the experimental database in Table 1. Figure 2c shows the continuous prediction of $V_D/V_E(\mu)$ given by the different models with the discrete experimental values in a range of ductility within 0 and 6. As can be observed all models can predict the $V_D/V_E(\mu)$ ratio with distinct accuracy at different ranges of $\mu$. Firslty, the prediction given by Akiyama clearly overestimates the ratio for small values of ductility. The reason is that Akiyama's expression was intended to be used with structures that undergo moderate to large plastic deformations. On the other hand, the modification proposed by Benavent-Climent et al. provides a lower bound of the experimental data accounting for the ductility of the system. The spectrum proposed by Decanini and Mollaioli follows a very similar trend and values to Benavent-Climent et al's modification of Akiyama's model. Lastly, the model proposed by Fajar seems to be the most accurate following closely the trend and average values of the experimental data.

## 4 Conclusions

This paper summarised the results of several shake table tests conducted by the authors on different types of structures. The results were curated and analysed in terms of the ratio of energy that contributes to damage against the total energy input and compared with the prediction given by several authors in the literature. The following conclusions can be put forth:

- There is an almost linear relation between the energy that contributes to damage and the total energy input in structures that undergo low levels of plastic deformations (i.e. for low levels of $V_D$ and $V_E$. As the level of inelastic deformations increases, the energy dissipation due to damping $W_\xi$ compared to the energy that contributes to damage $V_D$ decreases and this makes that $V_D$ tends to increase more than $V_E$.
- For realistic values of the damping ratio in RC structures (i.e. $\xi$ 5% the damping ratio does not seem to be influential on the $V_D/V_E$ ratio.
- For low levels of ductility demand ($\mu < 2$) the $V_D/V_E$ ratio is strongly dependent on $\mu$, but this dependency vanishes as $\mu$ increases, becoming irrelevant for design purposes.
- The model proposed by Fajfar and Vidic provides the most accurate prediction of the $V_D/V_E$ ratio estimated from the tests. The models proposed by Benavent-Climent and by Decanini and Mollaioli provide very similar results. The model proposed by Akiyama should be used only for moderate to large levels of plastic deformations (i.e.($\mu < 2$)) and provides a lower bound limit of the experimental results.

The conclusions presented here are based in a limited amount of data, therefore further validation would be required in the future, especially for systems with large values of damping.

**Acknowledgements.** This work has been supported by the Spanish Government under Project MEC PID2020120135RB-I00 and received funds from the European Union (Fonds Européen de Dévelopment Régional).

## References


1. George, W.H.: Limit design of structures to resist earthquakes. In: Proceedings World Conference of Earthquake Engineering, vol. 6 (1956)
2. Hiroshi, A.: Earthquake-Resistant Limit-State Design for Buildings. University of Tokyo Press (1985)
3. Benavent-Climent, A., Pujades, L.G., López-Almansa, F.: Design energy input spectra for moderate-seismicity regions: DESIGN ENERGY INPUT SPECTRA. Earthq. Eng. Struct. Dyn. **31**(5), 1151–1172 (2002)
4. Fajfar, P., Vidic, T.: Consistent inelastic design spectra: hysteretic andinput energy. Earthq. Eng. Struct. Dyn. **23**(5), 523–537 (1994)
5. Decanini, L.D., Mollaioli, F.: An energy-based methodology for the assessment of seismic demand. Soil Dyn. Earthq. Eng. **21**(2), 113–137 (2001)
6. Uang, C.M., Bertero, V.V.: Earthq. Eng. Struct. Dynamics **19**, 77 (1990)
7. Benavent-Climent, A., Morillas, L., Escolano-Margarit, D.: Seismic performance and damage evaluation of a reinforced concrete frame with hysteretic dampers through shake-table tests. Earthq. Eng. Struct. Dyn. **43**(15), 2399–2417 (2014)
8. Amadeo Benavent-Climent, Leandro Morillas, and Juan M Vico. A study on usingwide-flange section web under out-of-plane flexure for passive energy dissipation. *Earthquake Engineering & Structural Dynamics*, 40(5):473–490, 2011.
9. Benavent-Climent, A., Donaire-Avila, J., Oliver-Saiz, E.: Shaking table tests of a reinforced concrete waffle-flat plate structure designed following modern codes: seismic performance and damage evaluation: Shaking Table Tests of an RC Waffle-Flat Plate Structure. Earthq. Eng. Struct. Dyn. **45**(2), 315–336 (2016)
10. Benavent-Climent, A., Donaire-Avila, J., Oliver-Sáiz, E.: Seismic performance and damage evaluation of a waffle-flat plate structure with hysteretic dampers through shake-table tests. Earthq. Eng. Struct. Dyn. **47**(5), 1250–1269 (2018)
11. Benavent-Climent, A., Galé-Lamuela, D., Donaire-Avila, J.: Energy capacity and seismic performance of RC waffle-flat plate structures under two components of far-field ground motions: Shake table tests. Earthq. Eng. Struct. Dyn. **48**(8), 949–969 (2019)
12. Benavent-Climent, A., González-Sanz, G., Donaire-Avila, J., Galé-Lamuela, D., Escolano-Margarit, D.: Seismic response of a rc structure with new self-centering metallic dampers under biaxial near-field ground motions: shake table tests. J. Building Eng. **65**, 105753 (2023)
13. Gonzalez-Sanz, G., Escolano-Margarit, D., BenaventCliment, A.: A new stainless-steel tube-in-tube damper for seismic protection of structures. Appl. Sci. **10**(4), 1410 (2020)
14. Benavent-Climent, A., Escolano-Margarit, D., Yurkin, Y., Ponce-Parra, H., Arcos-Espada, J.: Shake table tests on a reinforced concrete waffle-flat plate structure with new hybrid energy dissipation devices. Earthq. Eng. Struct. Dyn. **52**, 727–749 (2022)
15. Benavent-Climent, A., Escolano-Margarit, D., Arcos-Espada, J., Ponce-Parra, H.: New metallic damper with multiphase behavior for seismic protection of structures. Metals **11**(2), 183 (2021). https://doi.org/10.3390/met11020183

# Elastic and Inelastic Input Energy Correlations: Effects of Damping, Pulse Type, and Significant Duration

Furkan Çalım[1](✉), Ahmet Güllü[2], and Ercan Yüksel[1]

[1] Faculty of Civil Engineering, Istanbul Technical University, Istanbul, Turkey
calimf@itu.edu.tr

[2] Ingram School of Engineering, Texas State University, San Marcos, TX 78666, USA

**Abstract.** In the seismic analysis and design of the structures, the energy-based approach has been considered a sought-after alternative to conventional design approaches in recent years. It accounts for the pulse type, frequency content, and duration-related cumulative damage potential of an earthquake. Precise computation of seismic input energy imparted into the structure during a seismic action is decisive in reaching a comprehensive energy-based seismic design method. Hence, based on the modal contributions, a methodology to determine the seismic input energy demands of multi-degree-of-freedom (MDOF) frames has been proposed. This paper evaluated the method through nonlinear response history analyses (NRHA) of four benchmark MDOF frames with three, six, nine, and twenty stories. The effects of the pulse type and significant duration of the ground motions, and the system's damping characteristics on the elastic and inelastic input energy correlations were investigated. Whereas the correlation was not significantly affected by the significant duration of the earthquake record or the period of the structure, better correlations were detected for the structural systems with a 5% damping ratio under non-pulse-like ground motion records. The correlation between the elastic and inelastic seismic input energies was nearly 98%, with a maximum mean difference of 18.9% for the non-pulse-like ground motion records.

**Keywords:** Energy-Based Design · Energy Balance Equation · Seismic Input Energy · Elastic and Inelastic Energy Spectra

## 1 Introduction

Current seismic design codes and analysis methods primarily use force- and displacement-based approaches for the earthquake-resistant design of structures. However, these approaches neglect the effect of some main parameters, e.g., duration and frequency content of the ground motion. Additionally, only the peak responses are considered by ignoring the cumulative damage potential of the ground motions to compute the seismic demands by conventional methods. Thus, a need for a more extensive seismic analysis and design approach has arisen. As an alternative to the abovementioned approaches, a new philosophy called the energy-based approach was introduced [1].


H. Varum et al. (Eds.): IWEBSE 2023, LNCE 236, pp. 138–153, 2023.
https://doi.org/10.1007/978-3-031-36562-1_11

The use of the energy concept in the overall performance analysis and seismic design of structures is considered a more rational approach with great potential since it not only covers the capabilities of the current approaches but also takes the hysteretic behavior of the structural members into account [2–7]. Precise calculation of the seismic input energy imparted into a structure during an earthquake action, whether a single-degree-of-freedom (SDOF) or a multi-degree-of-freedom (MDOF) system, is the first step and plays an essential role in the energy-based seismic design approach.

Akiyama [8] computed the input energy imparted by the 1940 El Centro ground motion record into a five-story structure and its equivalent one-story system with the same predominant period and total mass. It was concluded that the amount of imposed input energy mainly depends on the period and mass of the structures. It was also stated that it is possible to estimate the input energy of an MDOF system through its equivalent single-degree-of-freedom (E-SDOF) systems.

While the input energy imparted into an elastic SDOF system can easily be calculated through the dynamic equation of motion under an earthquake, some formulations based on modal contributions are required to be employed to compute the input energy demand of an elastic MDOF system [9, 10]. It can reliably be estimated using its E-SDOF systems by a series of theoretical formulations benefitting from the energy balance equations. It can also predict inelastic seismic demands for design purposes [11, 12]. For ease of calculation, the seismic input energy response of a nonlinear MDOF system can also be associated with the input energy imparted into the equivalent linear elastic MDOF system since it is the envelope [13].

The study's rationale is to evaluate the practical methodology to predict the nonlinear energy response of MDOF frames. To this extent, a parametric study was conducted to evaluate the correlation between the elastic and inelastic input energy responses of MDOF systems. The main parameters discussed were the significant duration and the pulse type of the seismic excitation, as well as the fundamental vibrational period and the damping ratio of the structure.

## 2 Formulation of the Seismic Input Energy

The equation of motion of a lumped-mass linear SDOF system under a seismic excitation is used to calculate the energy components. Multiplying all of the terms of the equation of motion with the relative velocity ($\dot{u}$) and integrating the product over the duration results in the energy balance equation, Eq. (1). In the equation, $u$ represents the relative displacement, and the dots stand for the derivatives of the displacement. Additionally, the subscript $g$ denotes the *ground*. While the right-hand side of the equation gives the relative input energy ($E_I$), the terms on the left-hand side of the equation represent the relative kinetic energy ($E_k$), damping energy ($E_d$), and absorbed energy ($E_a$), respectively [8].

$$m \int_0^t \ddot{u}(t) \cdot \dot{u}(t) dt + c \int_0^t \dot{u}(t) \cdot \dot{u}(t) dt + k \int_0^t u(t) \cdot \dot{u}(t) dt = -m \int_0^t \ddot{u}_g(t) \cdot \dot{u}(t) dt \tag{1}$$

The input energy induced by a linear elastic MDOF system can be calculated using the input energy of its E-SDOF systems. If the generalized displacement of mode $i$ is denoted as $D_i(t)$ and the modal displacement is given as $x_i(t) = \Gamma_i D_i(t)$, where $\Gamma_i$ is the modal participation factor, the input energy of the $i^{th}$ mode E-SDOF system can be calculated as Eq. (2).

$$(E_I/m)_{E-SDOF,i} = -\int_0^t \ddot{u}_g(t) \cdot \dot{D}_i(t)dt \tag{2}$$

Since the excitation factor of $i^{th}$ mode can be written as $L_i = \phi^T M\iota$, multiplying Eq. (2) by "$\phi^T M\iota$" and dividing by the term "$L_i$" will not change the result. In addition, replacing the "$D_i(t)$" term with "$x_i(t)/\Gamma_i$" in the expression yields Eq. (3).

$$(E_I/m)_{E-SDOF,i} = -\frac{1}{L_i \cdot \Gamma_i}\int_0^t \phi_i^T \cdot M \cdot I \cdot \ddot{u}_g(t) \cdot \dot{x}_i(t)dt \tag{3}$$

When the mode shapes are normalized with respect to the effective modal masses, the value of $L_i$ becomes equal to the value of $\Gamma_i$. Thus, Eq. (3) can be rewritten as Eq. (4) for normalized mode shapes. It is worth noting that the physical displacement is calculated as $u_i(t) = \phi_i x_i(t)$.

$$(E_I/m)_{E-SDOF,i} = -\frac{1}{\Gamma_i^2}(E_I)_{MDOF,i} \tag{4}$$

The final equation for the input energy calculations for linear elastic MDOF system through energy balance equations benefitting from its E-SDOF systems is obtained by taking the summation of the input energy contributions of each mode, Eq. (5) where $n$ is the number of modes to be used. Using up to the first three vibrational modes is adequate for this calculation [9, 10].

$$(E_I)_{MDOF} = \sum_{i=1}^{n} (E_I/m)_{E-SDOF,i} \cdot \Gamma_i^2 \tag{5}$$

For design purposes, the peak seismic input energy imposed on the MDOF systems by a ground motion can be calculated using its mass-normalized input energy spectrum. The spectral input energy values, which summarize the maximum values of the input energy response of the SDOF systems, are first obtained for each mode from the mass-normalized input energy spectrum. Then, they can be placed in Eq. (5) instead of using the input energy time series of E-SDOF systems to compute the elastic input energy for ease of calculation [14].

The study by Güllü et al. [15] proposed a methodology for the distribution of this energy to story levels of frame-type structures. Later, it was shown that the seismic input energy computed through modal contributions can be utilized for the accurate estimation of story responses.

## 3 Numerical Modeling

Four benchmark moment-resisting steel frames with three, six, nine, and twenty stories were selected to investigate the effects of structural and ground motion properties on the correlation between inelastic and elastic input energy responses [16, 17]. The inelastic input energy responses of the benchmark frames were computed through nonlinear response history analyses. A total of 44 ground motions grouped based on their impulsive characteristics and significant durations were selected for the analyses.

### 3.1 Benchmark Frames

Four benchmark steel MDOF moment-resisting systems previously studied in the literature were selected, and their geometric properties and load cases were identically adapted from these studies. Even though the nine- and twenty-story buildings originally had both perimeter and gravity frames, only the perimeter frames in the north-south direction were modeled in this study [18].

The three- and six-story systems are four-bay frames with a span length of 9.15 m, a ground story height of 4.27 m, and a normal story height of 3.51 m. The nine-story system is a five-bay frame with a span length of 9.15 m, a ground story height of 5.49 m, and a normal story height of 3.96 m. The twenty-story system is a five-bay frame with a span length of 6.10 m, a ground story height of 5.49 m, and a normal story height of 3.96 m. The columns of three- and six-story steel frames are fixed at the base, while the columns of nine- and twenty-story steel frames are pinned at the base, Fig. 1.

The total seismic masses above the ground level of the three-, six-, nine-, and twenty-story moment resisting frames are 515 tons, 1076 tons, 4500 tons, and 5526 tons, respectively. Modal properties of the frames, whose mode shapes are normalized for the effective modal masses, are given in Table 1.

**Table 1.** Modal properties of the benchmark frames.

| Mode | Period (sec) | | | | Modal Participation Factor (Γ) | | | |
|---|---|---|---|---|---|---|---|---|
| | 3-Story | 6-Story | 9-Story | 20-Story | 3-Story | 6-Story | 9-Story | 20-Story |
| 1 | 0.80 | 1.58 | 2.27 | 3.82 | 21.07 | 29.53 | 61.24 | 66.53 |
| 2 | 0.30 | 0.53 | 0.85 | 1.32 | 7.45 | -11.40 | 22.14 | 25.50 |
| 3 | 0.14 | 0.30 | 0.49 | 0.77 | 3.92 | -6.59 | 12.95 | 13.89 |

This study considered two damping ratios for the seismic energy response prediction of the frames by the modal contribution formulations and the nonlinear response history analyses (namely, 5% and 2%). The damping was defined as Rayleigh damping, in which α and β constants related to the mass and stiffness were calculated for the aforementioned damping ratios at $0.2T_1$ and $1.0T_1$, where $T_1$ is the predominant period of each frame [19].

### 3.2 Ground Motion Selection

The first parameter of this study in terms of the ground motion characteristics is the pulse type. The pulse-like type ground motions, having a velocity pulse in the fault-normal direction, carry lower frequency characteristics. Near-fault earthquakes are often called

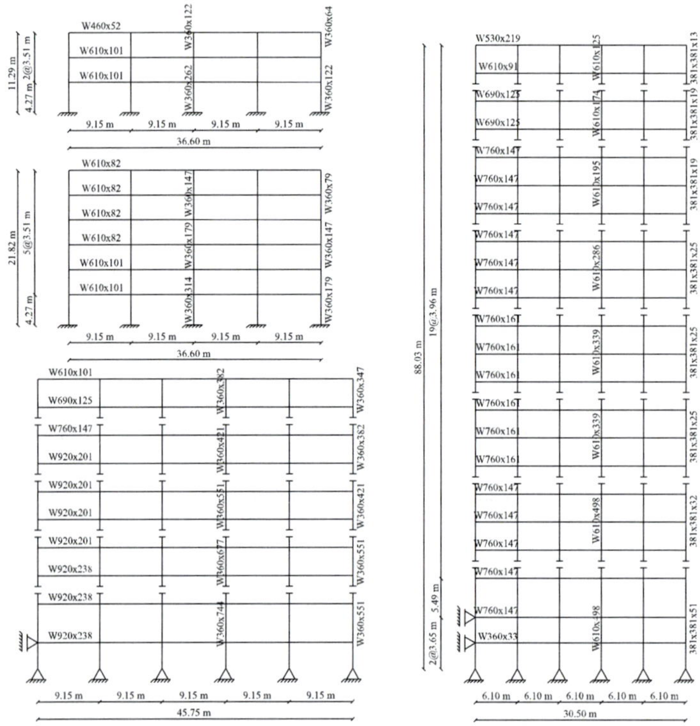


**Fig. 1.** Selected benchmark frames [14, 15].

pulse-like type ground motions, and they have relatively higher peak ground velocity (PGV) to peak ground acceleration (PGA) ratio. The energy of the pulse-like records is usually contained in a single or very few pulses. Thus, they tend to cause relatively higher seismic responses (e.g., shear force, displacement, inter-story drift, etc.) and higher potential damage to the structures than ordinary non-pulse-like ground motions. Therefore, typical seismic analysis and design methods have difficulty estimating the seismic demands from these types of ground motions [20, 21].

The second parameter of this study is the significant duration of the seismic excitation. The significant duration is frequently used as an intensity measure and an indicator of seismic demands. It is based on the normalized Arias intensity [22], which is a measure of the total energy content, Eq. (6), where $\ddot{u}_g(t)$ is the acceleration time history and $t_d$ is the total ground motion duration. The significant duration is then defined as the time interval corresponding to a specified amount of the normalized Arias intensity plot [23–25]. In this study, the significant duration is calculated using the 5%–95% interval of this plot ($D_{5-95}$), Fig. 2.

$$I_A = \frac{\pi}{2g} \cdot \int_0^{t_d} [\ddot{u}_g(t)]^2 dt \tag{6}$$

A total of 44 ground motions with different characteristics were selected among the actual recorded events from the PEER NGA database [26]. The ground motion

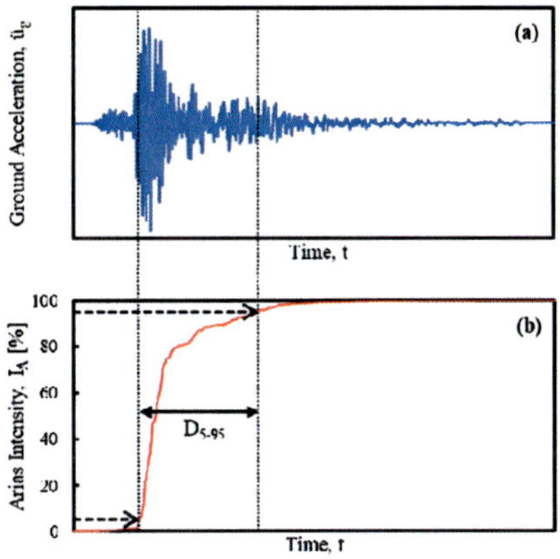


**Fig. 2.** Calculation of the significant duration of a sample ground motion record: (a) Acceleration time history, (b) Normalized Arias intensity.

records were grouped according to their pulse type and significant duration: 11 short-duration ordinary non-pulse-like records (Group #1), 11 long-duration ordinary non-pulse-like records (Group #2), 11 short-duration pulse-like records (Group #3), and 11 long-duration pulse-like records (Group #4), Table 2. Here, the ground motions with a significant duration ($D_{5\text{–}95}$) of up to 10 s were denoted as short-duration records. Those with a $D_{5\text{–}95}$ value longer than 10 s were denoted as long-duration records. For each earthquake group, acceleration and energy response spectra calculated for the 5% damping ratio and their mean spectra are presented in Figs. 3 and 4.

**Table 2.** Selected ground motions.

| *Group #* | *EQ #* | Event | Record | Total Dur [sec] | Significant Dur. [sec] | PGV [cm/sec] | PGA [cm/sec$^2$] |
|---|---|---|---|---|---|---|---|
| 1 | 1.01 | Loma Prieta | TRI090 | 40.0 | 4.5 | 33.2 | 157.0 |
| | 1.02 | Imperial Valley-06 | H-E08230 | 37.6 | 5.8 | 52.1 | 457.2 |
| | 1.03 | Kobe_ Japan | KBU090 | 32.0 | 6.2 | 30.4 | 305.9 |
| | 1.04 | Gazli_ USSR | GAZ000 | 13.6 | 6.4 | 66.3 | 688.4 |
| | 1.05 | Imperial Valley-06 | H-E08140 | 37.8 | 6.8 | 54.5 | 598.6 |
| | 1.06 | Gazli_ USSR | GAZ090 | 13.1 | 7.0 | 67.7 | 847.5 |
| | 1.07 | Kobe_ Japan | KBU000 | 32.0 | 7.1 | 55.5 | 270.5 |

*(continued)*

**Table 2.** (*continued*)

| *Group #* | *EQ #* | Event | Record | Total Dur [sec] | Significant Dur. [sec] | PGV [cm/sec] | PGA [cm/sec$^2$] |
|---|---|---|---|---|---|---|---|
| | 1.08 | Imperial Valley-06 | H-E11140 | 39.1 | 9.0 | 36.0 | 359.8 |
| | 1.09 | Imperial Valley-06 | H-BCR140 | 37.8 | 9.7 | 46.8 | 587.3 |
| | 1.10 | Imperial Valley-06 | H-BCR230 | 37.7 | 9.7 | 44.9 | 762.2 |
| | 1.11 | Imperial Valley-06 | H-AEP045 | 14.8 | 9.8 | 42.8 | 301.0 |
| 2 | 2.01 | Chi-Chi_ Taiwan-06 | CHY047W | 97.0 | 14.4 | 16.0 | 159.2 |
| | 2.02 | Chi-Chi_ Taiwan | HWA003-W | 65.0 | 25.9 | 10.4 | 48.3 |
| | 2.03 | Chi-Chi_ Taiwan | TAP021-E | 144.0 | 28.5 | 29.3 | 98.2 |
| | 2.04 | Chi-Chi_ Taiwan | CHY054-N | 90.0 | 36.1 | 18.3 | 93.8 |
| | 2.05 | Kocaeli_ Turkey | ATS000 | 150.4 | 36.7 | 36.5 | 247.9 |
| | 2.06 | Kocaeli_ Turkey | ATS090 | 150.4 | 37.2 | 35.1 | 182.3 |
| | 2.07 | Chuetsu-oki_ Japan | NIG014NS | 300.0 | 38.2 | 18.6 | 108.7 |
| | 2.08 | Chi-Chi_ Taiwan | CHY054-E | 90.0 | 39.8 | 17.8 | 93.1 |
| | 2.09 | Imperial Valley-06 | H-DLT352 | 100.2 | 50.5 | 33.0 | 343.1 |
| | 2.10 | Imperial Valley-06 | H-DLT262 | 100.2 | 51.4 | 26.3 | 231.2 |
| | 2.11 | Niigata_ Japan | NIG013NS | 299.8 | 73.6 | 12.9 | 95.7 |
| 3 | 3.01 | Christchurch_ NZ | PRPCS | 20.0 | 4.3 | 81.3 | 584.7 |
| | 3.02 | Northridge-01 | JEN292 | 28.6 | 6.2 | 97.4 | 605.4 |
| | 3.03 | Northridge-01 | LDM064 | 26.6 | 6.5 | 74.8 | 418.2 |

(*continued*)

**Table 2.** *(continued)*

| *Group* # | *EQ* # | Event | Record | Total Dur [sec] | Significant Dur. [sec] | PGV [cm/sec] | PGA [cm/sec$^2$] |
|---|---|---|---|---|---|---|---|
| | 3.04 | Northridge-01 | PAR--L | 22.2 | 6.6 | 76.1 | 546.9 |
| | 3.05 | Northridge-01 | JGB022 | 28.7 | 6.9 | 76.2 | 560.4 |
| | 3.06 | Parkfield-02_ CA | C02090 | 21.1 | 7.0 | 64.0 | 611.9 |
| | 3.07 | Parkfield-02_ CA | COW360 | 21.0 | 7.5 | 81.4 | 817.4 |
| | 3.08 | Northridge-01 | PAR--T | 22.2 | 8.0 | 54.1 | 296.6 |
| | 3.09 | Northridge-01 | WPI316 | 25.0 | 8.8 | 59.2 | 350.3 |
| | 3.10 | Parkfield-02_ CA | COW090 | 21.0 | 9.4 | 63.8 | 593.5 |
| | 3.11 | Kobe_ Japan | KJM090 | 150.0 | 9.5 | 76.1 | 617.7 |
| 4 | 4.01 | Northridge-01 | JEN022 | 28.6 | 12.5 | 111.5 | 402.7 |
| | 4.02 | Chi-Chi_ Taiwan | TCU068-N | 90.0 | 13.2 | 264.1 | 363.9 |
| | 4.03 | Chi-Chi_ Taiwan | TCU102-E | 90.0 | 14.9 | 91.7 | 298.2 |
| | 4.04 | Chi-Chi_ Taiwan | TCU052-N | 90.0 | 16.0 | 172.3 | 438.4 |
| | 4.05 | Tottori_ Japan | TTR008EW | 300.0 | 18.1 | 54.2 | 383.9 |
| | 4.06 | Chi-Chi_ Taiwan | TCU102-N | 90.0 | 19.6 | 66.4 | 168.5 |
| | 4.07 | El Mayor_ Mexico | CIWESHNE | 200.0 | 22.9 | 52.1 | 275.6 |
| | 4.08 | Darfield_ New Zealand | RHSCS04W | 136.9 | 23.9 | 62.7 | 229.2 |
| | 4.09 | Darfield_ New Zealand | REHSN02E | 150.0 | 24.8 | 62.2 | 257.9 |
| | 4.10 | Darfield_ New Zealand | REHSS88E | 150.0 | 30.5 | 26.6 | 233.4 |
| | 4.11 | Chi-Chi_ Taiwan | TCU063-E | 90.0 | 33.1 | 44.0 | 179.0 |

### 3.3 Modeling Strategy for Nonlinear Systems

The plastic hinge theory was used to model the inelastic behavior of steel frame elements. While the curvature type plastic hinges with elastic-perfectly plastic (EPP) material models with no cyclic degradation were defined at both ends of each member, mid-sections were considered elastic under the excitation. P-M2-M3 type hinges were assigned to the column elements, whereas moment hinges were assigned to the beam elements. Self and additional masses were lumped at the joints of the benchmark frames. Supports restricting the out-of-plane movement were assigned at every joint to deal with the plane frame instead of the space frame.

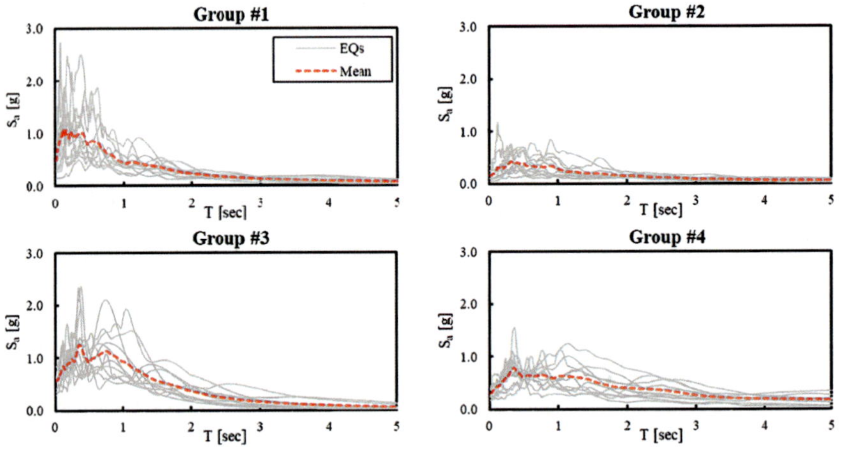


**Fig. 3.** Acceleration spectra for the selected earthquake records.

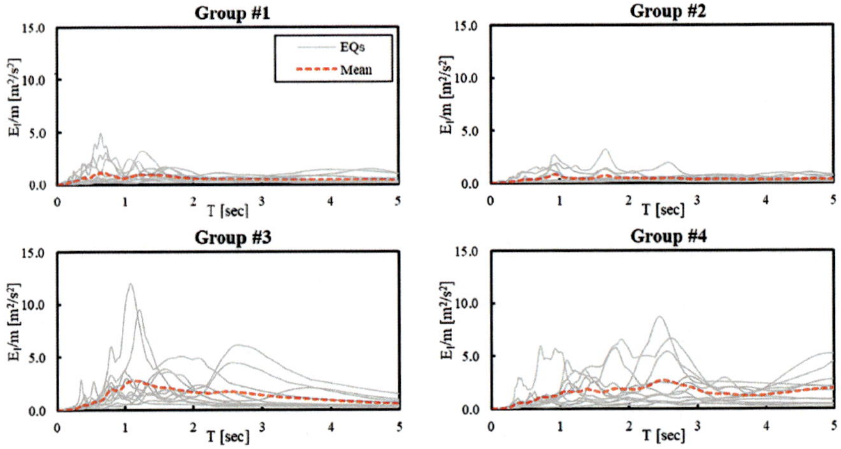


**Fig. 4.** Seismic input energy spectra for the selected earthquake records.

# 4 Results

In calculating the input energy response history of the linear elastic benchmark frames through theoretical formulations, seismic responses of their $i^{th}$ mode E-SDOF systems were first calculated through a MATLAB-based code [27] using Eq. (2).

The number of modes used in this process is the mode number satisfying a cumulative summation of the effective masses larger than 95% of the total mass. Thus, the energy responses of the E-SDOF systems corresponding to the first two modes of the three-story frame and the first three mode E-SDOF systems of the six-, nine-, and twenty-story frames were calculated, Table 3. Then, they were combined using Eq. (5) to calculate the elastic input energy responses of the MDOF systems.

**Table 3.** Effective masses of the benchmark frames (in %).

| Mode | 3-Story | | 6-Story | | 9-Story | | 20-Story | |
|---|---|---|---|---|---|---|---|---|
| | Modal | Cumul | Modal | Cumul | Modal | Cumul | Modal | Cumul |
| 1 | 86.23 | 86.23 | 81.00 | 81.00 | 83.10 | 83.10 | 80.00 | 80.00 |
| 2 | 10.78 | 97.01 | 12.07 | 93.07 | 10.90 | 94.00 | 11.80 | 91.80 |
| 3 | 2.99 | 100.00 | 4.03 | 97.10 | 3.70 | 97.70 | 3.50 | 95.30 |

Inelastic input energy imparted into the MDOF systems along with each energy component response was obtained through the performed nonlinear response history analyses. After they were normalized with respect to the total seismic mass of the structures, they were compared with the computed input energy response histories of each ground motion group. The comparisons for 5% and 2% structural damping cases are illustrated for Group #1 ground motion records in Figs. 5 and 6.

In the figures, straight lines represent the elastic input energy time series, while the dashed lines represent the inelastic input energy time series. For all types of ground motions and each MDOF frame type structure, it was observed that the elastic and inelastic response histories follow a similar trend.

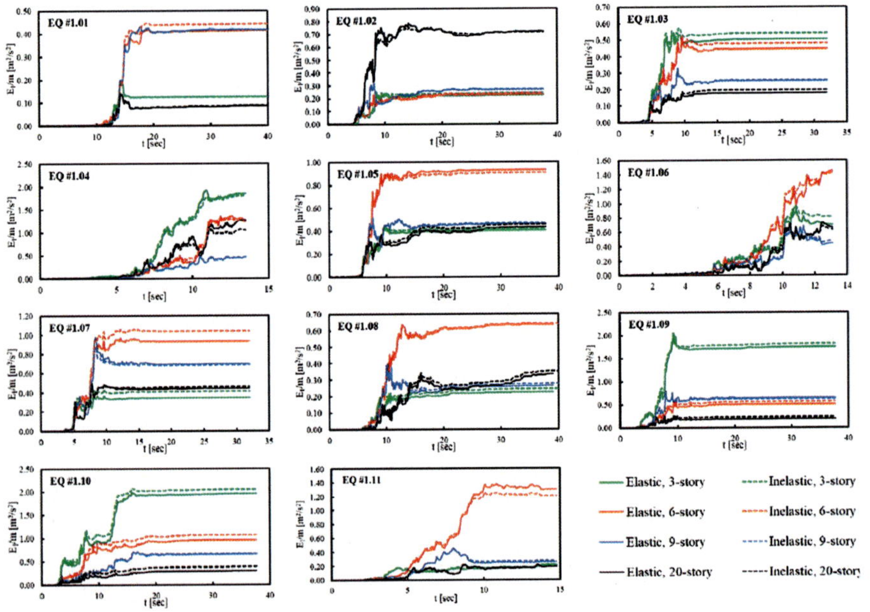


**Fig. 5.** Mass-normalized input energy response history of Group #1 records for a 5% damping ratio.

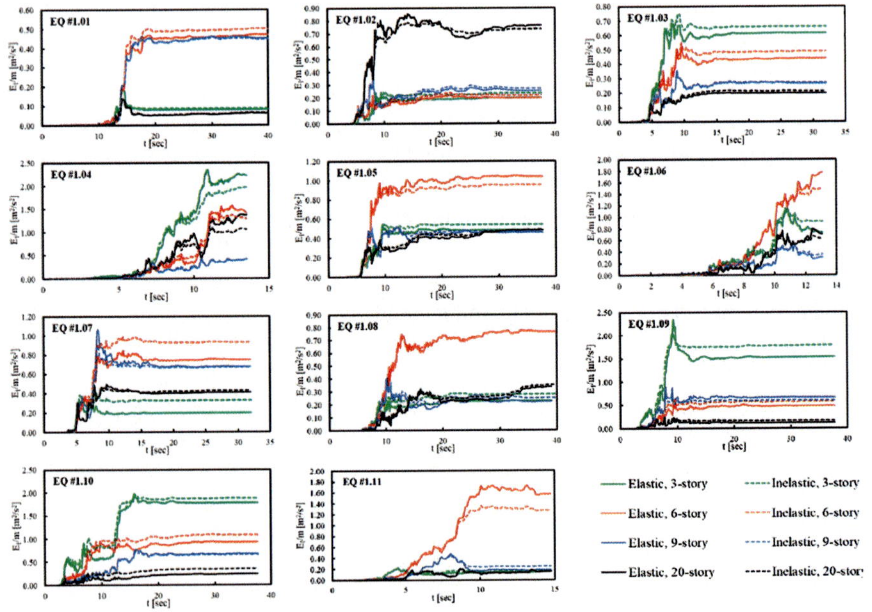


**Fig. 6.** Mass-normalized input energy response history of Group #1 records for a 2% damping ratio.

## 5 Discussions

The maximum values of the elastic input energies -$(E_I/m)_{el.}$- which were numerically calculated by the energy balance equations were compared with the inelastic input energies -$(E_I/m)_{in.}$- which were obtained after running a series of nonlinear time response analyses in for the benchmark frames with a damping ratio of 5% and 2% under the effect of each ground motion group. After they were plotted for each structure under the selected ground motion records in Fig. 7, no direct correlation was observed between the maximum amount of seismic energy input and the damping ratio. Accordingly, it can be stated that the damping must be considered separately in the analyses.

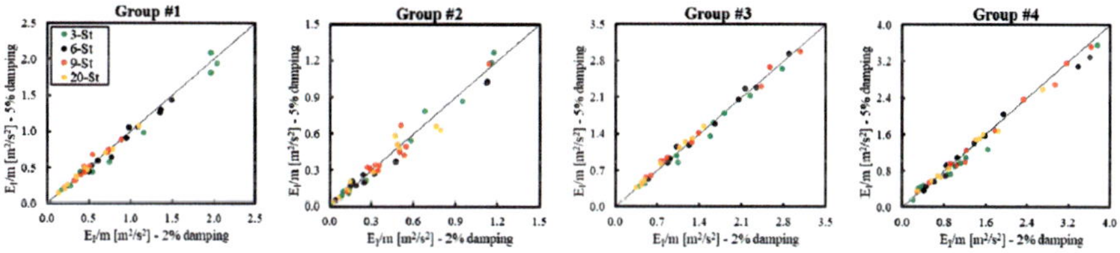


**Fig. 7.** Seismic input energy comparisons for different damping ratio values.

To make a judgment about the effect of the fundamental vibrational period of the structure, significant duration, and impulsive characteristics of the seismic excitation on the relationship between the peak $(E_I/m)_{el.}$ And $(E_I/m)_{in.}$ Values, relative percentage differences between them ($\varepsilon$) were calculated using Eq. (7). Also, the correlation coefficients ($r$) were computed by Eq. (8) where $x$ and $y$ represent the data sets (elastic and inelastic input energies) and $n$ is the total number of data (=11 in this case).

$$\varepsilon = \frac{|(E_I/m)_{el.} - (E_I/m)_{in.}|}{((E_I/m)_{el.} + (E_I/m)_{in.})/2} \tag{7}$$

$$r = \frac{n \cdot \left(\sum x \cdot y\right) - \left(\sum x\right) \cdot \left(\sum y\right)}{\sqrt{\left[n \cdot \sum x^2 - \left(\sum x\right)^2\right] \cdot \left[n \cdot \sum y^2 - \left(\sum y\right)^2\right]}} \tag{8}$$

Larger relative differences ($\varepsilon$) were obtained between the elastic and inelastic input energies when a 2% damping ratio was used, Table 4. The maximum $\varepsilon$ values for 5% and 2% damping ratios were obtained for the 20-story frame under long-duration pulse-like ground motion records, i.e., 17.2% and 30.1%, respectively. The arithmetic means of 176 $\varepsilon$ values (for 44 ground motions acted on four frames) were calculated as 8.0% and 12.5% for the 5% and 2% damping ratio cases, respectively.

Likewise, the correlation coefficients ($r$) between the elastic and inelastic input energy data series were computed to be less for the systems with a damping ratio of 2%, especially for the 20-story frame under Group #4 ground motion records, Table 5. The minimum $r$ values for 5% and 2% damping ratios were also obtained for the 20-story frame under long-duration pulse-like ground motion records, i.e., 0.823 and 0.410, respectively. The arithmetic mean of 176 $r$ values was calculated as 0.974 and 0.939 for the 5% and 2% damping ratio cases, respectively.

It was observed that the difference between the elastic and inelastic input energies majorly depends on the pulse type of the seismic excitation. While the relative percentage differences between them are relatively less and the correlation coefficients are higher (r = 1.0 means perfect relation between the investigated data sets) in ordinary-type non-pulse-like ground motions, higher dispersions were obtained in the pulse-like ground motions. When the period of the structure or the significant duration of the ground motion is considered a parameter, no direct relation between these variables and the values of ε or r was observed for the studied benchmark frames.

**Table 4.** Relative percentage differences for the models with different damping ratios (in %).

| Group # | 3-Story | | 6-Story | | 9-Story | | 20-Story | |
|---|---|---|---|---|---|---|---|---|
| | 5% damping | 2% damping | 5% damping | 2% damping | 5% damping | 2% damping | 5% damping | 2% damping |
| 1 | 6.1 | 9.5 | 5.8 | 11.4 | 2.5 | 5.3 | 8.3 | 11.1 |
| 2 | 2.7 | 6.7 | 0.5 | 2.2 | 0.9 | 3.1 | 2.6 | 4.6 |
| 3 | 13.8 | 18.9 | 17.0 | 18.3 | 9.0 | 14.0 | 10.0 | 13.7 |
| 4 | 9.7 | 16.9 | 10.6 | 18.8 | 11.9 | 16.1 | 17.2 | 30.1 |

**Table 5.** Correlation coefficients for the models with different damping ratios.

| Group # | 3-Story | | 6-Story | | 9-Story | | 20-Story | |
|---|---|---|---|---|---|---|---|---|
| | 5% damping | 2% damping | 5% damping | 2% damping | 5% damping | 2% damping | 5% damping | 2% damping |
| 1 | 0.997 | 0.992 | 0.989 | 0.969 | 0.994 | 0.991 | 0.992 | 0.990 |
| 2 | 0.999 | 0.996 | 1.000 | 1.000 | 1.000 | 0.994 | 0.999 | 0.995 |
| 3 | 0.984 | 0.975 | 0.981 | 0.975 | 0.978 | 0.974 | 0.973 | 0.949 |
| 4 | 0.980 | 0.985 | 0.967 | 0.924 | 0.934 | 0.910 | 0.823 | 0.410 |

It was seen that the energy response of the first mode E-SDOF system is dominant and captures a significant portion of the input energy of the MDOF system in low-story frames. However, as the number of stories increases, the effect of higher modes plays an important role as well, Fig. 8. The arithmetic means of percentile contribution of the first mode to the total elastic input energy were computed as 95.0%, 90.5%, 85.8%, and 78.1% for the three-, six-, nine-, and twenty-story frames, respectively.

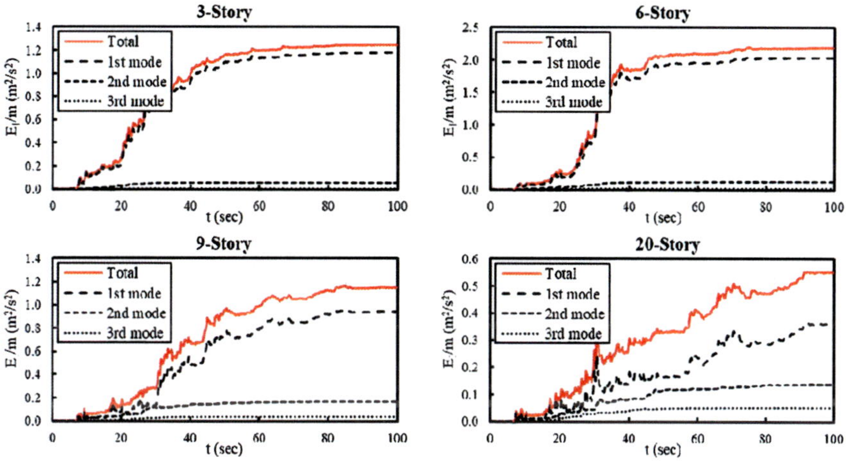


**Fig. 8.** The contribution of each mode to the total input energy for a sample ground motion.

## 6 Conclusions

The relationship between the elastic and inelastic response histories of steel MDOF systems was investigated in this study. The parameters were the earthquake characteristics, the structure's fundamental vibrational period, and the systems' damping ratio. Based on this, 22 pulse-like and 22 ordinary non-pulse-like type ground motions with different significant durations and four steel frame type structures with fundamental vibrational periods ranging from 0.80 s to 3.82 s were selected.

The elastic input energy responses were computed through a MATLAB-based code using the analogy between the input energy of MDOF and its E-SDOF systems. The first mode response was observed to cover a large portion of the total elastic input energy in the low-rise MDOF systems. However, the effects of other modes need to be considered in mid-rise and high-rise MDOF systems. In the elastic input energy computation, using only the first mode for the twenty-story frame results in a 28% and 33% average relative percentage difference. Thus, using as many modes as the mode number responsible for a cumulative summation of the effective masses larger than 95% of the total mass (generally up to the first three modes) is recommended in the precise computation of the elastic input energy an MDOF system is exposed to during an earthquake.

The computed elastic input energies were correlated with the inelastic input energy responses obtained from nonlinear response history analyses. It can be stated that the most influential parameter on the relationship between input energy responses of the linear elastic and nonlinear MDOF systems is the impulsive characteristics of the ground motion. The mean relative percentage differences between them ($\varepsilon$) were computed as 3.7% for the ordinary non-pulse-like records and 12.4% for the pulse-like records. Also, $\varepsilon$ generally tends to be larger for the ground motion records with relatively higher PGV values.

The analyses were performed for two different damping levels. The mean relative percentage differences between the seismic input energies imparted into the elastic and inelastic MDOF systems were computed as 12.5% and 8.0% for the 2% and 5% damping ratios, respectively. Thus, it can be stated for the range of selected benchmark frames that more precise seismic input energy predictions can be made for the structural systems with relatively higher damping levels. Therefore, the damping ratio of the system should be considered in the analyses.

The relationship between the elastic and inelastic input energy responses was not significantly affected by the earthquake record's significant duration or the structure's period in the studied vibrational period range.

Instead of performing time-taking nonlinear response history analyses, computation of the input energy time series benefitting from the energy responses of linear elastic E-SDOF systems might be reasonable because of its convenience and accuracy, especially for the non-pulse-like ground motion records.

## References

1. Housner, G.W.: Limit design of the structures to resist earthquakes. In: Proceedings of the 1st World Conference on Earthquake Engineering (1956)
2. Uang, C.M., Bertero, V.V.: Evaluation of seismic energy in structures. Earthquake Eng. Struct. Dynam. **19**(1), 77–90 (1990). https://doi.org/10.1002/eqe.4290190108
3. Fajfar, P., Fischinger, M.A.: Seismic procedure including energy concept. In: Ninth European Conference on Earthquake Engineering. Moscow, Russia (1990)
4. Chou, C.C., Uang, C.M.: Establishing absorbed energy spectra—an attenuation approach. Earthquake Eng. Struct. Dynam. **29**(10), 1441–1455 (2000). https://doi.org/10.1002/1096-9845(200010)29:10%3C1441::aid-eqe967%3E3.0.co;2-e
5. Akbas, B., Shen, J.: Earthquake-resistant design (EQRD) and energy concepts. IMO Teknik Dergi **192**, 2877–2901 (2003)
6. Dindar, A.A., Yalçın, C., Yüksel, E., Özkaynak, H., Büyüköztürk, O.: Development of earthquake energy demand spectra. Earthq. Spectra **31**(3), 1667–1689 (2015). https://doi.org/10.1193/011212eqs010m
7. Güllü, A., Yüksel, E., Yalçın, C., Dindar, A.A., Özkaynak, H., Büyüköztürk, O.: An improved input energy spectrum verified by the shake table tests. Earthquake Eng. Struct. Dynam. **48**(1), 27–45 (2019). https://doi.org/10.1002/eqe.3121
8. Akiyama, H.: Earthquake-Resistant Limit-State Design for Buildings. University of Tokyo Press (1985)
9. Kalkan, E., Kunnath, S.K.: Effective cyclic energy as a measure of seismic demand. J. Earthquake Eng. **11**(5), 725–751 (2007). https://doi.org/10.1080/13632460601033827
10. Mezgebo, M.G., Lui, E.M.: A new methodology for energy-based seismic design of steel moment frames. Earthq. Eng. Eng. Vib. **16**(1), 131–152 (2017). https://doi.org/10.1007/s11803-017-0373-1
11. Ucar, T.: Computing input energy response of MDOF systems to actual ground motions based on modal contributions. Earthquakes Struct. **18**(2), 263–273 (2020). https://doi.org/10.12989/eas.2020.18.2.263
12. Çalım, F., Güllü, A., Yüksel, E.: Evaluation of different approaches to estimate seismic input energy and top displacement demand of moment resisting frames. In: Benavent-Climent, A., Mollaioli, F. (eds.) IWEBSE 2021. LNCE, vol. 155, pp. 71–83. Springer, Cham (2021). https://doi.org/10.1007/978-3-030-73932-4_6
13. Mollaioli, F., Bruno, S., Decanini, L., Saragoni, R.: Correlations between energy and displacement demands for performance-based seismic engineering. Pure Appl. Geophys. **168**, 237–259 (2011). https://doi.org/10.1007/s00024-010-0118-9

14. Çalım, F.: Estimations of seismic input energy and inelastic top displacement demand in moment resisting frame type structures. (M.Sc. Thesis). Istanbul Technical University, Istanbul (2020)
15. Güllü, A., Çalım, F., Yüksel, E.: Estimation of the story response parameters through the seismic input energy for moment-resisting frames. Soil Dyn. Earthq. Eng. **164**, 107636 (2023). https://doi.org/10.1016/j.soildyn.2022.107636
16. Akbas, B., Shen, J., Hao, H.: Energy approach in performance-based seismic design of steel moment resisting frames for basic safety objective. Struct. Design Tall Build. **10**(3), 193–217 (2001). https://doi.org/10.1002/tal.172
17. Gupta, A., Krawinkler, H.: Seismic demands for performance evaluation of steel moment-resisting frame structures (SAC Task 5.4.3). Report No. 132, The John A. Blume Earthquake Engineering Center, Stanford (1999)
18. Surmeli, M., Yuksel, E.: A variant of modal pushover analyses (VMPA) based on a non-incremental procedure. Bull. Earthq. Eng. **13**(11), 3353–3379 (2015). https://doi.org/10.1007/s10518-015-9785-3
19. Applied Technology Council, ATC-72: Modeling and Acceptance Criteria for Seismic Design and Analysis of Tall Buildings. ATC, Redwood City, CA (2010)
20. Moustafa, A., Takewaki, I.: Characterization and modeling of near-fault pulse-like strong ground motion via damage-based critical excitation method. Struct. Eng. Mech. **34**(6), 755–778 (2010). https://doi.org/10.12989/sem.2010.34.6.755
21. Kohrangi, M., Vamvatsikos, D., Bazzurro, P.: Pulse-like versus non-pulse-like ground motion records: spectral shape comparisons and record selection strategies. Earthquake Eng. Struct. Dyn. **48**(1), 46–64 (2018). Portico. https://doi.org/10.1002/eqe.3122
22. Arias, A.: Measure of Earthquake Intensity. Massachusetts Inst. of Tech., Cambridge. University of Chile, Santiago de Chile (1970)
23. Trifunac, M.D., Brady, A.G.: On correlation of seismoscope response with earthquake magnitude and Modified Mercalli Intensity. Bull. Seismol. Soc. Am. **65**(2), 307–321 (1975). https://doi.org/10.1785/bssa0650020307
24. Abrahamson, N.A., Silva, W.J.: Report to Brookhaven National Laboratory (1996)
25. Kempton, J.J., Stewart, J.P.: Prediction equations for significant duration of earthquake ground motions considering site and near-source effects. Earthq. Spectra **22**(4), 985–1013 (2006). https://doi.org/10.1193/1.2358175
26. PEER Ground Motion Database, NGA-West 2. http://ngawest2.berkeley.edu/
27. Mathworks Inc., MATLAB 2016 software

Check for updates

# Cumulative Damage of Degrading Structures Due to Soft Soil Seismic Sequences

Jorge Ruiz-García(✉) and José M. Ramos-Cruz

Universidad Michoacana de San Nicolas de Hidalgo, Morelia 58030, México
jruizgar@stanfordalumni.org

**Abstract.** Several studies have attributed to cumulative damage as the origin of the collapse of tens of buildings in Mexico City during the strong September 19, 2017 (Mw7.1) earthquake. However, some of the previous studies based their conclusions on results from nondegrading elastoplastic or only-stiffness degrading single-degree-of-freedom, SDOF, systems. Therefore, an examination of cumulative damage of buildings having degrading (i.e., stiffness-and-strength degradation) global behavior is pertinent. From the results of this investigation, it is shown that the evolution of normalized hysteretic energy, NHE, which is closely related to the cumulative damage, strongly depends on the hysteretic features and the yield strength coefficient of the system as well as the relationship between its period of vibration and the predominant period of the ground motion. Furthermore, energy-based damage assessment should be carried out with analytical models that include degrading stiffness-and-strength hysteretic features and post-peak negative slope rather than using nondegrading hysteretic features. Finally, it was shown that cumulative damage is negligible between the strong 1985 and 2017 earthquakes for the case-study buildings at two distinct soft soil sites in Mexico City.

**Keywords:** Cumulative damage · seismic sequences · degrading structures · weak first-storey

## 1 Introduction

Mexico City is a highly populated city with high-density building environment which is exposed to a high seismic hazard due to its unique soft soil conditions. Seismic hazard is attributed to local earthquakes due to seismogenic geological faults along the Transvolcanic belt, shallow crustal (continental) earthquakes, interplate subduction earthquakes, and intermediate-depth intraslab earthquakes. Coincidentally, the most damaging historical earthquakes occurred on September 19, 1985 (Mw8.0) and 2017 (Mw7.1), exactly 32 years apart. Particularly, it was noted that 91% of the recorded 44 collapsed buildings as a consequence of the 2017 event were designed with older seismic provisions released prior to 1985 [1], while 57% of the collapsed buildings experienced a soft-storey mechanism.

Following the earthquake, several studies attributed the collapse of the buildings to cumulative damage during their life span [2, 3, 5]. For instance, Quinde and Reinoso


H. Varum et al. (Eds.): IWEBSE 2023, LNCE 236, pp. 154–164, 2023.
https://doi.org/10.1007/978-3-031-36562-1_12

[2] showed the normalized hysteretic energy, $NHE_m$, evolution of a single-degree-of-freedom, SDOF, having elastoplastic (i.e., nondegrading) behavior with natural period, $T$, equal to 1.8s subjected to a seismic sequence ensembled from earthquake ground motions recorded at SCT station during four historical earthquakes, which is later shown in Fig. 4b. Similarly, Massone et al. [3] studied the cumulative damage of a prototype 6-storey building model having weak-first storey with $T = 1.5$ s [4], which included a low-cycle fatigue criterium in the steel reinforcing modelling, as well as a simplified equivalent SDOF system having elastoplastic behavior. It should be noted that the equivalent SDOF system also have the same $T$ and design strength capacity of the prototype building. Both models were subjected to a seismic sequence ensembled from the earthquake records gathered during the September 19, 1985 and 2017 events recorded at the SCT station. The authors found that the 2017 seismic event governed the structural damage of the building under consideration; that is, the 1985 event did not induce cumulative damage prior to the 2017 event. In another study, Rodríguez [6] computed the damage spectrum of stiffness-degrading SDOF systems subjected to the same seismic sequence employed by Massone et al. [3] using a energy-based damage index introduced previously by the author [7]. Based on their study, the author concluded that cumulative damage was the cause of collapse of buildings during the September 19, 2017 earthquake.

The objective of this paper is to examine the $NHE_m$ spectra from degrading and nondegrading (elastoplastic) SDOF systems located at two distinct soft soil sites of Mexico City during strong historical earthquakes. Additionally, this study examined the evolution of cumulative damage of case-study RC buildings having weak-first storey configuration designed with the out-of-date 1976 edition of seismic provisions for Mexico City characterized as degrading and nondegrading equivalent SDOF systems under seismic sequences.

## 2 Case-Study Weak First-Story Buildings

This investigation focused their attention on the cumulative damage of reinforced concrete, RC, buildings having weak-first story, denoted to as WFS, located on soft soil sites of Mexico City. For this purpose, building models representative of RC-WFS buildings designed with the out-of-date 1976 edition of seismic provisions for Mexico City, named as NTCS-76, were considered in this investigation [8]. It should be noted that multistorey buildings at that time were designed assuming moment frames as structural system, neglecting the participation of infilled masonry walls. The static method of analysis was employed considering an elastic design spectrum for soft soil sites reduced by the ductility reduction factor, $Q$. Case-study buildings were designed with two different values of $Q$, equal to 4 and 6, commonly used in the Mexico City seismic provisions of the 1970's for moment-resisting frames. Bidimensional building models were generated coupling one exterior and one interior frame due to symmetry of the RC buildings in plan view as shown in Fig. 1a. Participation of the masonry infills above the first story was considered through strut elements as shown in Fig. 1b. A detailed description of the design process and modelling assumptions can be found in Ruiz-García and Cárdenas [8].

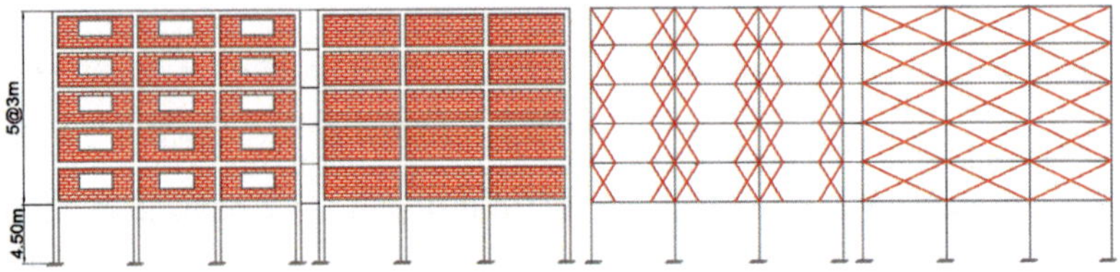


**Fig. 1.** a) Elevation view of the 6-story weak first-storey building considered in this investigation, b) modelling of the 6-story weak first-storey building.

It should be mentioned that prior studies have studied the cumulative damage of buildings through equivalent SDOF systems having a bilinear elastoplastic, EPP, envelope to account for their global behavior [2–4] as it is shown in Fig. 2a. However, as part of their investigation, Ruiz-García and Cárdenas [8] showed that RC-WFS buildings exhibit a negative slope in their capacity curve after reaching their strength capacity, which is attributed to the weak-first storey mechanism prior to collapse. Therefore, equivalent SDOF systems having a trilinear force-displacement envelope representative to the capacity curve of weak first-storey RC buildings, as shown in Fig. 2b, were considered to study the evolution of cumulative damage in this type of structures. It should be pointed out that the trilinear envelope exhibits cyclic strength and stiffness degradation as well as negative post-peak slope; that is, it exhibits degrading features as opposite to the nondegrading elastoplastic system.

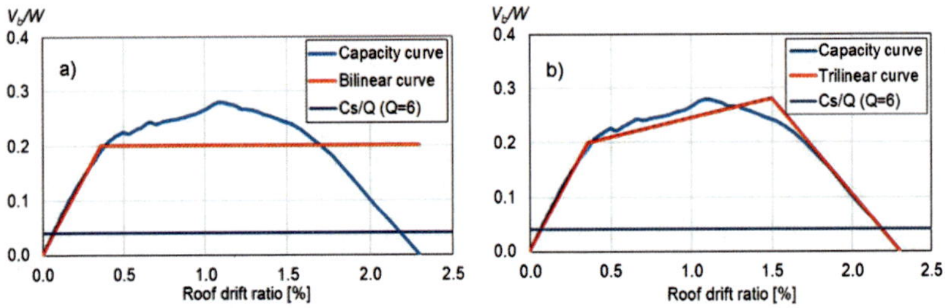


**Fig. 2.** Comparison of capacity curve of a 6-storey WFS building model with equivalent curves: a) bilinear (EPP) curve, b) trilinear curve.

For this purpose, the well-known Ibarra-Medina-Krawinker, IMK, hysteretic model [9], whose backbone envelope is shown in shown in Fig. 3, was considered in this investigation. Parameters $a_h$ (post-yield stiffness ratio), $m_c$ (displacement ductility at peak capacity), $a_c$ (negative slope coefficient), and $m_f$ (ultimate displacement ductility capacity) required in the IMK model were obtained from adjusting the trilinear curve to the capacity curve using an equal energy approach. Table 1 reports the main dynamic and mechanical properties (i.e., first-mode period of vibration, $T_1$, yield strength coefficient, $C_y$, and roof drift ratio at yielding, $q_y$) as well as adjusted parameters $a_h$, $m_c$, $m_f$, and $a_c$ for the case-study building models. Note that $C_y$ is greater than the seismic coefficient, Cs, prescribed in the NTCS-76 (Cs = 0.24 for soft soil sites) divided by $Q$ as shown in Fig. 2. Particularly, case-study buildings with the same number of stories but designed with $Q$

= 4 have greater $C_y$ than those designed with $Q = 6$ (for example, comparing 4NQ4 with 4NQ6). The equivalent degrading and nondegrading SDOF systems considered the same $T_1$ and $C_y$ for comparison purposes.

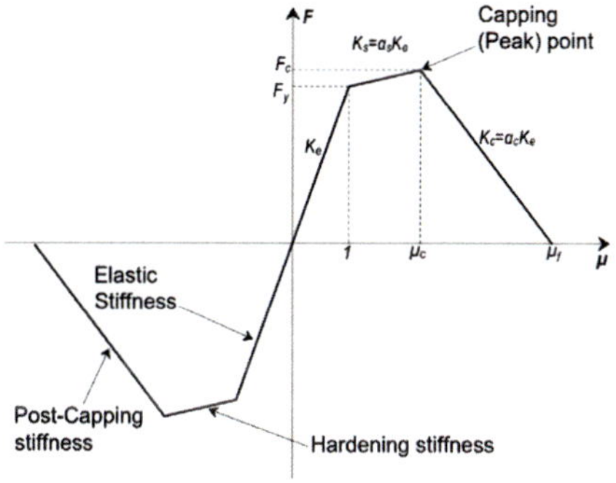


**Fig. 3.** Backbone curve of the Ibarra-Medina-Krawinkler, IMK, hysteretic model [9] considered in this investigation.

**Table 1.** Mechanical and dynamic properties of the case-study building models

| Model | $T_1$ [s] | $C_y$ | $q_y$ [%] | $a_h$ [%] | $m_c$ | $a_c$ [%] | $m_f$ |
|---|---|---|---|---|---|---|---|
| 4NQ4 | 0.92 | 0.30 | 0.59 | 9.7 | 2.7 | −20.6 | 8.3 |
| 4NQ6 | 0.96 | 0.22 | 0.46 | 11.1 | 3.0 | −23.5 | 8.1 |
| 6NQ4 | 0.96 | 0.25 | 0.45 | 15.8 | 3.8 | −43.2 | 7.1 |
| 6NQ6 | 1.03 | 0.20 | 0.36 | 12.6 | 4.2 | −63.0 | 6.4 |

Cumulative damage was quantified by means of the well-known Park-Ang damage index [5], $I_{P\&A}$, that take into account the cumulative normalized hysteretic energy, $NHE_m$, and the peak displacement ductility demand. For computing $I_{P\&A}$, $b$ value was set equal to 0.25, for a structure with nonductile elements, and $m_f$ was taken as the ultimate displacement ductility capacity of the structure.

## 3 Seismic Sequences

The seismic sequences consist of an ensemble of earthquake ground motions recorded at the same stations located on soft-soil sites during strong historical earthquakes that hit Mexico City. For instance, Fig. 4 shows two seismic sequences ensembled from earthquakes records gathered at SCT station from 4 historical subduction earthquakes, which was employed in Quinde and Reinoso [2], that includes the 1985 (Mw8.0) and 2017 (Mw7.1) earthquakes. Similarly, Rodriguez [6] employed a seismic sequence consisting only of the East-West component of the 1985 and 2017 seismic events. It should be mentioned that the East-West components recorded the highest peak ground acceleration.

In addition, two seismic sequences were ensembled from earthquakes records gathered at Culhuacan (CH84) station from six intraplate historical earthquakes, which includes the strong intraplate events occurred on June 15, 1999 (Mw7.0) and September 19, 2017 (Mw7.1). Both seismic sequences are shown in Fig. 5.

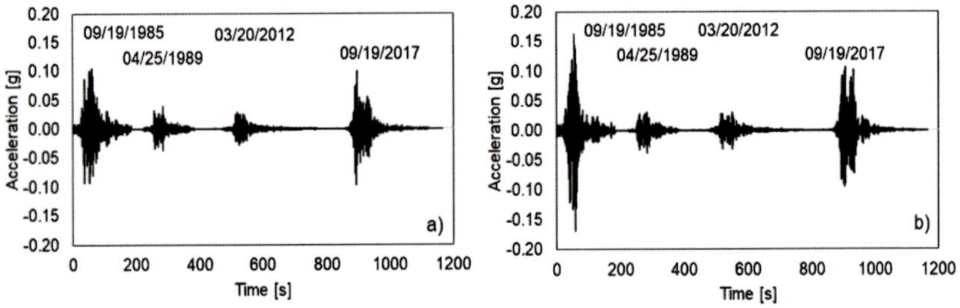


**Fig. 4.** Seismic sequence ensembled from earthquakes records gathered at SCT station from 4 historical earthquakes: a) component N00E (North-South), b) component N90W (East-West).

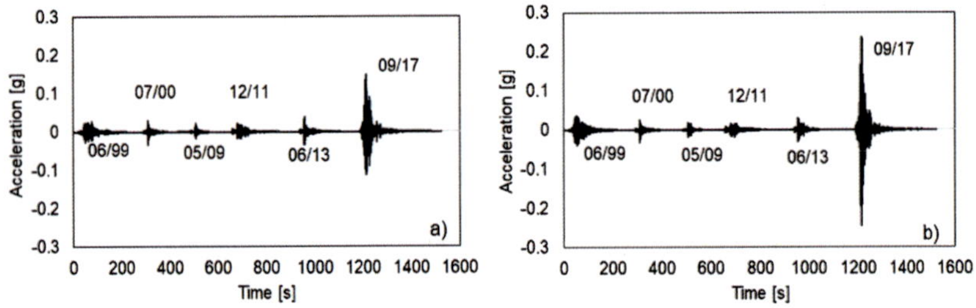


**Fig. 5.** Seismic sequence ensembled from earthquakes records gathered at CH84 station from six intraplate historical earthquakes: a) component N00E (North-South), b) component N90W (East-West).

## 4 Cumulative Damage Evaluation for Degrading and Nondegrading Structures

Figure 6 shows the normalized hysteretic energy, $NHE_m$ spectra computed for the EW component gathered at SCT station during the September 19, 1985 earthquake for SDOF systems with nondegrading (Fig. 6a) and degrading (Fig. 6b) hysteretic features, five levels of yield strength coefficient, $C_y$, and damping ratio equal to 5%. The degrading system have the following parameters of the trilinear envelope: $m_c = 3.0$, $a_h = 10\%$, $a_c = -20\%$. The $NHE_m$ spectra were computed for normalized periods of vibration with respect to the predominant period of the ground motion, $T/T_g$. Comparing both figures, it can clearly be seen that the $NHE_m$ demands depend on the type of hysteretic behavior, $C_y$ and $T/T_g$. It can also be observed that the limiting $T/T_g$ ratio were $NHE_m$ demands vanish (i.e., becomes negligible) depends on $C_y$. For instance, limiting $T/T_g$ ratios are about 1.5 and 2.0 for degrading systems with $C_y$ equal to 0.30 and 0.10, respectively. Interestingly, the peak of NHE$_m$ demand corresponds to $T/T_g$ ratios smaller than one, and $T/T_g$ decreases as the $C_y$ decreases, which is particularly true for degrading systems.

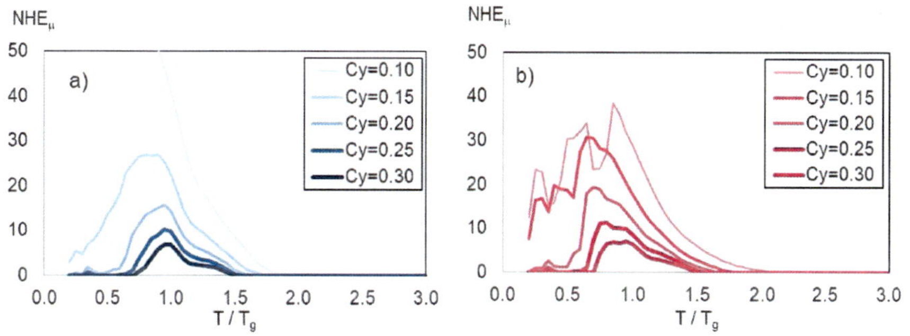


**Fig. 6.** Normalized hysteretic energy, NEH$_m$, computed for the EW component gathered at SCT station ($T_g = 2.04$ s) during the September 19, 1985 earthquake: a) nondegrading (EPP) behavior, b) degrading behavior.

Next, Fig. 7 displays a comparison of the NHE$_m$ demands for the same station computed from the 1985 event (solid line) and the seismic sequence (dashed line) showed in Fig. 4b. From the figure, it is evident that the NHE$_m$ demands increase when the seismic sequence is considered, although the amount of increment also depends on the type of hysteretic behavior, $C_y$ and $T/T_g$. Early observations of the NHE$_m$ trends noted from the 1985 event still hold for the seismic sequence.

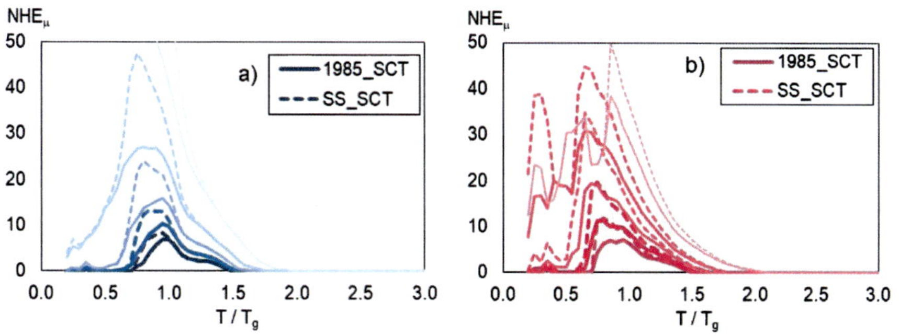


**Fig. 7.** Comparison of NEH$_m$ computed for the EW component gathered at SCT station ($T_g$ = 2.04s) during the September 19, 1985 earthquake (solid line) and the seismic sequence at SCT station shown in Fig. 4b: a) nondegrading (EPP) behavior, b) degrading behavior.

A similar NHE$_m$, spectra computed for the EW component gathered at CH84 station during six historical earthquakes, including the September 19, 2017 earthquake, revealed that the first five seismic events did not induce NHE$_m$ demands since the systems behaved elastically. Therefore, Fig. 8 displays the NHE$_m$ spectra corresponding to the 2017 seismic event. It can clearly be observed that the $NHE_m$ demands are larger for both degrading and nondegrading systems than those computed for the SCT station in 1985. Nevertheless, dependency of $NHE_m$ on $C_y$ and $T/T_g$ also applies for this station.

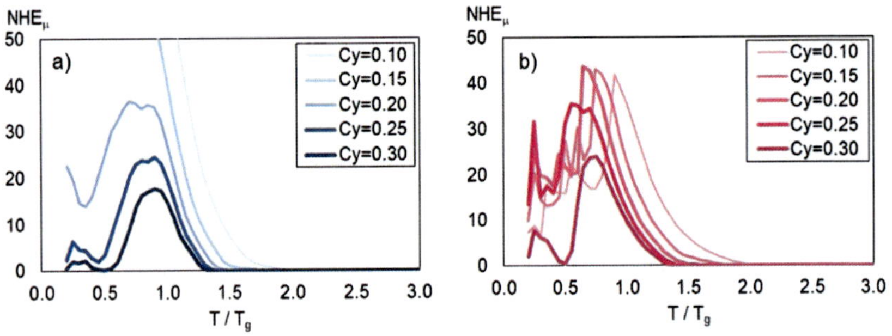


**Fig. 8.** Normalized hysteretic energy, $NHE_m$, computed for the EW component gathered at CH84 station ($T_g = 1.31$ s) during the September 19, 2017 earthquake: a) nondegrading (EPP) behavior, b) degrading behavior.

It is convenient to track the evolution of $NHE_m$ and $I_{P\&A}$ over time to identify possible cumulative damage in the case-study building. Figures 9a and 9b show the evolution of $NHE_m$ and $I_{P\&A}$ computed for degrading, with trilinear envelope defined with parameter given in Table 1, and nondegrading equivalent single-degree-of-freedom, ESDOF, systems of the 4-storey WFS building designed with $Q = 6$ (model 4NQ6), respectively, under the seismic sequence ensembled at SCT station shown in Fig. 4b.

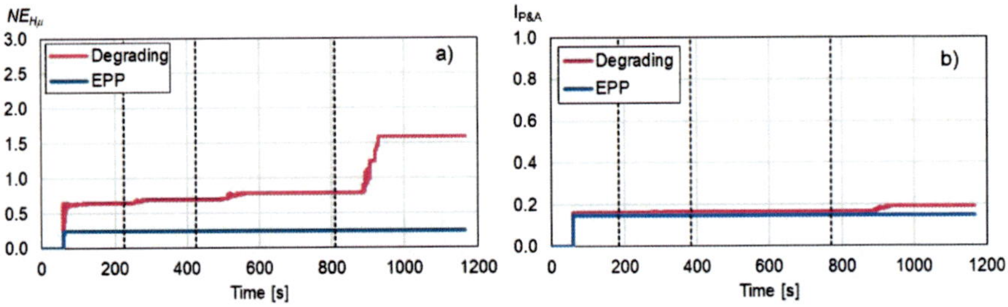


**Fig. 9.** Evolution of $NHE_m$ for degrading (Trilinear) and nondegrading (EPP) equivalent SDOF systems of a 4-storey ($Q = 6$) WFS building under the seismic sequence ensembled at SCT station (E-W component): a) $NHE_m$, b) $I_{P\&A}$.

It can clearly be seen that the evolution of $NHE_m$ and $I_{P\&A}$ depends on the type of hysteretic behavior. For instance, the $NHE_m$ demand for the degrading ESDOF slightly increases from the 1985 event to the 2017 event, while the $NHE_m$ demand for the nondegrading ESDOF system remains constant (i.e., no additional hysteretic energy is induced in the seismic events following the 1985 earthquake). In terms of the $I_{P\&A}$, this value slightly increases for the degrading ESDOF systems util the 2017 event. It should be mentioned that both the degrading and nondegrading ESDOF systems exhibited linear behavior under the seismic sequence shown in Fig. 4a.

A similar comparison corresponding to the 6-storey WFS building designed with $Q = 6$ (6NQ6 model) is shown in Fig. 10. In spite of the slightly larger $NHE_m$ demands than those for the 4NQ6 model, it can be seen that cumulative damage is negligible for the degrading system, and it does not occur for the nondegrading system.

To provide an explanation, Fig. 11 displays the displacement time-history along with the hysteretic response of both systems under the seismic sequence shown in Fig. 4b. It can be seen that the 1985 seismic event induce nonlinear behavior along with permanent lateral displacement at the end of the seismic excitation for both systems; however, $NHE_m$ does not grow for the EPP systems during the subsequent two seismic events since its hysteretic branch unload and reload in the elastic branch, without energy dissipation, while the degrading system slightly increase $NHE_m$ in small hysteretic cycles close to the origin. It should also be noted that the peak (maximum) displacement was triggered in the 1985 event for both systems. A similar examination for the North-South component of the seismic sequence revealed that both EPP and degrading systems behaved elastically.

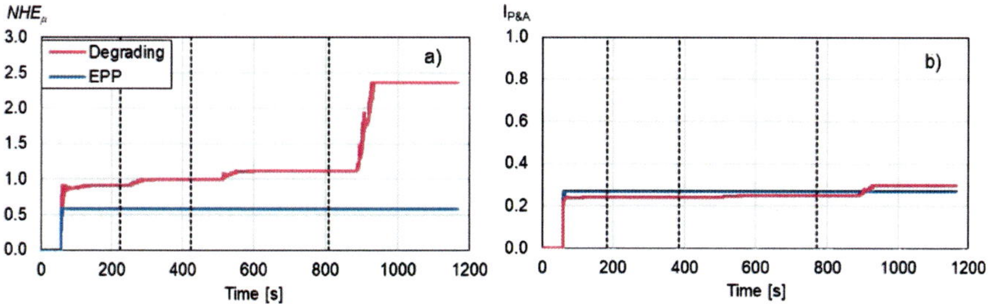


**Fig. 10.** Evolution of $NHE_m$ for degrading (Trilinear) and nondegrading (EPP) equivalent SDOF systems of a 6-storey ($Q = 6$) WFS building under the seismic sequence ensembled at SCT station (E-W component): a) $NHE_m$, b) $I_{P\&A}$.

Since $NEH_m$ demands vary from site to site, it is of interest to examine the evolution of $NEH_m$ and $I_{P\&A}$ for the site of CH84 station. Figures 12a and 12b show the evolution of $NEH_m$ and $I_{P\&A}$ computed for degrading and nondegrading, equivalent SDOF systems of the same 4-storey WFS building designed with $Q = 6$, respectively, under the seismic sequence ensembled at CH84 station.

It can be seen that the SDOF system exhibits energy demand until the 2017 seismic event and, therefore, cumulative damage does not occur due to the 1999 seismic event. Additionally, comparing Fig. 9a with Fig. 12a, it is evident that the earthquake record gathered at CH84 station site trigger significantly larger $NHE_m$ demands than those at the SCT station site during the 2017 seismic event. It should be noted that $T/T_g$ ratio is equal to 0.68, which anticipates large $NHE_m$ demands for this site, using Fig. 8, for a system with $C_y = 0.22$ It is evident that the EPP system significantly underestimate $I_{P\&A}$ with respect to the degrading system., which it can be explained since degrading systems trigger larger $NHE_m$ demands at this site for $T/T_g$ ratio about 0.68 than those for

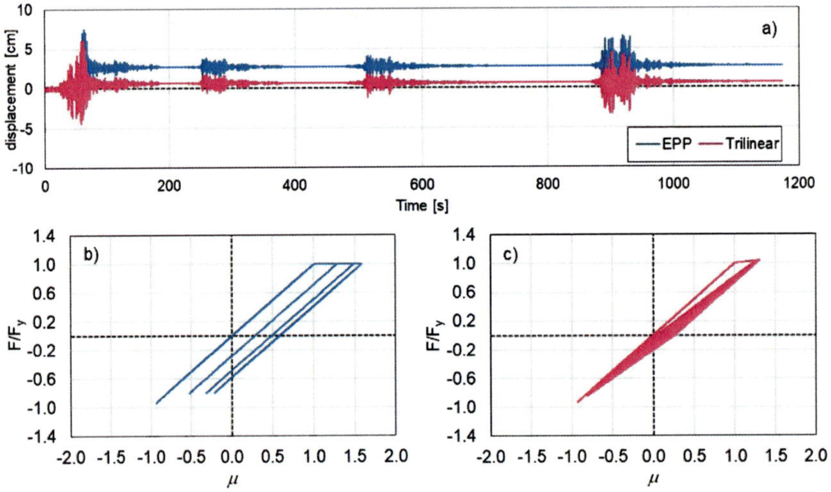


**Fig. 11.** a) Comparison of displacement time-history of a degrading (Trilinear) and nondegrading (EPP) equivalent SDOF systems of a 6-storey ($Q = 6$) WFS building under the seismic sequence ensembled at SCT station (E-W component), b) EPP hysteretic response, c) degrading hysteretic response.

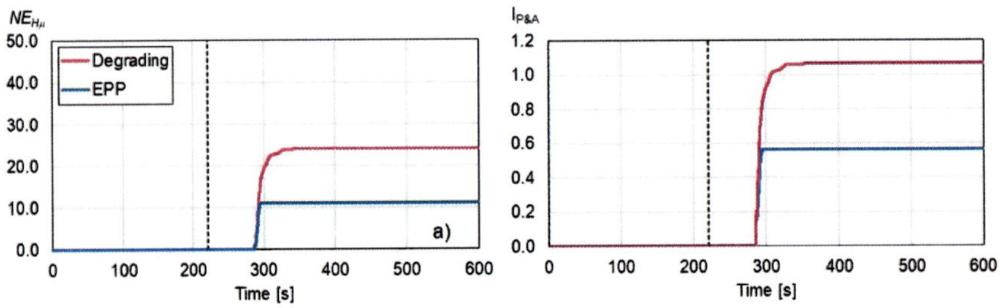


**Fig. 12.** Evolution of NHEm for degrading (Trilinear) and nondegrading (EPP) equivalent SDOF systems of a 4-storey ($Q = 6$) WFS building under the seismic sequence ensembled at CH84 station (E-W component): a) NHEm, b) IP&A.

EPP systems. For this reason, degrading equivalent SDOF systems of the 4NQ4, 6NQ4, and 6NQ6 also exhibited large $NHE_m$ demands and $I_{P\&A}$.

To explain the difference in $NHE_m$ demands and $I_{P\&A}$ for the degrading and non-degrading (EPP) systems, Fig. 13 displays the displacement time-history along with the hysteretic response of both systems. Both systems remain elastic under the 1999 seismic event and the 2017 seismic event trigger larger peak displacement demand (i.e., displacement ductility demand) in the degrading system than in the nondegrading EPP system, since the former exhibits in-cyclic strength degradation. Additionally, the degrading system exhibits wider inelastic cycles than the nondegrading system that explain the larger $NHE_m$ demands.

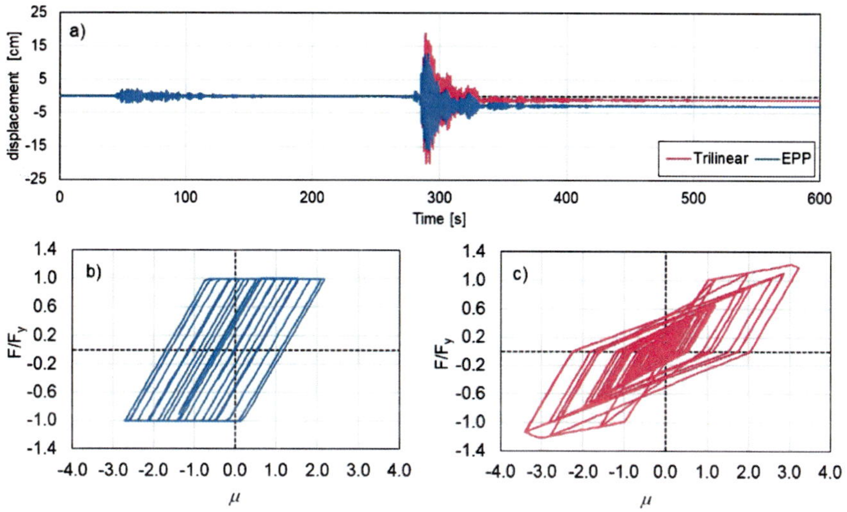


**Fig. 13.** a) Comparison of displacement time-history of a degrading (Trilinear) and nondegrading (EPP) equivalent SDOF systems of a 4-storey ($Q = 6$) WFS building under the seismic sequence ensembled at CH84 station (E-W component), b) EPP hysteretic response, c) degrading hysteretic response.

# 5 Discussion

Previous studies have claimed that cumulative damage was the cause of the collapsed buildings in Mexico City hit by the September 19, 2017 earthquake [2, 6]. However, slight cumulative damage (i.e., increment of $I_{P\&A}$ after a sequence of historical earthquakes) was only observed at the SCT site, while cumulative damage was not observed at the CH84 site even considering SDOF systems having degrading features. Examining the structural characteristics of collapsed buildings during the September 19, 2017 (Mw7.1) earthquake revealed that collapsed buildings had several structural pathologies [1, 10] such as soft story (i.e., vertical irregularities), corner asymmetry, flat slab (i.e., flexible diaphragm), and lack of ductile steel detailing that strongly influenced in their collapse more than only cumulative damage in their lifespan. Additionally, it was shown that cumulative damage was negligible between the strong 1985 and 2017 earthquakes for the case-study buildings at two distinct soft soil sites in Mexico City.

# 6 Conclusions

This study examined the possible cumulative damage on structures located at soft soil sites in Mexico City due to historical earthquakes. For this purpose, earthquake ground motions gathered at two distinctive sites (SCT and CH84 stations) were considered in this investigation. From the results of this study, it was shown that the evolution of normalized hysteretic energy, NHE, closely related to the cumulative damage, strongly depends on the hysteretic features of the SDOF system as well as the yield strength coefficient of the system, $C_y$, and the relationship between the period of vibration and the predominant period of the ground motion, $T/T_g$. Therefore, energy-based damage assessment should be carried out with analytical models that include degrading stiffness-and-strength hysteretic features and post-peak negative slope rather than using nondegrading hysteretic features.

## References


1. Galvis, F.A., Miranda, E., Heresi, P., Dávalos, H., Ruiz-García, J.: Overview of collapsed buildings in Mexico City after the 19 September 2017 (Mw7.1) earthquake. Earthquake Spectra **36**(2_suppl), 83–109 (2020). https://doi.org/10.1177/8755293020936694
2. Quinde, Pablo, Reinoso, Eduardo: Long duration and frequent, intense earthquakes: lessons learned from the 19 September 2017 earthquake for Mexico City's resilience. Earthquake Spectra **36**(2_suppl), 49–61 (2020). https://doi.org/10.1177/8755293020942548
3. Massone, L., Aceituno, D., Carrillo, J.: Cumulative damage in RC buildings-The case of the 2017 Puebla-Morelos earthquake. In: 17th World Conference on Earthquake Engineering, paper 2b-0082, Sendai, Japan (2020)
4. Arteta, C.A., Carrillo, J., Archbold, J., et al.: Response of mid-rise reinforced concrete frame buildings to the 2017 Puebla earthquake. Earthq. Spectra **35**(4), 1763–1793 (2019)
5. Park, Y.J., Ang, A.H.S.: Mechanistic seismic damage model for reinforced concrete. J. Struct. Eng. **111**(4), 722–739 (1985)
6. Rodriguez, M.E.: The interpretation of cumulative damage from the building response observed in Mexico City during the 19 September 2017 earthquake. Earthquake Spectra **36**(2_suppl), 199–212 (2020). https://doi.org/10.1177/8755293020971307
7. Rodríguez, M.E.: A measure of the capacity of earthquake ground motions to damage structures. Earthquake Eng. Struct. Dynam. **23**, 627–643 (1994)
8. Ruiz-García, J., Cárdenas, Y.: Seismic performance assessment of weak first-storey RC buildings designed with old and new seismic provisions for Mexico City. Eng. Struct. **232**, 111803 (2021)
9. Ibarra, L., Medina, R.A., Krawinkler, H.: Hysteretic models that incorporate strength and stiffness deterioration. Earthquake Eng. Struct. Dynam. **34**, 1489–1511 (2005)
10. Reinoso, E., Quinde, P., Buendía, L., Ramos, S.: Intensity and damage statistics of the September 19, 2017 Mexico earthquake: Influence of soft story and corner asymmetry on the damage reported during the earthquake. Earthq. Spectra **37**(3), 1875–1899 (2021)

# Reliability-Based Hysteretic Energy Spectra Using Modern Intensity Measures

E. Bojórquez(✉), J. Carvajal, J. Bojórquez, and A. Reyes-Salazar

Universidad Autónoma de Sinaloa, Mexico, Mexico
{eden,joelcarvajal.fic,juanbm}@uas.edu.mx

**Abstract.** This work is oriented to develop a reliability-based seismic design tool that account for cumulative demands. For this aim, dissipated hysteretic energy uniform annual failure rate (UAFR) spectra are obtained via the advanced spectral shape-based intensity measure $I_{Np}$. In order to obtain the new hysteretic energy spectra based on structural reliability, a set of 50 earthquake ground motions selected from soft soil of Mexico City, included the 1985 earthquake were selected due to their large energy amount demanded to the structures. Several nonlinear oscillators with elasto-perfectly plastic, bilinear and Takeda hysteretic behavior are subject to the set of records to compute the new tool for energy-based design. It is observed that the use of elasto-perfectly plastic models provided similar UAFR spectra in comparison with hysteretic models with different post-yielding stiffness. On the other hand, if stiffness degradation is considered, the UAFR energy spectra tend to be different. The new hysteretic energy UAFR spectra are the beginning toward energy-based earthquake resistant design that combine structural reliability and advanced ground motion intensity measures for degrading and/or no degrading structures.

**Keywords:** Hysteretic energy UAFR spectra · Intensity measures · Cumulative demands · Structural reliability

## 1 Introduction

Traditional response or seismic design spectra are based on the use of spectral acceleration at first mode of vibration of the structure $Sa(T_1)$ as intensity measure. However, the evolution of the intensity measures suggests that $Sa(T_1)$ has some limitations for seismic engineering, when it is used as an intensity measure. In particular, in some cases $Sa(T_1)$ has low efficiency in predicting nonlinear structural response. On the other hand, it does not take into account the effect of elongation of the vibration period of the structure when nonlinear behavior occurs. For this reason, recently modern intensity measures have been gained popularity; for example, Bojórquez and Iervolino [1] proposed the spectral shape-based intensity measure $I_{Np}$ were results a very good predictor of the structural response, and it has been used to develop new seismic hazard maps. In particular, it has been shown that $I_{Np}$ is a better indicator of ground motion intensity than the more commonly used measures such as peak ground acceleration ($PGA$), peak


H. Varum et al. (Eds.): IWEBSE 2023, LNCE 236, pp. 165–176, 2023.
https://doi.org/10.1007/978-3-031-36562-1_13

ground velocity (*PGV*), and in most cases $Sa(T_1)$ [2–6]. The results have demonstrated that the $I_{Np}$ measure can be used to accurately predict the performance of structural systems. Moreover, two additional issues may be relevant to achieve adequate seismic performance. The first is the inclusion of plastic deformation demands. The seismic behavior of the structure depends on its capacity to support plastic deformations during an earthquake. Therefore, knowing the accumulated demands of deformation of the structure is essential for designing seismic-resistant structures. Previous studies have shown that accumulated plastic deformation demands could be explicitly considered in seismic design through the concept of energy. The use of energy was initially proposed by Housner [7]. This concept has been used by various researchers to propose energy-based methodologies aimed at providing the structure with a dissipation capacity equal to or greater than that generated by the seismic load [8–18]. The capacity of a structure is determined by the energy it can absorb before failure. In particular, in this study, normalized hysteretic energy capacities has been selected as the performance parameter for energy-based design due to its close relationship with structural damage [19–23]. A second important issue is that the earthquake design spectra of most of the building codes are not associated to structural reliability levels, such the probability of exceeding a certain level of ground motion intensity given a specific failure rate [24–26]. Therefore, it is important to consider the use of seismic design spectra associated with levels of structural reliability to design buildings capable to support efficiently the loads produces by earthquakes with different magnitudes and intensities. For this reason, in this study, Uniform Annual Failure Rate (*UAFR*) spectra of normalized hysteretic energy are obtained and compared using $I_{Np}$ as the seismic intensity measure. The following section describes the procedure applied to obtain *UAFR* spectra.

## 2 Methodology

The structural response of the nonlinear single degree of freedom (*SDOF*) systems was estimated using Incremental Dynamic Analysis (*IDAs*) [27] in a first stage for several vibration periods and various lateral resistance coefficients at yielding (*Cy*) subjected to the ground motion records scaled in terms of $I_{Np}$. A total of 50 seismic records from Mexico City of lake zone have been used for this purpose. Table 1 shows the seismic records and summarizes some of their most relevant characteristics. The seismic records were scaled to represent different earthquake ground motion return periods [28]. The scaling factor was obtained for each of the studied ground motion earthquake and multiplied by the acceleration history of the record, so the spectral ordinate corresponding to the fundamental period of the structure being analyzed with a specific intensity is associated with a certain $I_{Np}$ value, depending on the case for which it is being scaled [29]. Different *SDOF* systems are analyzed in this work using structural reliability, taking into account the accumulated demands for the selected hysteretic behavior models, which in summary are represented as *UAFR* spectra.

### 2.1 Structural Reliability

Structural reliability is an important factor to consider in the design of earthquake-resistant structures. It can be described in terms of the annual exceedance probability [see Eq. (1)], which is the probability that a structure is capable of withstanding external loads or strains without significant damage or failure given a specific time period.

$$\nu_F = \int P(Q \geq 1|y)\left|\frac{d\nu_Y(y)}{d_y}\right|dy \tag{1}$$

In Eq. 1, $|d\nu_Y(y)/d_y|$ corresponds to the absolute value of the derivative of the seismic hazard curve of the specific site being evaluated; $Q$ expresses the ratio between demand and capacity; and $P(Q \geq 1|y)$ is the probability of failure given a seismic intensity ($y$). Note that structural failure occurs when the capacity of the structure is smaller than the demand exercised by the earthquake, in other words; demand/capacity $= Q \geq 1$. The seismic response or engineering demand parameter determined in this work is the normalized dissipated hysteretic energy by the strength and displacement at yielding.

## 2.2 Energy Demand

The most important consideration in structural design for earthquakes is to ensure that it has sufficient strength, ductility and energy dissipation capacity to withstand expected earthquakes without exceeding allowable deformations or stresses. This latter point is generally achieved through the use of seismic design techniques such as base isolation, damping systems, or energy dissipation systems. The use of these techniques can reduce a large portion of the total amount of energy that a structure must absorb during an earthquake and thus decrease the damage that can be inflicted on the structure. Energy-based methodologies focus on providing structures with energy dissipation capacities that are greater than or equal to the expected energy demands from the earthquake [8, 9]. The design criteria for a seismic-resistant structure can be formulated as follows in Eq. (2):

$$\textit{Energy Capacity} \geq \textit{Energy Demand} \tag{2}$$

The energy most related to structural damage is the hysteric energy $E_H$. It can be physically interpreted as the area enclosed by each of the hysteresis cycles that the structure undergoes during a seismic excitation. Although $E_H$ provides an approximate idea of the accumulated plastic deformation demands, this response parameter alone does not provide sufficient information to evaluate the structural behavior, so it is convenient to normalize it as shown in the following Eq. (3):

$$E_N \geq \frac{E_H}{F_y \delta_y} \tag{3}$$

where $F_y$ and $\delta_y$ are the strength and yield displacement, respectively. The normalized hysteresis energy ($E_N$) is a parameter that correlates better with structural damage [30, 31]. For this study, the normalized hysteresis energy is considered as a parameter to control accumulated damage.

In order to compute the energy demands, the SDOF systems were subjected to a group of seismic records described in the following sections.

## 2.3 Seismic Records

Seismic records are an important part of understanding how structures respond to earthquakes. Ground motion records can also be used to improve the design of new structures, providing engineers identify any potential weaknesses or vulnerabilities that could cause significant damage depending on the type of site. The soft soil of Mexico City requires large amounts of hysteretic energy capacities in constructed structures. In some cases, due to the nature of the soft soil and the high seismicity of the zone, many structures require significant repairs or demolition to ensure their safety and stability. Therefore, in this study, a set of earthquake ground motion records representative of the soft soil of Mexico City are analyzed, with magnitudes ranging from 5.9 to 8.1, as shown in Table 1. Notice that all the ground motion records were scaled to represent different intensity levels in terms of $I_{Np}$, which let compute the fragility and structural reliability of the selected nonlinear systems, finally, to assess *UAFR* spectra in terms of normalized hysteretic energy (Fig. 1).

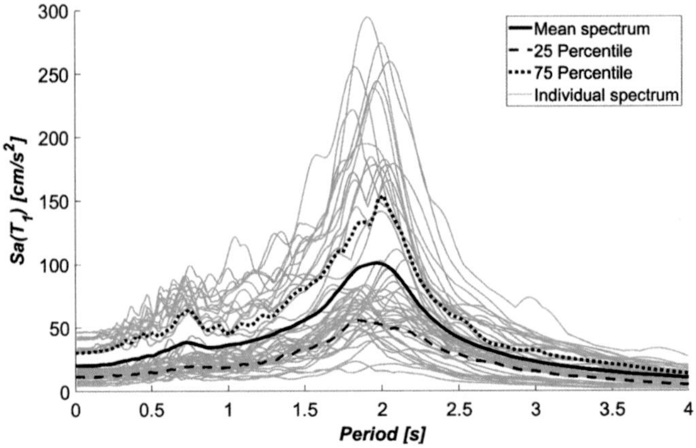


**Fig. 1.** Seismic response spectra corresponding to soft soil of Mexico City including mean, 25$^{th}$ y 75$^{th}$ percentile.

**Table 1.** Seismic records of soft soil of Mexico City

| Record no | Date | Station | Mw | PGA (cm/s$^2$) |
|---|---|---|---|---|
| 1 | 16-Sep-85 | S.C.T | 8.1 | 167.9 |
| 2 | 14-Sep-95 | Alameda | 6.4 | 253.44 |
| 3 | 9-Aug-00 | Alameda | 6.1 | 207.36 |
| 4 | 10-Dec-94 | Cibeles | 6.3 | 43.52 |
| 5 | 9-Oct-95 | Cibeles | 6.5 | 378.88 |

*(continued)*

**Table 1.** (*continued*)

| Record no | Date | Station | Mw | PGA (cm/s$^2$) |
|---|---|---|---|---|
| 6 | 30-Sep-99 | Cibeles | 6.5 | 340.48 |
| 7 | 22-Jan-03 | Cibeles | 6.5 | 267.52 |
| 8 | 1-Jan-04 | Cibeles | 6.3 | 193.28 |
| 9 | 15-May-93 | C.U | 5.9 | 186.88 |
| 10 | 10-Dec-94 | C.U | 6.3 | 293.12 |
| 11 | 9-Oct-95 | C.U | 6.5 | 384 |
| 12 | 11-Jan-97 | C.U | 6.9 | 273.92 |
| 13 | 22-May-97 | C.U | 6 | 203.52 |
| 14 | 30-Sep-99 | Centro | 6.5 | 360.96 |
| 15 | 9-Aug-00 | C.U | 6.1 | 236.8 |
| 16 | 20-Mar-12 | Centro | 7.4 | 429.4 |
| 17 | 15-May-93 | CUPJ | 5.9 | 21.87 |
| 18 | 10-Dec-94 | CUPJ | 6.3 | 152.96 |
| 19 | 9-Oct-95 | CUPJ | 6.5 | 194.21 |
| 20 | 11-Jan-97 | CUPJ | 6.9 | 156.25 |
| 21 | 30-Sep-99 | Multifamiliar | 6.5 | 279.75 |
| 22 | 9-Aug-00 | CUPJ | 6.1 | 188.75 |
| 23 | 20-Mar-12 | Multifamiliar | 7.4 | 347 |
| 24 | 9-Oct-95 | Sec | 6.5 | 243.07 |
| 25 | 10-Dec-94 | Garibaldi | 6.3 | 168.43 |
| 26 | 14-Sep-95 | Garibaldi | 6.4 | 273.12 |
| 27 | 9-Oct-95 | Garibaldi | 6.5 | 236.83 |
| 28 | 1-Jan-04 | Garibaldi | 6.3 | 130.25 |
| 29 | 9-Oct-95 | Liverpool | 6.5 | 233.03 |
| 30 | 3-Feb-98 | Liverpool | 6 | 143.25 |
| 31 | 9-Aug-00 | Liverpool | 6.1 | 208.25 |
| 32 | 22-Jan-03 | Liverpool | 6.5 | 286.25 |
| 33 | 13-Apr-07 | Esc | 6.3 | 182.25 |
| 34 | 18-Apr-14 | Esc | 7.2 | 351.6 |
| 35 | 15-May-93 | Sector | 5.9 | 213.76 |
| 36 | 10-Dec-94 | Sector | 6.3 | 312.32 |

(*continued*)

**Table 1.** (*continued*)

| Record no | Date | Station | Mw | PGA (cm/s$^2$) |
|---|---|---|---|---|
| 37 | 14-Sep-95 | Sector | 6.4 | 422.4 |
| 38 | 9-Oct-95 | Sector | 6.5 | 375.04 |
| 39 | 11-Jan-97 | Sector | 6.9 | 321.28 |
| 40 | 15-Jun-99 | Sector | 6.4 | 270.08 |
| 41 | 30-Sep-99 | Sector | 6.5 | 291.84 |
| 42 | 9-Aug-00 | Sector | 6.1 | 227.84 |
| 43 | 20-Mar-12 | Sector | 7.4 | 442.5 |
| 44 | 8-May-14 | Sector | 6.4 | 388.2 |
| 45 | 10-Dec-94 | Tlatelolco | 6.3 | 157.97 |
| 46 | 9-Oct-95 | Tlatelolco | 6.5 | 234.04 |
| 47 | 1-Jan-04 | Tlatelolco | 6.3 | 175.75 |
| 48 | 13-Apr-07 | Deportivo | 6.3 | 175.75 |
| 49 | 18-Apr-14 | Deportivo | 7.2 | 329.2 |
| 50 | 14-Sep-95 | Tlatelolco | 6.4 | 355.84 |

### 2.4 Structural Models

This study considers two hysteretic behavior models: elasto-plastic and modified Takeda. Different degrees of post-yield behavior and the effect of energy capacities are analyzed for the selected *SDOF* systems. The first model includes the elasto-perfectly plastic hysteresis behavior (EPP), which exhibits linear elastic behavior up to a resistance value called the elastic limit, and then behaves plastically until reaching a maximum displacement. In addition, three elastoplastic models with different post-yield stiffness coefficients corresponding to 3%, 5%, and 10% have been selected (see Fig. 2a). The second model corresponds to the Takeda-modified model which is commonly used to represent reinforced concrete structures that exhibit stiffness degradation effects (Fig. 2b). Notice that for the present study values of $\alpha = 0.4$, $\beta = 0.6$ have been used [32, 33]. The analyzed *SDOF* systems have a period ranging from 0.1 to 4.0 s.

## 3 Numerical Results

This section presents the results of the numerical analysis conducted in this study. The results of the *UAFR* spectra are discussed in terms of the ground motion intensity measure $I_{Np}$. The spectra shown in the next section were obtained from 50 typical soft soil records from Mexico City. A range of 40 *SDOF* systems were studied for periods between 0.1 to 4.0 s and by using hysteresis behavior models with different degrees of post-yielding stiffness (hysteretic behavior), energy dissipation capacities, uniform annual failure rates and stiffness degradation.

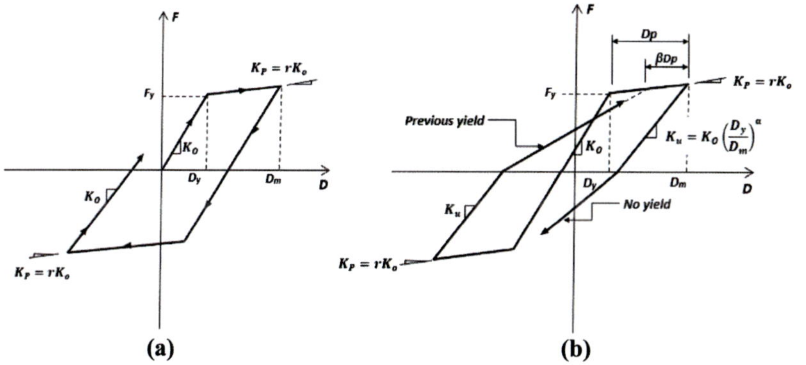


**Fig. 2.** Hysteretic model and cyclic responses of the *SDOF* systems: (a) Elasto-plastic and (b) Takeda-modified models.

## 3.1 Influence of Post-yield Behavior on $E_N$ *UAFR* Spectra

The effect of post-yield behavior on *UAFR* spectra for soft soil ground motions in Mexico City is analyzed. The spectra were determined for the elasto-perfectly plastic model (EPP) and the bilinear model with 3% (BL03), 5% (BL05), and 10% (BL10) post-yield behavior, with a uniform failure rate and different energy capacities. Figure 3a and 3b show the *UAFR* spectra for the normalized hysteretic energy capacities equal to 6 and 9, respectively, and the intensity measure $I_{Np}$. It is observed that for the post-yielding levels in the studied elastoplastic models, no significant influence is observed on the different *UAFR* spectra for the group of seismic records that represent soft soil. Additionally, it was observed that in both cases of normalized hysteretic energy capacity, the spectra are identical in the different cases analyzed in this study. For this reason, the results suggest that the lateral resistance required to control a specific energy and structural reliability is similar for hysteretic behaviour with different post-yielding stiffness.

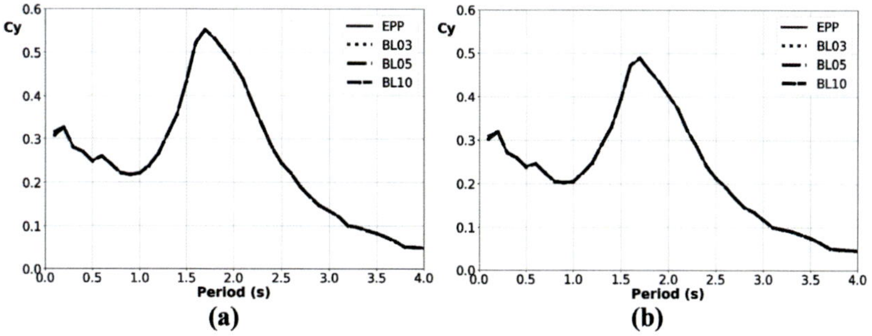


**Fig. 3.** The *UAFR* spectra for $\nu_F = 0.004$, different post-yielding stiffness and normalized hysteretic energy capacities of: (a) $E_N = 6$ and (b) $E_N = 9$.

### 3.2 Effect of the *UAFR* on the Hysteretic Energy Spectra

In this section the *UAFR* spectrum is compared for the three failure rates (0.0025, 0.004, and 0.008). Figure 4 illustrates that the most significant differences occur near the soil's fundamental period, which is close to 2.0 s. For example, When the annual failure rate is 0.004, the required lateral strength exceeds the normalized hysteretic energy of 0.0025 by approximately 11%. In the case of a failure rate of 0.008, the necessary lateral strength is even more significant, increasing by up to 35%. These results suggest that as failure rates decrease, the required lateral strength increases. In other words, if the return period ($T_R$) of earthquakes increases, the required lateral strength is significantly higher. In conclusion, the results show that the failure rates have a significant influence on the *UAFR* spectra in terms of normalized hysteretic energy. This is important for understanding the behavior of buildings under different failure conditions and for designing suitable structures for a specific structural reliability level.

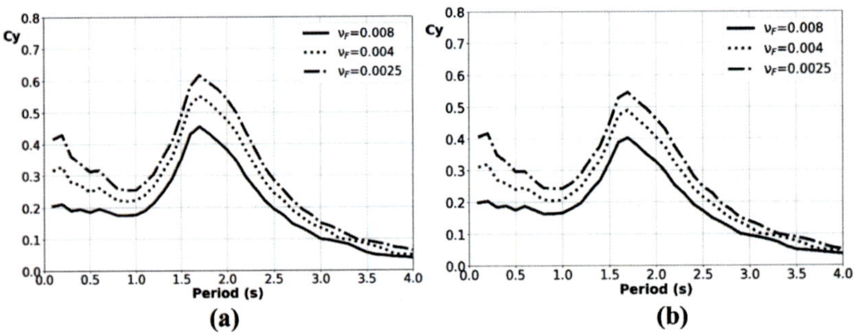


**Fig. 4.** The *UAFR* spectra for the *EPP* model with different failure rates and normalized hysteretic energy capacities of: (a) $E_N = 6$ and (b) $E_N = 9$.

### 3.3 Influence of Normalized Hysteretic Energy Capacity on *UAFR* Spectra

The analysis of a structure's ability to dissipate energy is essential to prevent the accumulation of plastic deformations that may lead to fatigue failure. The approach proposed by Terán-Gilmore and Jirsa [19] is based on the structure's capacity to dissipate hysteresis energy during repeated seismic loading, and utilizes a hysteresis cycle model to analyze how energy is dissipated and distributed within the structure. In this study, $E_N$ capacities of 3, 6, 9 and 12 were proposed to calculate the *UAFR* spectra, which are within the range of values obtained in previous studies by Bojórquez et al. [34]. The Fig. 5 compares the *UAFR* spectra with different normalized hysteretic energy capacities obtained for the intensity measure $I_{Np}$, the elasto-perfectly plastic model and the same failure rate of occurrence. The results show that the required lateral resistance is significantly higher for periods close to the soil period. On the other hand, when comparing *UAFR* spectra with different levels of normalized hysteretic energy capacity, it is observed that as the energy capacity in structures increases, the required lateral resistance decreases, particularly in the region of periods close to the soil period.

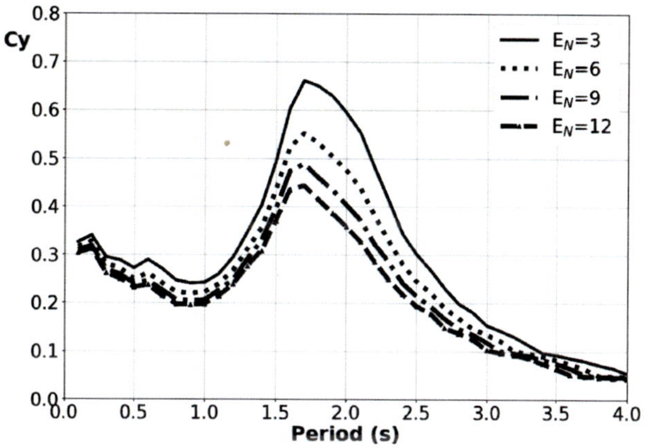


**Fig. 5.** Normalized hysteretic energy UAFR spectra with $\nu_F = 0.004$ for the nonlinear *EPP* model by capacities: $E_N = 3$, $E_N = 6$, $E_N = 9$ y $E_N = 12$.

### 3.4 Effect of Structural Degradation on *UAFR* Spectra

Seismic events can produce a reduction in the structural capacity of a building due to the accumulated damage over the nonlinear cycles. This phenomenon can affect the behavior of structures during their lifetime, which is crucial for the design of seismic-resistant buildings. In addition, the evaluation of structural degradation is a relevant issue in the analysis of existing buildings that have been affected by earthquakes and to observe the damage that occurred after an earthquake. In this study, an analysis is carried out to evaluate the influence of structural degradation on the performance of structures using the Takeda-modified model. It focuses on the normalized hysteresis energy dissipation capacity as a parameter of structural performance. Figure 6 shows the *UAFR* spectra corresponding to systems with elasto-perfectly plastic behavior (EPP) and Takeda-modified to represent degradation models. It can be observed that for short and intermediate structural periods less than 2.0 s, the effect of degradation in the systems is more evident in the required lateral resistance, reaching up to 0.3 times larger for periods oscillating around the soil period. Figure 7 clearly illustrates that the lateral resistance required is superior when the stiffness degradation is taking into account (values larger than one) via the lateral resistance ratio of the Takeda-modified behavior divided by the elasto-perfectly plastic model when $\nu_F = 0.004$ and $E_N = 9$. For this reason, the structural degradation can have a significant impact on the lateral resistance of seismic structures built on soft soils. Therefore, structural degradation should be considered when designing structures on soft soils.

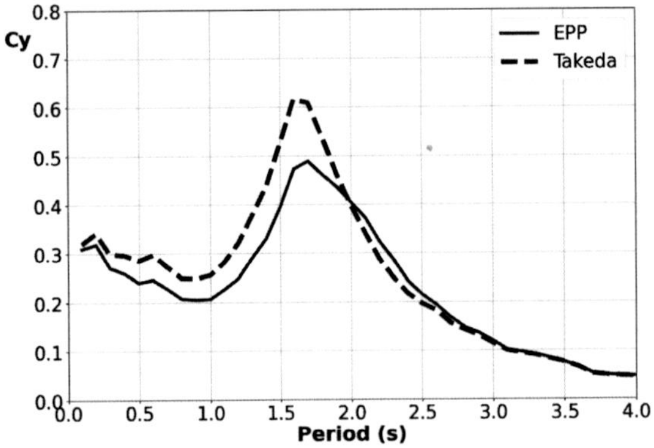


**Fig. 6.** The *UAFR* spectra for the elasto-perfectly plastic and Takeda-modified behavior models with $\nu_F = 0.004$ and a capacity $E_N = 9$.

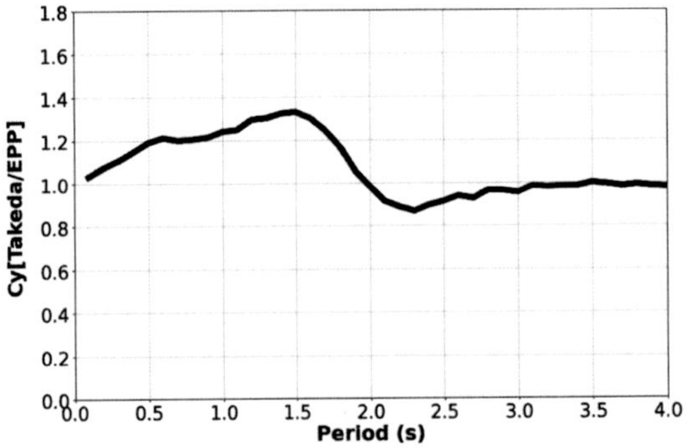


**Fig. 7.** Lateral resistance ratio of the Takeda-modified behavior divided by the elasto-perfectly plastic model by considering $\nu_F = 0.004$ and a capacity $E_N = 9$.

# 4 Conclusion

Based on the analysis of *UAFR* spectra for the seismic intensity measure $I_{Np}$, similar results were observed for the selected hysteresis models with different degrees of post-yield stiffness, suggesting that the *EPP* model can provide reasonable results for estimating the response with bilinear behavior for different grades of soft soil structures in Mexico City. On the other hand, the influence of spectra with different levels of energy capacity was observed, where the lateral resistance required for structures oscillating around the soil period is significantly higher, and as energy capacities increase, the spectra decrease. Finally, for structures experiencing structural degradation on soft soil, up to 30% higher lateral resistance is required compared to structures without degradation (see Fig. 7). Structural degradation can have a significant impact on the resistance of seismic structures designed on soft soils. Therefore, the use of seismic intensity measures representative of the spectral shape ($I_{Np}$) is recommended to obtain more accurate results when designing earthquake-resistant structures. A more detailed analysis is recommended to evaluate the behavior of different degrees of stiffness degradation in nonlinear models. This will allow observing the trend of resistance and the effects of structural degradation in obtaining *UAFR* spectra with normalized hysteresis energy capacity.

**Acknowledgments.** The authors express their gratitude to the CONACYT in Mexico for funding the research reported in this study under Grant Ciencia de Frontera CF-2023-G-1636 and for the scholarships given to the PhD student Joel Carvajal. The financial support given by the Universidad Autónoma de Sinaloa under Grant PROFAPI is appreciated.

## References

1. Bojórquez, E., Iervolino, I.: A spectral shape-based scalar ground motion intensity measure for maximum and cumulative structural demands. In: Proceedings of 14th European Conference on Earthquake Engineering (2010)
2. Buratti, N.: Confronto tra le performance di diverse misure di intensitá dello scuotimento sísmico. Congreso Nacional de Ingeniería de Italia, ANIDIS, Bari Italia (2011)
3. Buratti, N.: A comparison of the performances of various ground–motion intensity measures. DICAM – Structural Engineering, University of Bologna, Italy (2012)
4. Jamshidiha, H.R., Yakhchalian, M., Mohebi, B.: Advanced scalar intensity measures for collapse capacity prediction of steel moment resisting frames with fluid viscous dampers. Soil Dyn. Earthq. Eng. **109**, 102–118 (2018)
5. Javadi, E., Yakhchalian, M.: Selection of optimal intensity measure for seismic assessment of steel buckling restrained braced frames under near-fault ground motions. J. Rehabil. Civil Eng. **7–4**, 1–20 (2018)
6. Minas, S., Galasso, C.: Accounting for spectral shape in simplified fragility analysis of case-study reinforced concrete frames. Soil Dyn. Earthq. Eng. **119**, 91–103 (2019)
7. Housner, G.W.: Limit design of structures to resist earthquakes. In: The First World Conference on Earthquake Engineering. Berkeley, CA, USA (1956)
8. Akiyama, H.: Earthquake-Resistant Limit-State Design for Buildings. University of Tokyo Press, Tokyo, Japan (1985)
9. Uang, C.M., Bertero, V.V.: Evaluation of seismic energy in structures. Earthquake Eng. Struct. Dynam. **19**(1), 77–90 (1990)
10. Akbas, B., Shen, J., Hao, H.: Energy approach in peformance-based seismic design of steel moment resisting frames for basic safety objective. Fe Struct. Design Tall Build. **10**(3), 193–217 (2001)
11. Decanini, L.D., Mollaioli, F.: An energy-based methodology for the assessment of seismic demand. Soil Dyn. Earthq. Eng. **21**(2), 113–137 (2001)
12. Choi, H., Kim, J.: Energy-based seismic design of buckling-restrained braced frames using hysteretic energy spectrum. Eng. Struct. **28**(2), 304–311 (2006)
13. Bojorquez, E., Reyes-Salazar, A., Teran-Gilmore, A., Ruiz, S.E.: Energy-based damage index for steel structures. Steel Compos. Struct. **10**, 331–348 (2009)
14. Lieping, Y., Guangy, C., Zhe, Q.: Study on energy-based seismic design method and the application for steel braced frame structures. In: The Sixth International Conference on Urban Earthquake Engineering, Tokyo Institute of Technology, Tokyo, Japan (2009)
15. Mezgebo, M.G., Lui, E.M.: A new methodology for energy-based seismic design of steel moment frames. Earthq. Eng. Eng. Vib. **16**(1), 131–152 (2017). https://doi.org/10.1007/s11803-017-0373-1

16. Quinde, P. Terán-Gilmore, A. and Reinoso, E.: Cumulative structural damage due to low cycle fatigue: an energy-based approximation. J. Earthquake Eng. **25**(12), 2474–2494 (2021)
17. Yang, H. Wang, H., Jeremić, B.: An energy-based analysis framework for soil structure interaction systems. Comput. Struct. **265** (2022)
18. Zhai, C., Ji, D., Wen, W., Li, C., Lei, W., Xie, L.: Hysteretic energy prediction method for mainshock-aftershock sequences. Earthq. Eng. Eng. Vib. **17**(2), 277–291 (2018). https://doi.org/10.1007/s11803-018-0441-1
19. Teran-Gilmore, A., Jirsa, J.O.: Energy demands for seismic design against low-cycle fatigue. Earthquake Eng. Struct. Dynam. **36**(3), 383–404 (2007)
20. Bojorquez, E., Ruiz, S.E., Teran-Gilmore, A.: Reliability-based evaluation of steel structures using energy concepts. Eng. Struct. **30**(6), 1745–1759 (2008)
21. Salkhordeh, M., Govahi, E., Mirtaheri, M.: Seismic fragility evaluation of various mitigation strategies proposed for bridge piers. Structure **33**, 1892–1905 (2021)
22. Rodríguez-Castellanos, A., Ruiz, S.E., Bojórquez, E., Orellana, M.A., Reyes-Salazar, A.: Reliability-based strength modification factor for seismic design spectra considering structural degradation. Nat. Hazard. **21**, 1445–1460 (2021)
23. Bojórquez, J., et al.: Structural reliability of reinforced concrete buildings under earthquakes and corrosion effects. Eng. Struct. **237** (2021)
24. Inoue, T., Cornell, C.A.: Seismic hazard analysis of MDOF structures. In: Proceedings of the International Conference on Acoustics, Speech and Signal Processing, vol. 6, Issue 1, pp. 437–444, Ciudad de Mexico, Mexico (1991)
25. Esteva, L., Ruiz, S.E., Rivera, J.L.: Reliability and performance-based design of structures with energy-dissipating devices. In: Proceedings of the 9th World Seminar on Seismic Isolation, Energy Dissipation and Active Vibration Control of Structures, Kobe, Japan (2005)
26. Bojorquez, E., Bojorquez, J., Ruiz, S.E., Reyes-Salazar, A., Velazquez-Dimas, J.: Response transformation factors for deterministic-based and reliability-based seismic design. Struct. Eng. Mech. **46**, 755–773 (2013)
27. Shome, N., Cornell, A.: Probabilistic seismic demand analysis of nonlinear structures. Reliability of Marine Structures Program, Report No. RMS-35, Department of Civil Engineering, Stanford University (1999)
28. Chan, M., Ruiz, S.E., Montiel, A.: Escalamiento de acelerogramas y número mínimo de registros requeridos para el análisis de estructuras. Revista de Ingeniería Sísmica **7** (2005). https://doi.org/10.18867/ris.72.123
29. Vamvatsikos, D., Cornell, C.A.: Incremental dynamic analysis. J. Earthquake Eng. Struct. Dyn. **31**, 491–514 (2002)
30. Krawinkler, H, Nassar, A.: Seismic design based on ductility and cumulative damage demands and capacities. In: Krawinkler, H., Fajfar, P. (eds.) Nonlinear Seismic Analysis and Design of Reinforced Concrete Buildings. Elsevier Applied Science, London, U.K. (1992)
31. Teran-Gilmore, A., Simon, R.: Use of constant cumulative ductility spectra for performance-based seismic design of ductile frames. In: Proceedings of the 8th U.S. National Conference on Earthquake Engineering, San Francisco, CA, USA (2006)
32. Takeda, T., Sozen, M.A., Nielson, N.N.: Reinforced concrete response to simulated earthquakes. J. Struct. Div. ASCE **96**, 2557–2573 (1970)
33. Takeda, T.: Study of the Load-Deflection Characteristics of Reinforced Concrete Beams Subjected to Alternating Loads (In Japanese). Transactions, Architectural Institute of Japan, vol. 76, (1962)
34. Bojórquez, E., Reyes-Salazar, A., Terán-Gilmore, A., Ruiz, S.E.: Energy-based damage index for steel structures. Steel Compos. Struct. **10**(4), 331–348 (2010)

# A Displacement-Based Seismic Design Procedure with Local Damage Control, Dissipated Energy and Resilience

J. Abraham Lagunas(✉), A. Gustavo Ayala, Juan Gutiérrez, and Cristhian Montiel

Instituto de Ingeniería, Universidad Nacional Autónoma de México, Ciudad Universitaria, 04510 CDMX, México
jlagunasm@iingen.unam.mx

**Abstract.** This paper proposes a seismic design procedure for reinforced concrete buildings based on local displacements including energy and damage control. The procedure is based on the assumption that the performance of a multi-degree of freedom non-linear system can be approximated from the performance of a single-degree-of-freedom reference system generally associated to the fundamental mode of the building. The behaviour curve of this system is constructed from the frequency results of two spectral modal analyses corresponding to an undamaged (elastic) structure and to the same structure with the damage distribution accepted to occur under design conditions. The inelastic displacement is determined based on a rotation level proposed in the design at the ends of the beams for the level where the maximum interstory drift is presented. Finally, through a damage index defined by the ratio between the hysteretic energy and the input energy, the dissipated energy and the resilience indices are related for design decision-making, *e.g.* repair cost. It is shown that with this design method, it is possible to guarantee the design objectives and to control not only the distribution and level of damage expected for a given design demand but also the time and costs associated to recovery of functionality.

**Keywords:** displacement-based design · local displacements · damage index · dissipated energy control · damage control

## 1 Introduction

In the past, moderate and intense earthquakes such as Northridge (1994), Kobe, (1995), and the 1985 and 2017 Mexico earthquakes affected the built infrastructure of communities throughout the world, causing considerable human and economic losses, because existing seismic-resistant structures have not exhibited satisfactory seismic performances despite having been designed using current regulations and accepted design methods, highlighting their limitations. The current design approach of most modern codes is force-based, in which it is required to calculate, by means of an elastic modal spectral analysis, a load pattern associated to a regulatory design demand given by a spectrum reduced by a behaviour factor, function of ductility and over-strength.


H. Varum et al. (Eds.): IWEBSE 2023, LNCE 236, pp. 177–191, 2023.
https://doi.org/10.1007/978-3-031-36562-1_14

The seismic behaviour factor used in the design is defined at the designer's discretion, depending on the structural system being used and the level of detailing to be provided to ensure a certain regulatory deformation capacity. Finally, to demonstrate the validity of a particular design, the inelastic interstory distortions obtained at the design forces are compared with the prescribed allowable limits, and, if they do not meet the specifications, the stiffness and/or strength of the structural elements are increased to ensure that they do not exceed the specified limits. A limitation of the force-based seismic design philosophy is that the methods consider force reduction factors to obtain the design resistances of the structural elements, representing the inelastic behaviour of the elements inadequately. These reduction factors attempt to represent the global damage but the forces that cause it are not consistent with the expected damage.

The above mentioned has motivated the search and development of alternative seismic analysis and design procedures that guarantee the fulfillment of a performance objective, which have received the name of displacement-based methods. These methods seek to approximate the non-linear behaviour of the structure and, in this way, guarantee a performance objective through a transparent and easy-to-apply formulation. One of the best known is that proposed by Priestley *et al.* [1] and more recently [2].

The analysis of earthquake-induced displacements in a structure, according to several researchers, is the most direct way to measure its performance and the intensity of damage, because displacements are directly related to deformations and these in turn to structural damage. However, this design approach, despite limiting damage to a structure, preventing its collapse and protecting human life, does not take into account the consequences of such damage on the post-earthquake socio-economic development of an affected community or communities. This fact highlights the need for a change in the design philosophy used in practice so that engineers include in their design objectives aspects such as the loss of functionality caused by an earthquake and correspondingly its recovery. Thus, for an assumed distribution of damage under design conditions, having the control of local displacements in the structural elements of a building, also allows the control of the consequences of this damage. This design approach is of relevance in the implicit control of the seismic resilience and recovery of functionality of a structure with minimum economic and social impact to the affected communities.

In this paper, keeping in mind the current importance of the seismic resilience, a new performance-based design procedure is proposed and validated. The proposed procedure measures performance in terms of maximum local displacements in damaged structural elements, damage which is controlled in distribution and intensity during the design process to guarantee a certain level of resilience according to the established needs and design objectives.

## 2 Description of Proposed Design Method

The basic principle of the design method proposed in this paper is that it is possible to estimate the performance of a structure from the performance of a simplified reference system corresponding to a single degree-of-freedom (SDOF) oscillator whose behaviour curve is established from the capacity curve of the multiple-degree-of-freedom (MDOF) structure. Several authors show that it is possible to approximate the real capacity curve by a bilinear curve considering equivalence of deformation energies between the real curve and the bilinear curve, Fig. 1a. Using concepts of structural dynamics and the procedure included in [3], among other references, it is possible to extract from the bilinear capacity curve of the MDOF system, the bilinear capacity curve, also known as the behaviour curve of the SDOF reference system defined in the space of pseudo-spectral accelerations, $S_a$, *vs.* spectral displacements, $S_d$, Fig. 1b. This curve corresponds to the dominant mode of vibration of the structure, which in the case of the buildings being considered corresponds to the fundamental.

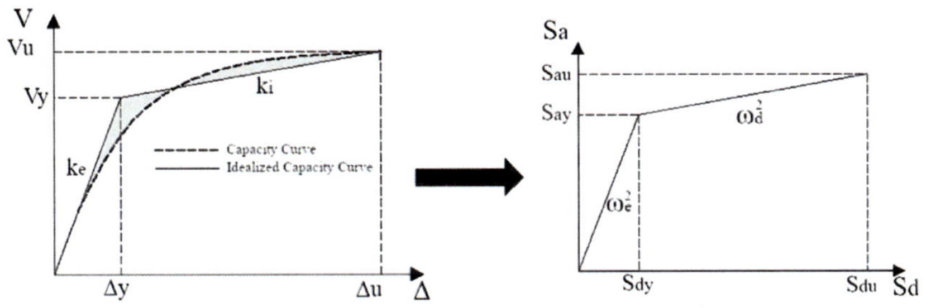


a) b)

**Fig. 1.** Transformation of the capacity curve to the fundamental mode reference system behaviour curve, a) real and idealized capacity curve, b) reference system behaviour curve.

The first slope of the bilinear capacity curve shown in Fig. 1a represents the lateral stiffness of the structure in the elastic behaviour range, $k_e$, and the second slope in the inelastic range corresponding to the structure with damage, $k_i$. The bilinear curve in Fig. 1b corresponding to the reference system behaviour curve (RSBC), defined by two characteristic points, the yield ($S_{dy}$, $S_{ay}$) and the ultimate ($S_{du}$, $S_{au}$); where the first branch represents the elastic behaviour of this reference system and the second its inelastic behaviour, associated with a state of damage. The slopes of the branches of the RSBC are the square of the corresponding circular frequencies, $\omega_e^2$ and $\omega_d^2$.

The proposed displacement-based design method with local and global damage control consists of obtaining the approximate behaviour curve of the structure, where its non-linear behaviour is approximated by the response of a reference system of a non-linear SDOF associated to the fundamental mode of vibration of the structure. The behaviour curve obtained with this method is represented in the space of pseudo-spectral accelerations *vs.* spectral displacements.

The capacity curve of the MDOF system is defined by two linear elastic models to approximate the behaviour of the structure, one is to represent the elastic stage, without damage, and the other to represent approximately the inelastic stage, with damage. The elastic model of the structure is defined from its geometry and the effective stiffnesses of the structural elements that compose it, Fig. 2a. The inelastic model, corresponding to the structure with damage, is a replica of the elastic model in which the damage in elements is introduced by means of hinges with stiffnesses corresponding to the levels of local inelastic rotation (damage) considered as a design objective, *e.g.* plastic joints.

The distribution and intensity of damage assumed by the designer are established from building displacement demands suggested in the literature, *e.g.* [4] y [5] following the strong column-weak beam criterion, *e.g.,* see Fig. 2b. The slopes of the branches of the behaviour curve of the reference system are extracted from the results of the spectral modal analyses of both models. This approach allows the proposed method to be applied using theoretical concepts known to the generality of design professionals and available in commercial structural analysis programs. Therefore, this design approach consists of defining the behaviour curve of the reference system in order to determine the strength required to meet the desired performance objective.

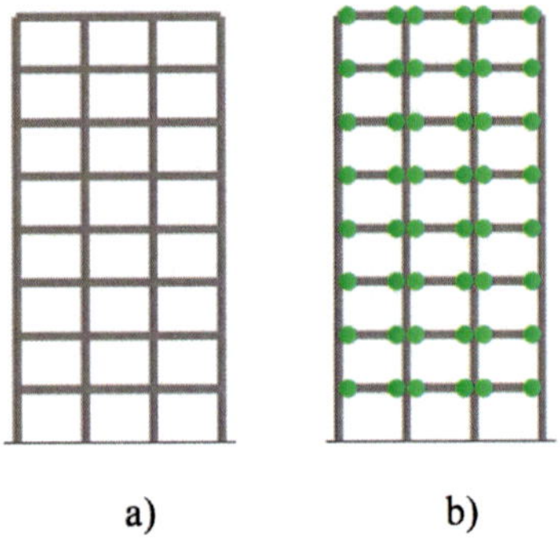


**Fig. 2.** Simplified model representation, a) Elastic, b) Damaged

## 2.1 Ultimate Displacement

In previous versions of this method [2], the ultimate displacement of the reference system was obtained using the maximum allowable values of distortions or displacements established by current seismic design standards. However, here it is proposed to find the design interstory drift as a function of the maximum inelastic rotation that the frame beams are allowed to develop in accordance with the design limit state, including the contribution of the node deformation and the slip of the longitudinal reinforcement bars at the ends of the beams.

With the aforementioned process, the distribution of maximum displacements is obtained, and therefore, the interstory drifts, with which it is possible to know the expected damage in structural elements, but also in non-structural elements and in contents according to documents such as FEMA 273 [6] or FEMA P-58 [7] and of course the global damage of the structure. This information is used to verify if the structural system satisfies its design objectives. This is intended to serve as a basis for a design based on resilience, since the method using as design objectives information about the distribution and intensity of the damage that the structural system will be allowed to present in the event of a seismic event. The performance results used and obtained with this design method together the cost of repairing the expected damage and the suspension of activities of a building thus designed can be used by the stakeholders/designers/users to make financial decisions on deciding which damage distribution an intensity is the best for their purposes. Damage to the structural and non-structural elements through the allowable deformations established by international design codes, *e.g.*, ASCE 41-17 [8].

In reinforced concrete frames, assuming that the inflection point of the beams or columns are located at mid-span length or at mid- height, respectively, the interstory displacement can be represented by the deformation of a beam-column assembly of an interstory of a reinforced concrete frame, as shown in Fig. 3.

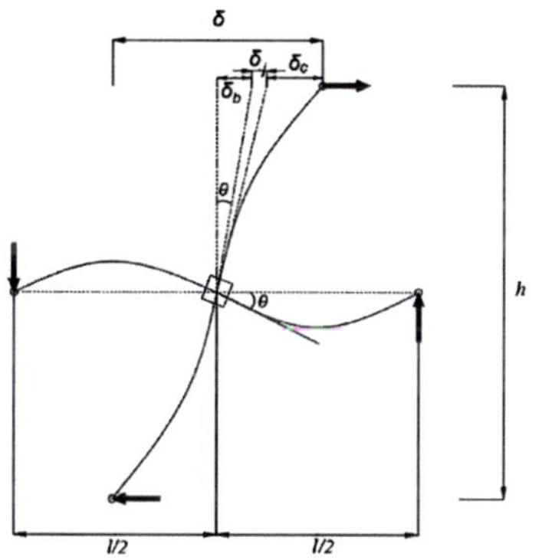


**Fig. 3.** Decomposition of the beam-column joint's interstory drift Wenbin and Jiaru [9].

Figure 3 illustrates that the interstory displacement of a regular frame may be defined as the sum of the displacements associated with the rotations of the beams, columns and nodes. This interstory displacement is composed of an elastic and inelastic displacement, $\delta_y$ and $\delta_p$ defined according to Eqs. 1, 2 respectively:

$$\delta_y = \delta_{yb} + \delta_{yc} + \delta_{vj} \tag{1}$$

$$\delta_p = \delta_{pb} + \delta_{pc} + \delta_{sj} \tag{2}$$

where $\delta_{yb}$ and $\delta_{yc}$ represent the contribution of beam and column deformation to yield interstory displacement, $\delta_y$; $\delta_{pb}$ and $\delta_{pc}$ represent the contribution of beam and column inelastic deformation to inelastic interstory displacement and $\delta_p$; $\delta_{vj}$ and $\delta_{sj}$ represent the contribution of node shear deformation and longitudinal reinforcement yield at the beam-column joint, respectively.

To define the yield displacement of the interstory, this work is based on the result of the analysis of an extensive experimental database of reinforced concrete beam-column joints by Priestley *et al.* [1], where it is suggested the yield rotation, $\theta_y$, may be calculated with the following expression:

$$\theta_y = 0.5\varepsilon_y\left(\frac{l_b}{h_b}\right) \tag{3}$$

where $\varepsilon_y$ is the axial strain of the reinforcing steel, $l_b$ is the length of the beam measured from the column axis, and $h_b$ is the depth of the beam.

Therefore, with the data of the height of the interstory and the yield distortion, it is possible to define the yield interstory drift. In this work, a change of variable from the yield distortion ($\theta_y$) defined by Priestley *et al.* [10] to the $DIR_y$ is done.

The inelastic displacements of beams and columns can be obtained from the ductilities of each type of element, as follows:

$$\mu_{\delta c} = \frac{\delta_{yc} + \delta_{pc}}{\delta_{yc}} \rightarrow \delta_{pc} = (\mu_{\delta c} - 1)\delta_{yc} \tag{4}$$

$$\mu_{\delta b} = \frac{\delta_{yb} + \delta_{pb}}{\delta_{yb}} \rightarrow \delta_{pb} = (\mu_{\delta b} - 1)\delta_{yb} \tag{5}$$

by defining the following:

$$\alpha_b = \frac{\delta_{yb}}{\delta_y}; \alpha_c = \frac{\delta_{yc}}{\delta_y}; \alpha_{sj} = \frac{\delta_{sj}}{\delta_p} \tag{6}$$

where the factors $\alpha_b$ and $\alpha_c$ are the ratios of the contribution of beam and column deformation to the yield displacement of the interstory respectively and $\alpha_{sj}$ is the ratio of the contribution of the deformation due to the sliding of the longitudinal reinforcement of the beam-column joint to the inelastic displacement of the interstory.

Substituting Eqs. 4 to 6 into Eq. 2 gives the following equation for calculating the inelastic displacement of the interstory in terms of the local ductilities and the contributions of the elements to the yield displacement:

$$\delta_p = \frac{(\mu_{\delta b} - 1)\alpha_b + (\mu_{\delta c} - 1)\alpha_c}{1 - \alpha_{sj}}\delta_y \tag{7}$$

Note that the first term of the product of Eq. 7 is an increment to the yield displacement of the interstory.

To calculate the displacements due to the deformations of the beams and columns, the following equations are used:

$$\delta_{yb} = \theta_{yb}\frac{h_c}{2} \tag{8}$$

$$\delta_{yc} = \frac{\varphi_{yc}}{3}\left(\frac{h_c}{2}\right)^2 \tag{9}$$

where $h_c$ represents the column height, $\varphi_{yc}$ the yield curvature of the column, which may be obtained with the following equation [10]:

$$\varphi_{yc} = 2.1\left(\frac{\varepsilon_y}{h_c}\right) \tag{10}$$

In Eq. 8, the expression of Priestley [11] is used to find the curvature and yield rotation of the beam.

$$\varphi_{yb} = 1.7\left(\frac{\varepsilon_y}{h_b}\right) \tag{11}$$

$$\theta_{yb} = \frac{\varphi_{yb} l_b}{6} \tag{12}$$

where $\varepsilon_y$ is the unit strain of the reinforcing steel and $h_b$ is the span of the beam. In Eq. 12, $l_b$ is the length of the beam.

Once the yield displacements of beams and columns are known, the contribution factors to the yield displacement may be calculated. Wenbin and Jiaru [9] suggest that the factor $\alpha_{sj}$ representing the contribution that the displacement due to longitudinal member slip deformation makes to the inelastic displacement may be taken to be approximately 10% for beam-column joints of normal proportions.

In order to have the control of local damage in beams and columns, the value of local ductilities is estimated. According to a simplified elastic analysis of the evolution of the displacements of a node of the structure, a design rotational ductility of the beams may be estimated as a function of the values of the allowable rotations established in the design codes. This is to achieve the desired local damage in the structural elements. Therefore, a design rotation is established to meet some performance limit state, for example, those established in Table 10-7 of ASCE 41-7 [8]. According to the above, the displacement ductility of a beam is a function of a design rotation, $\theta_{Dis}$, and can be calculated with the following equation.

$$\mu_{\delta_b} = 1 + \frac{\theta_{Dis}}{2\theta_{yb}} \tag{13}$$

As it is well known, that to guarantee and stable behaviour of a structure, undesirable mechanisms such as weak floors should not be allowed to occur, so the strong column-weak beam criterion is used, *i.e.*, columns are not allowed to develop inelastic behaviour. Therefore, in Eq. 7 the column ductility will be taken equal to one ($\mu_{\delta c}$=1).

Once the above is established, the inelastic distortion may be calculated with the following equation.

$$DIR_p = \frac{2\delta_p}{H} \tag{14}$$

where $\delta_p$ is the inelastic interstory drift, and $H$ is the interstory height.

Knowing the $DIR_p$, the inelastic displacement of the reference system can be estimated with the following equation.

$$S_{dp} = \frac{DIR_p H_k}{\Gamma_1^D \left( \phi_{k,1}^D - \phi_{k-1,1}^D \right)} \tag{15}$$

where $\Gamma_1^D$ is the modal participation factor of the first mode of vibration of the damage system, $\phi_{k,1}^D$ is the first modal shape ordinate at level $k$ which happens to be the level at which the maximum interstory drift occurs.

Therefore, the ultimate displacement of the reference system is:

$$S_{du} = S_{sy} + S_{sp} \tag{16}$$

## 3 Design Procedure

The displacement-based design procedure for reinforced concrete frames considers the following steps.

1. Preliminary design of the structure. A preliminary design is performed based on the designer's experience or a conventional analysis for gravity and seismic loads, *i.e.,* a force-based design. Since only a preliminary notion of the dimensions of the elements is required, it is not necessary to perform a detailed analysis and design.
2. Modal analysis of the elastic structure. With the pre-design of the previous step, an eigen analysis is performed to obtain the dynamic properties of the structure associated to the fundamental mode of vibration of the structure such as the modal shapes ($\phi_1^E$), the period ($T_{E1}$) and the participation factor ($\Gamma_{1E}$). Once these properties are obtained, the slope of the first branch of behaviour curve of the reference system is defined
3. Determination of the yield displacement of the behaviour curve of the reference system. According to the previous steps, the yield displacement of the reference system can be calculated by the following equation.

$$S_{dy} = \frac{DIR_y H_z}{\Gamma_1^E \left( \phi_{z,1}^E - \phi_{z-1,1}^E \right)} \tag{17}$$

   where $DIRy$ is the yield distortion calculated with Eq. 3, $H_z$ is the critical interstory height $z$ where the maximum modal distortion occurs, $\Gamma_1^E$ is the modal participation factor of the elastic model and $\phi_{z,1}^E$ are the modal coordinates of the interstory where the maximum distortion occurs in the elastic range.
4. Construction of the damage model. A target damage distribution is proposed according to the strong column-weak beam criterion, and a model of the damaged structure is constructed, which consists of a replica of the elastic model, except that in this model perfect plastic hinges are placed at the ends of the beams where inelastic behaviour is accepted to occur.
5. Modal analysis of the damaged mode A modal analysis of the damaged model is performed from which the modal shape $\phi_1^D$, the period ($T_{D1}$) and the participation factor, $\Gamma_{D1}$, associated with the fundamental mode, are obtained. With these values, the slope of the second branch of the behaviour curve can be known.
6. Determination of the ultimate displacement of the behaviour curve of the reference system. To calculate this displacement, the deformation obtained from the modal analysis of the damaged model is considered. The inelastic displacement of the interstory can be estimated with Eq. 7. Then the ultimate displacement is obtained with Eq. 16.
7. Determination of the ductility and inelastic stiffness ratio of the reference system. The ductility ($\mu_G$) is determined as a function of the yield and ultimate displacements. According to Eq. 18. Whereas the inelastic stiffness ratio ($\alpha$) is determined by Eq. 19.

$$\mu_G = \frac{S_{du}}{S_{dy}} \tag{18}$$

$$\alpha = \left(\frac{\omega_E}{\omega_D}\right)^2 = \left(\frac{T_E}{T_D}\right)^2 \tag{19}$$

8. Inelastic displacement estimation according to the ultimate displacement of the reference system. an inelastic displacement spectrum is constructed according to the found values of $\mu_G$ and $\alpha$, where the spectral displacement ($S_d$) corresponding to the period of the elastic model ($T_{E1}$) is obtained, if $S_d$ is different from the ultimate displacement ($S_{du}$) the initial structure needs to be modified such that the stiffness of the structure satisfies the required displacement. Alternatively, if on the first iteration the displacements are not equal, the required period ($T_{Req}$) of the inelastic displacement spectrum is obtained to satisfy the ultimate displacement. Once $T_{Req}$ is known, the initial stiffness of the structure is modified to reach that period, so that the stiffness distribution does not change significantly along the height. The process is repeated until a satisfactory design is achieved.
9. Computation of yield strength from the behaviour curve of the reference system. Once the previous step is satisfied, the yield strength ($R_y/m$) is obtained from an inelastic spectrum of resistances per unit mass, associated to the ductility value ($\mu_G$) and post-flow stiffness (α) calculated earlier. The value of yield strength ($R_y/m$) is found according to the period of the structure found that satisfies the ultimate displacement ($T_{Req} = T_{E1}$).
10. Ultimate resistance of the reference system behaviour curve is obtained. The ultimate resistance of the reference system ($R_u/m$) is obtained by the following equation:

$$R_u/m = R_y/m(1 + \alpha(\mu_G - 1)) \tag{20}$$

11. Reference system behaviour curve. By obtaining the characteristic points, the yield ($S_{dy}$, $R_y/m$) and ultimate ($S_{du}$, $R_u/m$), the behaviour curve of the reference system can be obtained, Fig. 4.

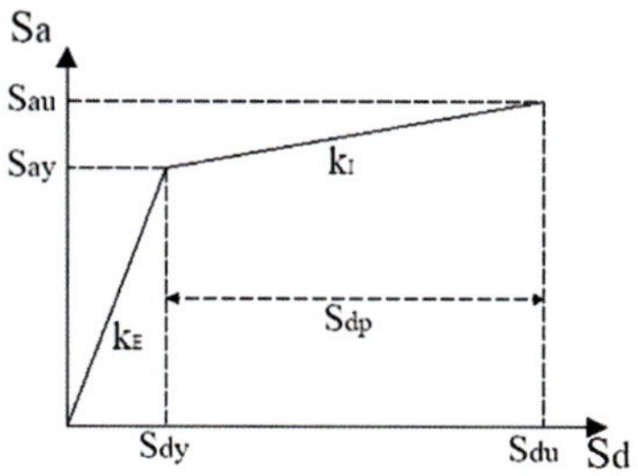


**Fig. 4.** Reference system behaviour curve

12. Calculation of design forces. The design forces of the structural elements are determined by means of two spectral modal analyses. The first one for the elastic model, *i.e.*, the undamaged model, which includes gravity loads and a seismic demand. The seismic demand is defined by the elastic design spectrum scaled by a factor $\lambda_E$ (Eq. 21) this factor is the ratio between the strength per unit mass of yield and the

pseudo-acceleration corresponding to the elastic period of the structure.

$$\lambda_E = \frac{R_y/m}{S_{a,TE}} \tag{21}$$

The second analysis is performed on the damaged model using an elastic design spectrum scaled by a factor $\lambda_D$ (Eq. 22) this factor is the ratio between the differences of the ultimate strength and the yield strength (inelastic strength) and the pseudo-acceleration corresponding to the inelastic period of the structure.

$$\lambda_D = \frac{R_u/m - R_y/m}{S_{a,TD}} \tag{22}$$

The sum of the above results lead to the design forces for the elements that accept and those that do not accept damage. As could be observed, the behaviour curve only represents the properties of the fundamental mode of a structural system. In order to take into account the contribution of the higher modes, they are implicitly considered when performing the modal analysis, since the maximum responses of the structure of the two simplified models are superimposed.

13. Structural elements design. The design of the elements that allow damage should be done covering the demand of the results of the previous step and complying with what is established for ductile structures in the design codes. In the case of the elements that do not accept damage, for example, columns, they should be designed to ensure a minimum strength criterion, in which the column strengths are greater than the beam strengths, in order to ensure a strong column-weak beam behaviour.

## 4 Application of the Design Methodology

The design methodology described in this paper was applied to the models of two reinforced concrete frames, one 8- and the other 10-storys, in order to show that with this methodology it is possible to approximate the inelastic response of the structures with a large margin of acceptance.

As a seismic design demand, a spectrum obtained from the average of several synthetic records associated to a return period of 250 years was used, which corresponds to the collapse prevention design level in Mexico City [12]. The buildings were assumed to be located at the soft soil site of the SCT accelerometric station where the acceleration records were simulated.

To validate the methodology, the results obtained are compared with the non-linear dynamic analyses (NLDA) corresponding to the records used to construct the design demand.

Different levels of damage were proposed in order to evaluate the resilience indices based on the costs and repair times involved in each of these designs. These damage levels refer to both spatial distribution of damage and maximum beam rotation levels for different performance criteria, i.e., immediate occupancy (IO), life safety (LS), and collapse prevention (CP) [8].

### 4.1 Structural Models

The structures used consist of 8- and 10-story reinforced concrete moment resisting frames, with 7.00 m bays and heights of 3.50 m in all floors. The preliminary design was carried out in accordance with NTC-CDMX [12]. The total dead load and instantaneous live load of the floors is 536 kgf/m$^2$ and 180 kgf/m$^2$, respectively. Figure 5 shows the geometric configuration of the frames as well as an example of the proposed damage distribution design. The rotation at the ends of the beams for design were assign according to the performance levels defined by ASCE 41–17, *i.e.,* IO 0.01, LS 0.02 CP 0.03 [8].

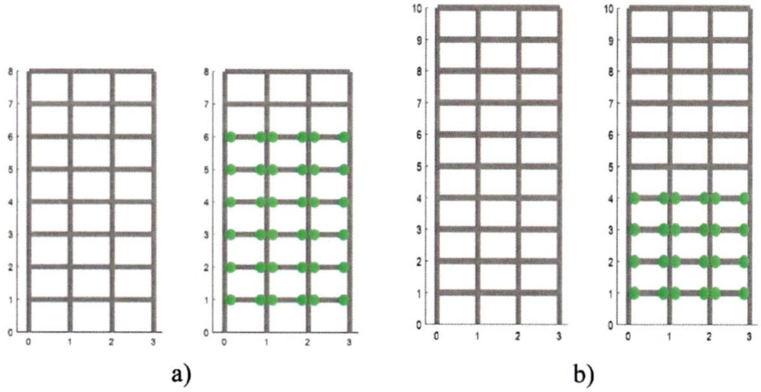


**Fig. 5.** Models and damage level distributions examples, a) 8 story, b) 10 story

It is assumed in the analysis of the frames that the floor system is a rigid diaphragm, *i.e.,* the slab and the beams work together in such a way that the properties of the beams are taken as T-type sections. The supports have been idealized as fixed. The moments of inertia of the columns are taken as 70% of that corresponding to their thick section and for beams 50% as indicated by the NTCDS [12].

### 4.2 Results and Evaluation

With the purpose of validating the design methodology, NLDA were carried out corresponding to each record that was used to create the design spectrum. The results are compared with those obtained from the application of the design method proposed in this paper. The analyses were carried out in Opensees program [13].

Figure 6 shows the displacement profile and distortions obtained with the design method with a blue line and with the mean of the non-linear dynamic analyses with a red line for an example of the 8 and 10 level model. For this example, the design rotation corresponds to life safety ($\theta_{Dis} = 0.02$), and as damage distribution, hinges were assigned to the first four stories in both cases. As can be seen, these results show a good approximation of the design methodology with NLDA profiles in both cases.

Figure 7 shows the average performance levels obtained from the NLDA for both models considered. As can be seen, the distribution of the damage obtained in both cases corresponds to the distribution of damage defined in the design, hinges in the end of beams in the first four stories for both cases.

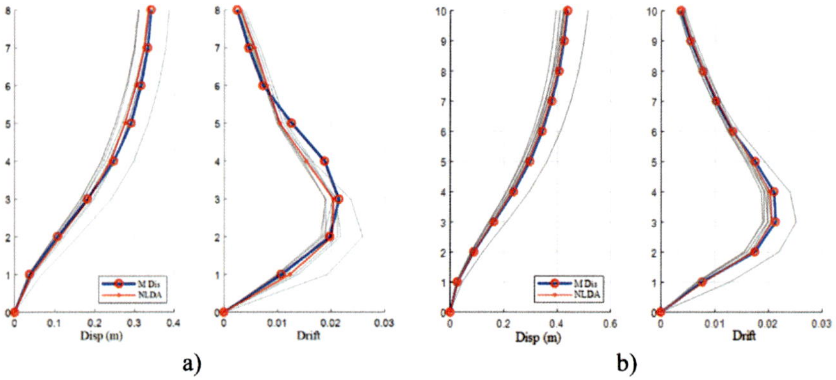


**Fig. 6.** Displacement and drift profile results, a) 8 story $\theta_{Dis} = 0.02$ (LS) damage on beams for the first four stories, b) 10 story $\theta_{Dis} = 0.02$ (LS) damage on beams for the first for stories.

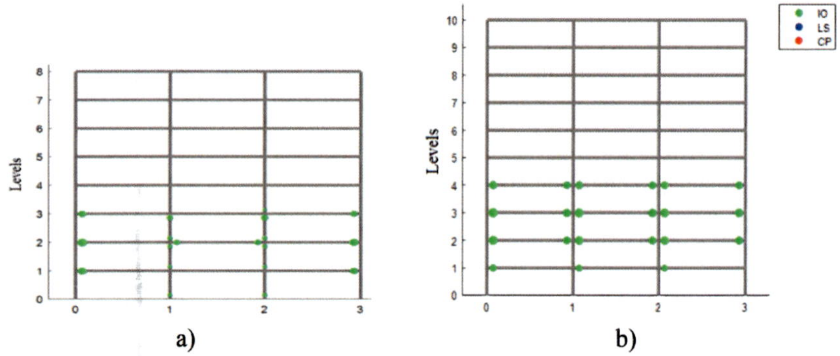


**Fig. 7.** Distribution of damage obtained from the average of the NLDA, a) 8 story, b) 10 story

### 4.3 Consequence Assessment

With the purpose of examining how the results obtained with this methodology are related to the resilience indices, initial costs and repair costs were evaluated, repair times are also included to quantify the loss of functionality. The fragility data from FEMA P-58 were used [7].

Figure 8 shows the relationship between ductility and total cost, considering initial cost and repair cost. The vertical axis shows the normalized total cost with the highest cost obtained from all the models. As can be seen in the graph, for a higher ductility corresponds to a higher total cost.

If we define the coefficient between the hysteretic energy and the input energy as a damage index ($E_H/E_I$), a similar trend can be observed by relating this energy index to the total cost, as shown in Fig. 9a, for a higher energy index corresponds a higher cost. This is because the repair costs increase as this energy damage index increases, as seen in Fig. 9b.

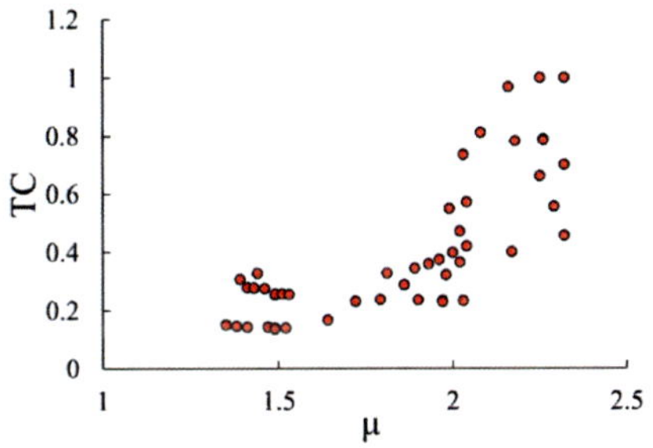


**Fig. 8.** Ductility ($\mu$) *vs.* Normalized Total Cost (TC)

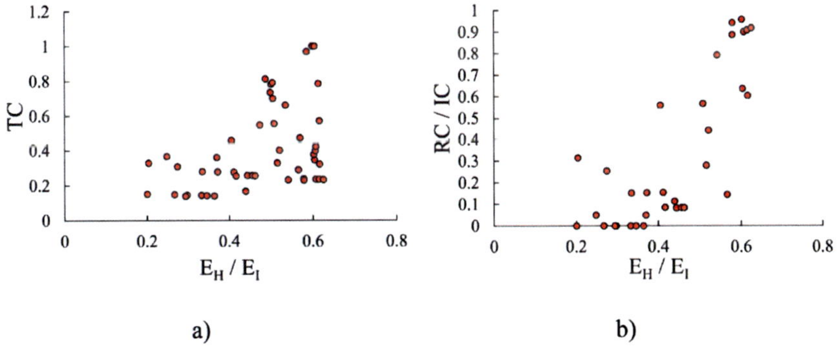


**Fig. 9.** a) Ratio Energy damage index ($E_H/E_I$) *vs.* Normalized Total Cost (TC), b) Ratio Energy damage index ($E_H/E_I$) *vs.* Repair Cost-Initial Cost ratio (RC/IC)

## 5 Conclusions

This paper presented a displacement-based seismic design method with local and global damage and dissipated energy control for reinforced concrete frame structures. This method is based on concepts of structural dynamics and on the methodology developed by López and Ayala [2], where the performance of a non-linear MDOF system can be approximated by the performance of a non-linear SDOF reference system. In this work, it is proposed to adapt a new equation to determine the maximum inelastic interstory drift of the damaged structure to find the ultimate displacement of the reference system.

The methodology is applicable for the design of regular, low and medium height reinforced concrete frames that are controlled by their fundamental mode of vibration. To validate the design procedure, the performance of the models designed with the proposed method are verified with non-linear dynamic analysis (NLDA) corresponding to the records that were used to construct the design spectrum from the mean of these was compared. The following conclusions are drawn from the results obtained.

- The floor distortions and displacements obtained with the design method have a good approximation with respect to those obtained with the average of the NLDA. Taking into account that the proposed method considers the effects of the local deformation of the elements that make up the floor where the critical distortion occurs.

- The equation of the inelastic displacement of the interstory proposed in the design method, which is a function of the allowable local deformation capacity of the reinforced concrete frame beams (plastic rotations), allows determining the ultimate displacement to obtain a more transparent ductility of the system than that recommended in the design codes.
- It is observed that the code recommended design interstory drifts do not guarantee local damage control since, as shown in the results, the distortions are larger than those prescribed in design codes, *e.g.*, 0.015 [12], if we want to have an adequate performance in the local inelastic behaviour complying with an international code such as the ASCE 41-17 [8].
- In all the models, the beams were the elements that were damaged with inelastic rotations below those established as design rotations in the proposed procedure. In most of the cases, the columns remained elastic, except in some situations, where it was observed that some end-inelastic rotations were formed, but they did not have significant values. Therefore, the structures designed with this method complied with the strong column-weak beam mechanism.
- It is observed that the damage of the structural elements can be controlled by keeping the local demands (plastic rotations) below the allowable ones, as well as keeping the interstory distortions below the thresholds established by design codes, according to the desired nonstructural performance. With this, it can be evaluated whether the structural system meets the design objectives.
- The damage distribution proposed for the design of the frames complied acceptably with the results obtained from the non-linear dynamic analyses.
- For the analysis of the consequences, an energy-based damage index is proposed, which indicates the amount of hysteretic energy that must be dissipated by the structural system. The results obtained show that this index is an indirect indicator of the total and repair costs, resilience indexes, and as a consequence of the estimated total costs, for a higher value of the index a higher cost is expected.

**Acknowledgments.** This research was partially sponsored by DGAPA-UNAM through the project PAPIIT IN104821 and the Institute for Safety of Constructions of Mexico City through the project "1536", the authors are grateful for their sponsorship. The first author also thanks the CONACyT for the scholarship granted to carry out his doctoral studies.

## References

1. Priestley, M., Calvi, M., Kowalski, M.: Displacement-Based Design of Structures. IUSS PRESS, Pavia, Italy (2007)
2. López, S.: Diseño por desplazamientos de estructuras de concreto. Tesis de maestría en Inge-niería (Estructuras). México: Posgrado de Ingeniería, UNAM. México (2013)
3. ATC: Applied Technology Council (ATC): Seismic Evaluation and Retrofit of Concrete Buildings. Volumes 1 and 2, Report No. ATC-40, Applied Technology Council, Redwood City, CA (1996)
4. Miranda, E.: Approximate seismic lateral deformation demands in multistory buildings. J. Struct. Eng. **125**(4), 417–425 (1999)

5. Reyes, C.: The serviceability limit state in the seismic design of buildings. (in Spanish) Doctoral Thesis. Facultad de Ingenieria, UNAM, Mexico (1999)
6. FEMA: FEMA 273 - Guidelines for the seismic rehabilitation of buildings. Federal Emergency Management Agency, Washington D.C. (1997)
7. FEMA: FEMA P 58 - Seismic Performance Assessment of Buildings Volume 3-Supporting Electronic Materials and Background Documentation. Federal Emergency Management Agency, Washington D.C. (2018)
8. ASCE.: ASCE/SEI, 41-17 - Seismic Evaluation and Retrofit of Existing Buildings. American Society of Civil Engineers, Reston, VA. (2017)
9. Wenbin, L., Jiaru, Q.: Rules of drift decomposition and drift-based design of RC frames. Advances in Building Technology, pp. 311–318 Elsevier (2002)
10. Panagiotakos, T., Fardis, M.: Deformation of reinforced concrete members at yielding and ultimate. Struct. J. **98**(2), 135–1488 (2001)
11. Priestley, M.: Brief comments on elastic flexibility of reinforced concrete frames, and significance to seismic design. Bull. NZEE **31**(4), 246–259 (1998)
12. GCDMX, NTCDS: Complementary Technical Standards for Seismic Design, (in Spanish) Official Gazette of the Government of Mexico City. México (2020)
13. McKenna, F., Fenves, G.L., Scott, M.H.: Open System for Earthquake Engineering Simulation. University of California, Berkeley CA (2000)

# An Energy-Based Seismic Design Procedure with Damage Control for Frames with Linear and Non-linear Fluid Viscous Dampers

Francisco Bañuelos-García[1](✉), A. Gustavo Ayala[2], and Marco A. Escamilla[2]

[1] Facultad de Ingeniería, Universidad Autónoma del Estado de México, Cerro de Coatepec S/N, Ciudad Universitaria, 50110 Toluca, Mexico
hbanuelosg589@profesor.uaemex.mx

[2] Instituto de Ingeniería, Universidad Nacional Autónoma de México, Ciudad Universitaria, 04510 CDMX, Mexico
{gayalam,mescamillag}@iingen.unam.mx

**Abstract.** Recent tendencies in earthquake engineering have as one of their main goals for the rehabilitation of existing and the design of new buildings to guarantee acceptable level of resilience are when they are subject to design demands during their life cycle. To fulfil this goals and guarantee the seismic performance of structures by controlling the dissipation energy the use of fluid viscous dampers is an excellent design alternative. Most of the seismic design procedures currently used in practice accept that a small or no part of the dissipated energy must be is due to hysteresis with the larger part must be absorbed by protective devices. This may not be the most appropriate, as the seismic demands of current design codes have increased by up to 25%, implicitly implying that larger and/or more devices would be required, which would directly impact the cost of structure. To overcome this problem, this paper presents an approximate seismic energy-based design procedure for new and existing structures equipped with linear and non-linear viscous dampers, in which a balance between supplementary damping and damage to the structure is considered. The framework of the design procedure proposed is based on the concept that the performance of a multiple degree of freedom structure may be approximated from the performance of a reference single degree of freedom system, with properties generally those of the fundamental mode of the structure. To illustrate the application of the procedure proposed four regular plane frames: 8, 12, 17, and 20-storey were designed. From the analysis and discussion of the results obtained, it is concluded that the procedure proposed allows the design of structures equipped with viscous dampers that satisfy a design objective.

**Keywords:** design procedure · energy dissipation · non-linear fluid viscous dampers


H. Varum et al. (Eds.): IWEBSE 2023, LNCE 236, pp. 192–204, 2023.
https://doi.org/10.1007/978-3-031-36562-1_15

# 1 Introduction

The main objective of seismic design is to guarantee adequate behaviour of a building by accomplishing given design performance levels, PL, when subjected to earthquake scenarios which may occur during its life cycle. Based on this objective, during the recent past the design methods based on forces have been used, however these methods cannot guarantee acceptable performance under seismic demands associated with those used in the seismic design conditions. For this reason, diverse performance-based design methods have been developed, particularly those based on displacements as these parameters are the most relevant to performance, *e.g.*, [4, 14, 15]. Nevertheless, there are many situations where it is not possible to guarantee the PLs considered due to architectural restrictions and/or code related issues. For this reason, it is necessary to consider other strategies to satisfy the PLs, such as the use of passive energy dissipation devices *e.g.*, viscous dampers, which dissipate part of the input earthquake energy and therefore reduce the displacements of the structure and the corresponding earthquake induced damage. Moreover, in cases when it is required to add a large amount of supplementary damping to satisfy the PLs (*i.e.*, $\xi_{\text{Dampers}} > 15\%$), and this amount is neither practical nor reasonable [8], the alternative is to accept damage in the structure, by using performance-based design procedures adapted to include passive energy dissipation devices.

The most appropriate procedure to analyse structures with viscous dampers is the dynamic step-by-step analysis; nevertheless, current design practices and most building codes recommend for the seismic evaluation and design of structures the use of modal spectral analysis with seismic demands given by design spectra. Due this situation, in the majority of cases in which viscous dampers are used as passive energy dissipation devices, the associated damping is non-proportional leading to non-classical eigen-values and vectors, making the conventional modal spectral analysis used in the practice of the seismic design of structures, not feasible to apply. To overcome this limitation Constantinou and Symmans [5] propose an approximate simplified procedure, in which the effect of the viscous dampers is considered as an equivalent supplemental modal viscous damping. In this procedure, the approximate amount of damping, associated to the viscous dampers, is easily determined in terms of the dynamic properties of the structure such as the fundamental mode shape and period. Using this approximation, Lopez-García and Soong [11], propose a simplified sequential search algorithm to calculate the distribution of dampers and damping coefficients according to interstorey velocities. In the case of Hwang *et al.* [8] propose a damper distribution based on storey shear strain energy, in this lines there is a large number of procedures [7, 9, 10]. The assumption of non-proportional damping as proportional is not always appropriate because it can produce significant errors in the response of structures [17]. Therefore, diverse studies focusing on the development of simplified procedures to use the modal spectral analysis in structures with viscous dampers have been carried out.

This paper presents a simplified energy-based seismic evaluation and design procedure based on a displacement-based design procedure previously developed [4]. This procedure considers explicitly a combination of damage control and added non-linear fluid viscous dampers [12] to comply with target design displacements. To illustrate the application of this procedure and to evaluate the performance of structures designed with the procedure proposed, four structures are designed. The additional damping required by the structure under design conditions with the accepted damage distribution is calculated using the approximation proposed. To distribute the nominal damping coefficients of the devices that produce the design amount of damping, assuming that they are located at the central spans of each floor, a procedure using as relative weights of these damping coefficients the drifts in the structure. As design seismic demand, the EW component of the SCT record of the 1985 Michoacan earthquake in Mexico is considered. The design obtained, is evaluated by comparing the interstorey drifts and overall structural performance. The accuracy of the results is assessed by comparing the performances extracted from the results of the step-by-step non-linear dynamic analysis of the structure designed with the target design performance. Finally, some conclusions about the design method and the results obtained are presented stressing the most relevant advantages of the displacement-based design of structures with viscous dampers and damage control.

## 2 Fundamentals of the Procedure Proposed

The procedure proposed is based on the assumption that the performance of a non-linear multi-degree of freedom (MDOF) structure may be approximated from the performance of a reference simplified non-linear single-degree of freedom (SDOF) system, normally associated to the fundamental mode of the structure [3]. The principle of this evaluation/design method is that the non-linear capacity curve of a MDOF structure may be approximated by a bilinear curve, using the equivalence of deformation energies corresponding to the real capacity curve and its bilinear approximation, and that, in accordance with basic principles of structural dynamics, the bilinear capacity curve of the reference SDOF system, also referred to as the behaviour curve of the reference system, may be directly extracted from this capacity curve. The behaviour curve of the reference system is obtained from the results of two conventional modal spectral analyses, one for the elastic phase of behaviour, *i.e.,* structure without damage, and other for the inelastic phase, *i.e.,* structure with the assumed distribution of damage.

### 2.1 Added Damping Consideration

As mentioned above, most seismic analysis procedures for structures with viscous dampers consider the effect of these devices in the response of the structure by assuming a proportional damping matrix, which is representative of the actual non-proportional damping matrix. This assumption, however, is not strictly valid, as the response of the structure may show significant errors when compared against that obtained using a step-by-step dynamic analysis of the structure with non-proportional damping matrix. To minimize the errors involved in this approximation, this paper proposes the following:

The equation that describes the dynamic equilibrium of a MDOF structure can be written as:

$$[\mathrm{M}]\{\ddot{\mathrm{u}}(\mathrm{t})\} + [\mathrm{C}]\{\dot{\mathrm{u}}(\mathrm{t})\} + [\mathrm{K}]\{\mathrm{u}(\mathrm{t})\} = \{\mathrm{p}(\mathrm{t})\} \tag{1}$$

where

[M] = Mass matrix

$[C]$ = Damping matrix
$[K]$ = Stiffness matrix
$\{p(t)\}$ = Force vector
$\{u(t)\}, \{\dot{u}(t)\}, \{\ddot{u}(t)\}$ = Displacement, velocity, and acceleration vectors

For a structure with viscous dampers, the damping matrix can be represented as:

$$[C] = [C_0] + [C_D] \tag{2}$$

where

$[C_0]$ = Inherent damping matrix (assumed as proportional)
$[C_D]$ = Damping matrix associated to the viscous dampers (non-proportional)

Since the damping matrix associated to the viscous dampers is non-proportional, neither to the mass nor the stiffness matrices, it is possible to approximate it as the sum of a proportional damping matrix ($[C_{DP}]$) and a residual damping matrix, ($[C_{DR}]$), *i.e.*,:

$$[C_D] = [C_{DP}] + [C_{DR}] \tag{3}$$

The first order approximation of the proportional matrix, also referred as Rayleigh damping may be written as:

$$[C_{DP}] = a_{0D}[M] + a_{1D}[K] \tag{4}$$

To uncouple the equilibrium equations given by Eq. 1 the following change of coordinates is required: $\{u(t)\} = [\Phi]\{x(t)\}, \{\dot{u}(t)\} = [\Phi]\{\dot{x}(t)\}, \{\ddot{u}(t)\} = [\Phi]\{\ddot{x}(t)\}$, to give:

$$[M][\Phi]\{\ddot{x}(t)\} + [C][\Phi]\{\dot{x}(t)\} + [K][\Phi]\{x(t)\} = \{p(t)\} \tag{5}$$

where

$[\Phi]$: Normalized modal matrix

Premultiplying each term in Eq. 5 by $\{\phi\}_i^T$ gives:

$$\{\phi\}_i^T[M]\{\phi\}_i\{\ddot{x}(t)\} + \{\phi\}_i^T[C]\{\phi\}_i\{\dot{x}(t)\} + \{\phi\}_i^T[K]\{\phi\}_i\{x(t)\} = \{\phi\}_i^T\{p(t)\} \tag{6}$$

Assuming that the inclusion of viscous dampers in the structure does not significantly modify the modal characteristics of the structure, a set of uncoupled dynamic equilibrium equations expressed in terms of modal coordinates $x_i(t)$ is obtained:

$$\ddot{x}_i(t) + 2(\xi_{0i} + \xi_{DPi})\omega_i\,\dot{x}_i(t) + \boldsymbol{\{\phi\}_i^T[C_{DR}]\{\phi\}_i\dot{x}}(t) + \omega_i^2 x_i(t) = \{\phi\}_i^T\{p(t)\} \tag{7}$$

where

$\omega_i$ = Frequency for mode i
$\xi_{0i}$ = Inherent viscous damping ratio for mode i
$\xi_{DPi}$ = Proportional viscous damping ratio of devices for mode i

$\Gamma_i$ = Modal participation factor

Even though the term $\{\boldsymbol{\phi}\}^T[\mathbf{C_{DR}}]\{\boldsymbol{\phi}\}$ is not a diagonal matrix, experimental evidence has shown that as the damping ratio of a structure is increased, the contribution of the higher modes of the structure to its total response may be ignored. Therefore, for practical applications in the approximate method, propose only the contribution of the fundamental mode of the MDOF system is usually considered [8]. Based on this consideration, the damping ratio associated to the residual damping matrix can be expressed as: $\xi_{DRi} = \frac{\{\phi\}_i^T[C_{DR}]\{\phi\}_i}{2\omega_i}$, and Eq. 7 can be rewritten as:

$$\ddot{x}_i(t) + 2(\xi_{0i} + \xi_{DPi} + \xi_{DRi})\omega_i\dot{x}_i(t) + \omega_i^2 x_i(t) = \{\phi\}_i^T\{p(t)\} \tag{8}$$

where

$\xi_{DRi}$ = Residual viscous damping ratio of devices for mode i

The proportional viscous damping ratio associated to the added devices can be calculated using the energy based approximation proposed by [5]:

$$\xi_{DPi} = \frac{T_i \sum_{j=1}^{nd}(C_j)(\mathrm{Cos}\,\theta_j)(\phi_i - \phi_{i-1})^2}{4\pi\left(\sum_{i=1}^{n} m_i\phi_i^2\right)} \tag{9}$$

where

$T_i$= Natural period of vibration of mode i
$\theta_j$ = Angle of the viscous damper j
$\phi_i$ = Horizontal modal displacements of mode i
$C_j$ = Damping coefficient of the damper at storey j

The residual viscous damping ratio of the devices for mode i can be defined as:

$$\xi_{DRi} = \frac{\{\phi_i^T\}[C_{DR}]\{\phi_i\}}{2\omega_i} = \frac{\{\phi_i^T\}[C_D - C_{DP}]\{\phi_i\}}{2\omega_i} \tag{10}$$

Thus, ignoring the response contribution of higher modes, the total damping ratio for the first mode is defined as:

$$\xi_{Design1} = (\xi_{01} + \xi_{DP1} + \xi_{DR1}) \tag{11}$$

In the order to considering non-linear viscous dampers in the procedure proposed, it can be used the following equation:

$$C_{\alpha_j} = \frac{(2)^{(2+\alpha_j)}\cdot(\pi)^{(2+\alpha_j)}\cdot\left(\phi_{rj,1}\right)^{(1+\alpha_j)}\cdot(D_{roof})^{(1-\alpha_j)}\cdot\left(C_{L_k}\right)\cdot\left(f_j\right)^{(1-\alpha_j)}}{\lambda_k + \left(\frac{1}{T_1}\right)^{(\alpha_j)}} \tag{12}$$

## 3 Energy-Based Design Procedure

In accordance with the concepts, the application of the displacement-based design method proposed to structure with non-linear fluid viscous dampers intended to satisfy the life safety limit state, LSLS, can be summarized in the following steps:

1. Preliminary configuration and dimensioning of the elements of the structure according to engineering judgement and/or designer experience. The objective of this step is to define a realistic stiffness distribution of the structural elements throughout the height of the structure, so that design displacement shape is in accordance with those of real structures.
2. Modal analysis of the elastic bare structure designed in the previous step. From this analysis, the modal participation factor PF, the fundamental period of the structure, $T_E$, and the displacement shape $\phi$ of the fundamental mode are obtained. From this displacement shape, the spectral yield displacement of the reference SDOF, $Sd_y$, may be calculated using the following equation:

$$Sd_y = \frac{IDT_y\, H_k}{PF_1^E\left(\phi_{k,1}^E - \phi_{k-1,1}^E\right)} \tag{13}$$

where

$IDT_y$ = Yield interstorey drift
$H_k$ = Height of the critical storey k (where maximum drift occurs)
$PF_1^E$ = Modal participation factor of the fundamental mode (elastic structure)
$\phi_{k1}^E$ = Modal shape ordinate of the critical interstorey, k

The yield interstorey drift for a reinforced concrete structure may be calculated using the Eq. (14) [2]

$$\Delta_y = \frac{0.5\varepsilon_y L_1}{h_{v1}} \tag{1}$$

where

$\boldsymbol{\varepsilon_y}$ = yield strain of the reinforcing steel
$\boldsymbol{L_1}$ = beam length
$\boldsymbol{h_{v1}}$ = beam depth

3. Definition of a design damage distribution for the PL in accordance with the characteristics of the structure and the design demands, using a strong-column weak-beam strategy and considering the contribution of the viscous dampers added to the structure. Structural damage is introduced at the ends of the elements, where damage is accepted to occur under design conditions by adding hinges with zero or residual rotational stiffness, equal to a reduced bending stiffness of the damaged element sections.
4. Modal analysis of the damaged bare structure to obtain the fundamental modal shape, participation factor and period, $T_D$, and, from this period, the slope of the second branch of the idealized bilinear behaviour curve of the reference SDOF system. From

this modal shape, the target spectral displacement of the reference SDOF, $Sd_{PL}$, is obtained using the following equation:

$$Sd_{PL} = \frac{IDT_{PL}\, H_k}{PF_1^D\left(\phi_{k,1}^D - \phi_{k-1,1}^D\right)} \tag{15}$$

where

$IDT_{PL}$ = Interstorey drift of PL

5. Calculation of the target yield and ultimate spectral displacements of the reference SDOF system, $Sd_{PL}$ and $Sd_y$ respectively, corresponding to the fundamental mode using the results of modal analysis. Its ductility, $\mu$, and post-yielding to initial stiffness ratio, $\alpha$, are obtained using Eqs. (16) and (17)

$$\mu = \frac{Sd_{PL}}{Sd_y} \tag{16}$$

$$\alpha = \left(\frac{T_E}{T_D}\right)^2 \tag{17}$$

6. Modification of the effective viscous damping ratio from the inelastic displacement spectrum for the given $\mu$ and $\alpha$, until the spectral displacement associated to the fundamental period is equal to the target spectral displacement of the structure (step 4). See Fig. 1:

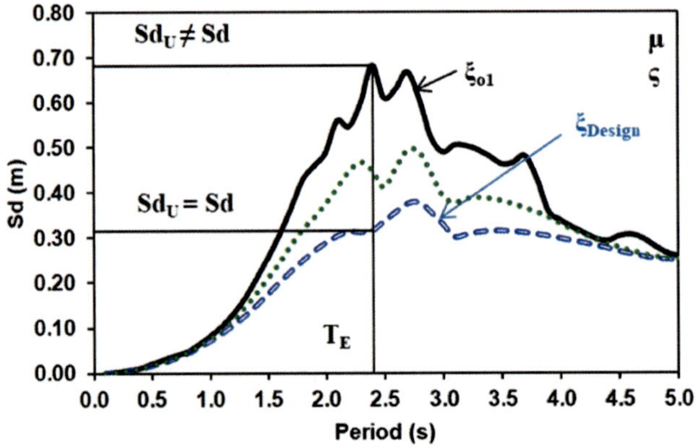


**Fig. 1.** Inelastic displacement spectrum

7. Calculation of the damping coefficients of the viscous dampers, using Eq. (9), where the proportional part ($\xi_{DP1}$) may be defined using the Constantinou and Symmans [5] proposal Eq. (18), and the residual part $\xi_{DR1}$ as defined by Eq. 10. The damping coefficient for each interstorey is calculated in proportion to the relative modal

displacement drifts normalized with the interstorey height of the building.

$$C_t = \frac{4\pi\xi_{DPi}\left(\sum_{j=1}^{nd}\frac{\phi_i-\phi_{i-1}}{H_j}\right)\left(\sum_{i=1}^{n} m_i\phi_i^2\right)}{T\sum_{j=1}^{nd}(\mathrm{Cos}\,\theta_j)\left(\frac{(\phi_i-\phi_{i-1})^3}{H_j}\right)} \tag{18}$$

$$C_j = \left(\frac{\left(\frac{\phi_i-\phi_{i-1}}{H_j}\right)}{\sum_{i=1}^{nd}\left(\frac{\phi_i-\phi_{i-1}}{H_j}\right)}\right)C_t \tag{19}$$

where

$H_j$= Height of the storey j

8. Determination of the yield strength, $Sa_y$, using the period of elastic model $T_E$ in the inelastic strength spectrum, corresponding to the values of $\mu$ and $\varsigma$ previously calculated, Fig. 2:
9. Calculation of the ultimate strength, $Sa_u$, of the reference system using the following equation:

$$Sa_u = Sa_y[1 + \varsigma(\mu - 1)] \tag{20}$$

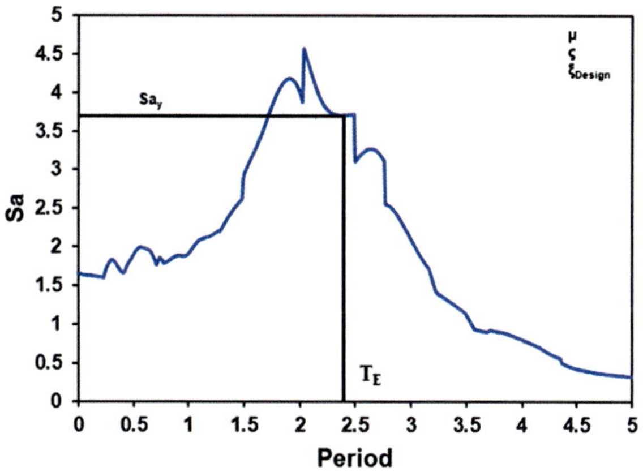


**Fig. 2.** Strength per unit mass spectrum for $\mu$ and $\varsigma$, associated to the PL

10. Determination of the design forces of the elements using the results of three different analyses: a gravity load analysis of the undamaged structure, a modal spectral analysis of the undamaged structure, using the elastic design spectrum scaled by the ratio of the strength per unit mass at the yield point of the behaviour curve and the elastic pseudo-acceleration for the initial period, $\lambda_E$, and a modal spectral analysis of the damaged structure, using the elastic spectrum scaled by the ratio of the difference of ultimate and yield strengths per unit mass and the pseudo-acceleration

for the period of the damaged structure, $\lambda_D$. Each modal spectral analysis considers the total viscous damping ratio ($\xi_{Design}$). The design forces are obtained by adding the forces due to gravity loads and the forces of the modal spectral analyses of the undamaged and damaged structure.

11. Determination of the design of every structural element in accordance with the forces obtained from the analysis of the simplified models and using the applicable design rules. The design process must be carried out in such a way that the design criteria of the code do not significantly alter the expected performance.

## 4 Application Examples

To illustrate the application of the design procedure developed in this paper, four structures, was designed. The nominal properties of the materials used in the design are: for the concrete, a compressive strength f'c = $3.00 \times 10^4$ kN/m$^2$, a modulus of elasticity Ec = $27.00 \times 10^6$ kN/m$^2$, and a weight density $\gamma = 23.53$ kN/m$^3$, and for the steel reinforcement a yield stress fy = $4.50 \times 10^5$ kN/m$^2$ and a modulus of elasticity Es = $2.00 \times 108$ kN/m$^2$. Based on the results of the preliminary design of the frame, the sections of the structural elements were defined.

To validate the results, obtained from the displacement-design method proposed, the seismic design demand considered was the response spectra of the EW component of the 1985 the earthquake in Mexico recorded at the SCT site. To validate the seismic performance of the designed structure, the drifts, obtained from the non-linear step-by -step analysis of the structure, were compared with the maximum drift considered as design target. The non-linear step-by-step analysis was carried out with the Perform 3D V5 [6], program. The analysis considerations were: 1) nearly elasto-plastic bilinear stable hysteretic behaviour for all beams and columns, 2) non-classical damping matrix due to the incorporation the viscous dampers, 3) nominal yield moments for beam and columns obtained from the design method proposed and 4) the drift considered as design target was 0.015, as specified by the current NTC-DS-2020 [13] to satisfy the LSLS. Figure 3 shows, for each of the frames considered, their geometry, location of dampers and accepted damage distribution.

## 5 Design Results

Figure 4 shows the maximum interstorey drifts calculated using step-by-step analyses for four different cases: (a) bare structure (SA), (b) structure with viscous damping (AV), (c) structure design with the damping coefficients calculated by procedure proposed (PP) and (d) structure design with the damping coefficients calculated by procedure Constantinou and Symmans procedure [5], which is used for some technical documents *e.g.*, ASCE 7-10 [2], ASCE/SEI 41-17 [1], among others. It may be observed that when the bare structure is subjected to the design seismic demand corresponding to the LSLS, some interstoreys lightly exceed the target drifts for this limit state (IDT LSLS). However, when the viscous dampers are added to the structure, the interstorey drift prescribed by the code is not exceeded at any interstoreys. In addition, the procedure proposed gives

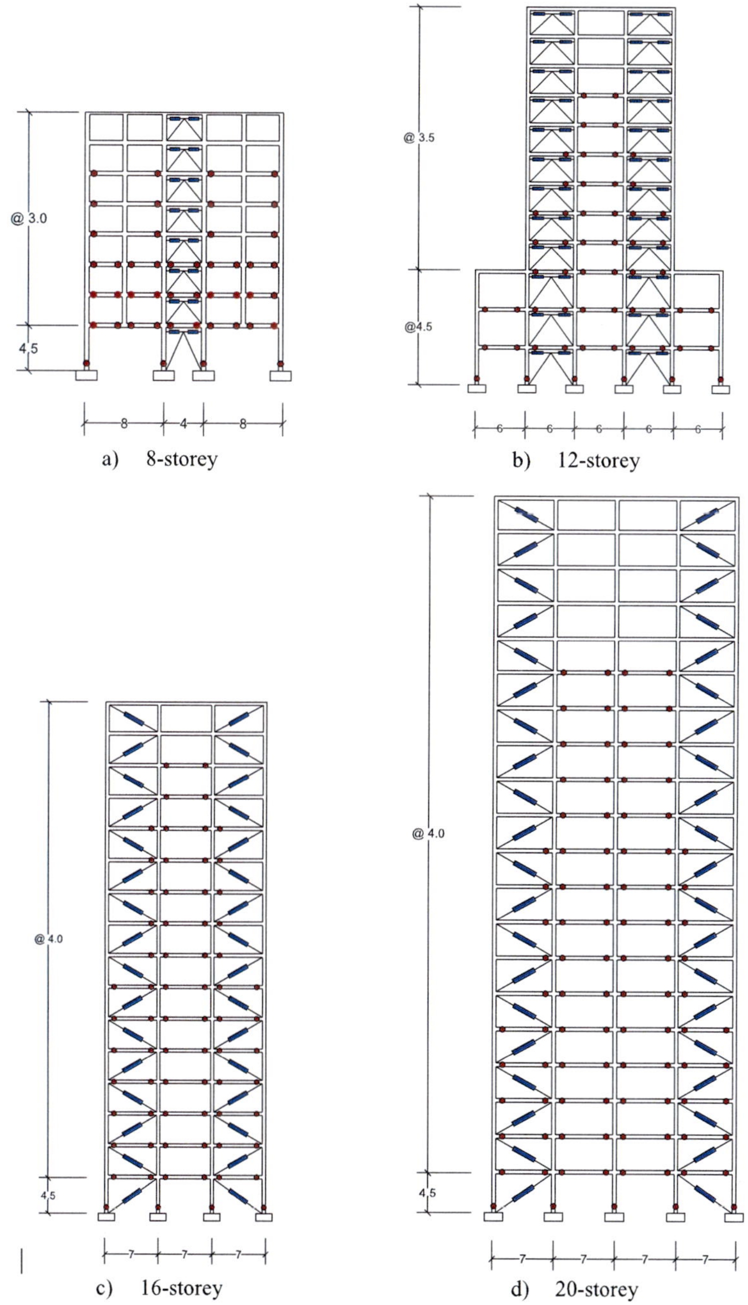


**Fig. 3.** Geometry, location of dampers and accepted damage distribution for the 8-, 12- 16- and 20- storey example frames

a better approximation when is compared with the results of the non-linear step-by-step analysis (see Fig. 4). With regards to the damage distributions, the results of the design procedure proposed show the same damage distribution than that obtained from the results of the non-linear step-by-step analysis. Moreover, the overall size of these devices was between 10 and 20% smaller compared to the procedure recommended in ASCE 7-10 [2]. This can have a direct impact on the total cost of fluid viscous dampers.

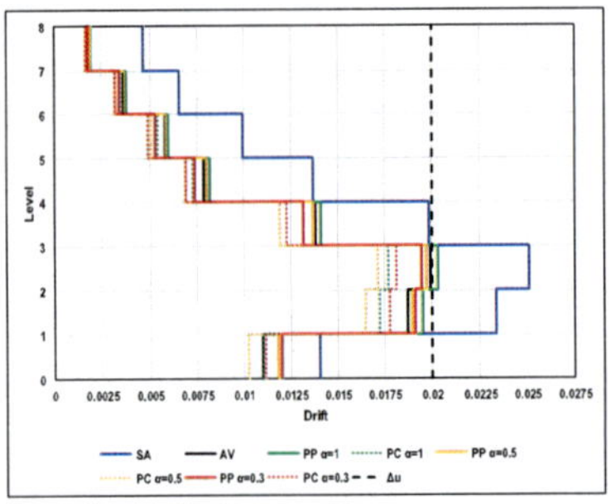


a) 8-storey

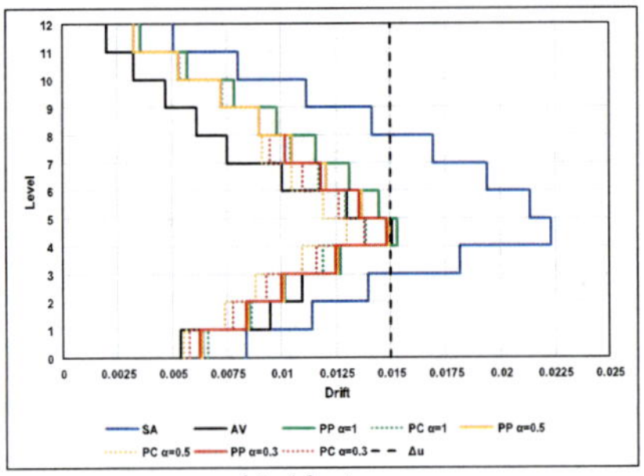


b) 12-storey

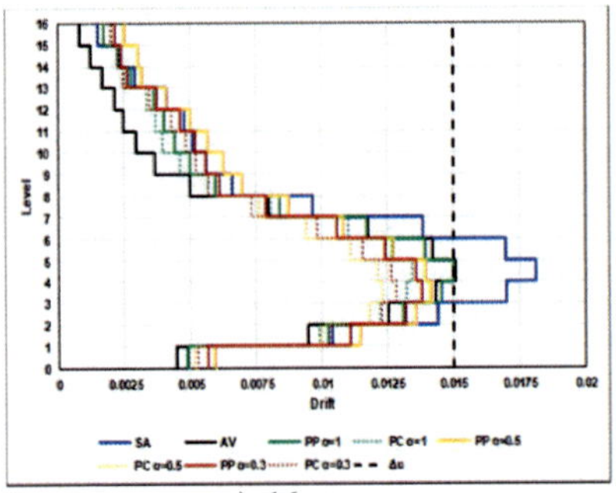


c) 16-storey

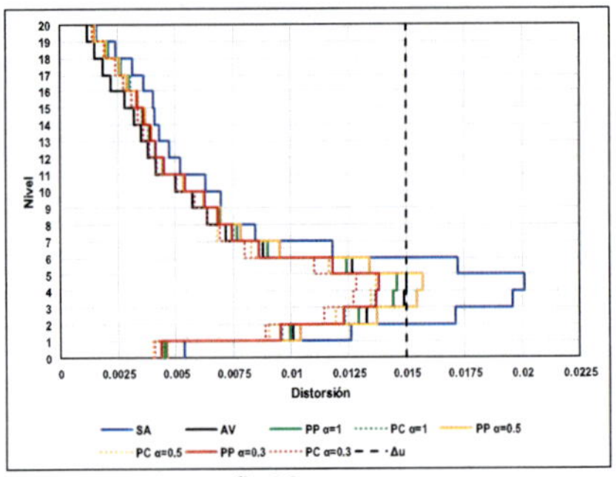


d) 20-storey

**Fig. 4.** Geometry, location of dampers and accepted damage distribution for the 8-, 12- 16- and 20- storey example frames

## 6 Conclusions

This paper presented the formulation and a practical application of a new energy - based seismic evaluation/design method for structures equipped with non-linear fluid viscous dampers and/or when these devices are used as a retrofit measure. This procedure seeks a balance between energy dissipation by supplementary viscous damping and inelastic deformations. To validate the design procedure proposed, the performance of the designed structure was obtained using non-linear step-by-step dynamic analysis considering as seismic demand the same demand used for its design. From the analysis of the results obtained the following conclusions may be extracted:

- For each model designed, a damage state consistent with the design objective corresponding to the LSLS was proposed, and then supplementary damping by non-linear fluid viscous dampers was added to satisfy the design objective. In no case the maximum value of supplementary damping (*i.e.* 25%) was exceeded results that makes economically feasible to add this type of devices in a structure [8], so it can be said that a balance was achieved between the energy dissipation due to the damage in the structure, as well as the energy dissipated by these devices.
- For the structures designed with the procedure proposed, the maximum interstorey drifts, produced by the design demand, are closer to those obtained from the non-linear step-by-step analysis of the structure, considering a non-proportional damping matrix, than the procedure recommended in technical documents such as: ASCE 7-10 [2], ASCE/SEI 41-17 [1], which can significantly reduce the cost of these devices. Nevertheless, considering that the design interstorey drift is recommended to limit the damage of the LSLS, both results can be considered satisfactory.
- The distribution of damage, used as target within the design procedure proposed was reproduced in the results of non-linear step-by-step analysis of the structure with a non-proportional damping matrix. This result was due to the correction to the added damping coefficient in the procedure, something that guaranteed that the response of the structure was reduced, and the accomplishment of the design objective of controlling the distribution of damage in the structural elements.
- The comparison of the effort involved in the application of this method using the computational tools available in most design offices, *e.g.*, SAP 2000 [16] and the quality of results obtained compared with those of other design methods, place it as an excellent design tool, since requires only two modal spectral analyses.

**Acknowledgments.** This research was partially sponsored by DGAPA-UNAM through the project PAPIIT IN104821, the authors are grateful for their sponsorship. The first author also thanks the National Council of Science and Technology for the scholarship granted to carry out his doctoral studies.

## References

1. ASCE: ASCE/SEI 41-17 Seismic Evaluation and Retrofit of Existing Buildings, ASCE/SEI 41-17. American Society of Civil Engineers, Reston (2017)
2. ASCE: ASCE 7-10 Minimum Design Loads for Buildings and Other Structures. American Society of Civil Engineers, Reston (2010)
3. Ayala, A.G.: Evaluation of the seismic performance design of structures, a new approach. Revista Internacional de Métodos Numéricos para Cálculo y Diseño en Ingeniería 17 (2001)
4. Ayala, A.G., Castellanos, H., Lopez, S.: A displacement-based seismic design method with damage control for RC buildings. Earthq. Struct. **3**, 413–434 (2012). https://doi.org/10.12989/EAS.2012.3.3_4.413
5. Constantinou, M.C., Symans, M.D.: Experimental and analytical investigation of seismic response of structure with supplemental fluid viscous dampers. National Center for Earthquake Engineering Research, State University of New York at Buffalo, N.Y. (1992)

6. CSI: Non-linear Analysis and Performance Assessment for 3D Structures. Computers and Structures, Inc. Perform 3D V5, Berkeley, CA (2007)
7. Diotallevi, P.P., Landi, L., Dellavalle, A.: A Methodology for the direct assessment of the damping ratio of structures equipped with nonlinear viscous dampers. J. Earthq. Eng. **16**, 350–373 (2012). https://doi.org/10.1080/13632469.2011.618521
8. Hwang, J.-S., Lin, W.-C., Wu, N.-J.: Comparison of distribution methods for viscous damping coefficients to buildings. Struct. Infrastruct. Eng. **1–14** (2010). https://doi.org/10.1080/15732479.2010.513713
9. Landi, L., Conti, F., Diotallevi, P.P.: Effectiveness of different distributions of viscous damping coefficients for the seismic retrofit of regular and irregular RC frames. Eng. Struct. **100**, 79–93 (2015). https://doi.org/10.1016/j.engstruct.2015.05.031
10. Levy, R., Lavan, O.: Fully stressed design of passive controllers in framed structures for seismic loadings. Struct. Multidiscip. Optim. **32**, 485–498 (2006). https://doi.org/10.1007/s00158-005-0558-5
11. Lopez Garcia, D., Soong, T.T.: Efficiency of a simple approach to damper allocation in MDOF structures. J. Struct. Control **9**, 19–30 (2002). https://doi.org/10.1002/stc.3
12. Makris, N., Constantinou, M.C.: Fractional-derivative Maxwell model for viscous dampers. J. Struct. Eng. **117**, 2708–2724 (1991). https://doi.org/10.1061/(ASCE)0733-9445(1991)117:9(2708)
13. GCDMX: Complementary Technical Standards for Seismic Design. Official Gazette of the Government of Mexico City, Mexico (2020). (in Spanish)
14. Panagiotakos, T.B., Fardis, M.N.: Deformation-controlled earthquake-resistant design of RC buildings. J. Earthq. Eng. **3**, 495–518 (1999). https://doi.org/10.1080/13632469909350357
15. Priestley, M.J.N., Calvi, G.M., Kowalsky, M.J.: Displacement Based Seismic Design of Structures, 1st edn. IUSS Press, Pavia (2007)
16. CSI: SAP2000 Integrated Software for Structural Analysis & Design. Computers and Structures Inc., Berkeley (2004)
17. Veletsos, A.S., Ventura, C.E.: Modal analysis of non-classically damped linear systems. Earthq. Eng. Struct. Dyn. **14**, 217–243 (1986). https://doi.org/10.1002/eqe.4290140205

# Effect of Bi-directional Ground Motion on the Response of Base-Isolated Structures with U-Shaped Steel Damper

Satoshi Yamada[1(✉)], Shoichi Kishiki[2], and Yu Jiao[3]

[1] The University of Tokyo, Tokyo, Japan
s_yamada@arch1.t.u-tokyo.ac.jp
[2] Tokyo Institute of Technology, Yokohama, Japan
[3] Tokyo City University, Tokyo, Japan

**Abstract.** Current design practices for base-isolated structures generally conduct in plane time-history analysis using a single horizontal seismic component of the ground motion. However, base-isolated structures are subjected to bi-directional horizontal ground motion. So, some error included in the evaluation of the maximum displacement of the seismic isolation layer and damage to the dampers during an earthquake. This research focuses on the damage on U-shaped steel dampers, widely used in Japan as seismic isolation devices.

A quantitative comparison between the damage on damper obtained using: (1) a one-directional (1D) model (SDOF system with single seismic component), and (2) a bidirectional (2D) model (Multiple Shear Springs with 2DOFs and both seismic components) is conducted. From analytical results, damage on damper up to fracture were compared. Then, the safety factor that should be considered in the current design based on 1D analysis was examined. Moreover, the values of the maximum displacement of the isolation layer were also compared and the safety factor that should be considered in the current design method was also examined.

**Keywords:** Base-isolation · Steel damper · Bi-directional ground motion · Maximum displacement · Damage evaluation

## 1 Introduction

Base isolated structure is a structural system suitable for energy based seismic design because the damper installed in the isolated layer absorbs the input energy by earthquake. The energy attribute to damage $E_D$ is the energy that exclued dissppated enegy by damping from input energy [1]. In a base isolated structure, $E_D$ is absorbed by dampers installed in the isolated layer. Currently, in-plane structural design is generally performed. It is same with base isolated structures. However, many types of dampers installed in the isolated layer are designed to absorb input energy from two directions. So, it must absorb more energy than in-plane condition. Performance of damper is generally expressed assuming in-plane conditions, such as in-plane fatigue characteristics [2]. For


H. Varum et al. (Eds.): IWEBSE 2023, LNCE 236, pp. 205–215, 2023.
https://doi.org/10.1007/978-3-031-36562-1_16

example, it is known that the performance of U-shaped steel dampers, which are generally used in Japanese base isolated structures, deteriorates when subjected to random external forces in two directions [3]. This research focuses on the damage evaluation methods for U-shaped steel dampers subjected to bi-directional excitation. A quantitative comparison between the damage on damper obtained using: (1) a one-directional (1D) model (SDOF system with single seismic component), and (2) a bidirectional (2D) model (Multiple Shear Springs with 2DOFs and both seismic components) is conducted. From analytical results, damage on damper up to fracture were compared. Then, based on the energy absorption capacity of damper and input energy brought by earthquake, the safety factor that should be considered in the current design based on 1D analysis was examined. Moreover, the values of the maximum displacement of the isolation layer were also compared and the safety factor that should be considered in the current design method was also examined.

## 2 Damage Evaluation Method for U-shaped Steel Dampers

Deformation capacity of the U-shaped steel dampers is affected by the loading direction. So far, low-cycle fatigue curves have been proposed based on the results of cyclic loading test conducted under the condition that the loading direction is fixed at 0° and 90° as shown in Fig. 1 [2]. By applying these evaluation formulas to the results of response analysis under unidirectional excitation, which is generally conducted at present, damage index of the U-shaped steel dampers installed in the $x$ and $y$ directions, as shown in Fig. 2, $D_{x,0}$ and $D_{y,90}$ can be obtained. The larger of these values is the damage index of the U-shaped steel damper due to the earthquake. When D calculated by Minors rule reaches 1.0, it is evaluated as damper has fractured.

$$D = Max\left(D_{x,0}, D_{y,90}\right) \tag{1}$$

Since the isolated layer deforms in two directions during an earthquake, it will be also damaged due to deformation in the perpendicular direction. Therefore, cumulative damage to the U-shaped steel damper under bidirectional excitation is the sum of damage in 0° and 90°. The damage index of the damper under bi-directional excitation $D_2$ is obtained by Eq. (2).

$$D_2 = Max\left(D_{x,0} + D_{x,90}, D_{y0} + D_{y,90}\right) \tag{2}$$

Furthermore, it has been clarified that the plastic deformation capacity of U-shaped steel dampers subjected to bi-directional external forces decreases due to the torsional deformation (Fig. 3). The effect of torsional deformation is expressed by the sway-motion index $J_f$, which is a function of the increment of the shear deformation angle $\gamma$ and incremental in-plane rotation angle $\Delta\phi$ as shown in Fig. 4. The limitation of damage $D_{2,lim}$ up to fracture is given as Eq. (4) in relation with the sway-motion index $J_f$, (Fig. 5) [3].

$$J_f = \sum_{i=1}^{n-1}\left(\gamma_i \cdot \gamma_{i+1} \cdot \frac{\gamma_i + \gamma_{i+1}}{2} \cdot |\Delta\phi_i|\right) \tag{3}$$

$$D_{2,lim} = \begin{cases} 1.0 & (J_f \leqq 15) \\ 1.6 - 0.04 \cdot J_f & (15 < J_f \leqq 30) \\ 0.4 & (30 < J_f) \end{cases} \quad (4)$$

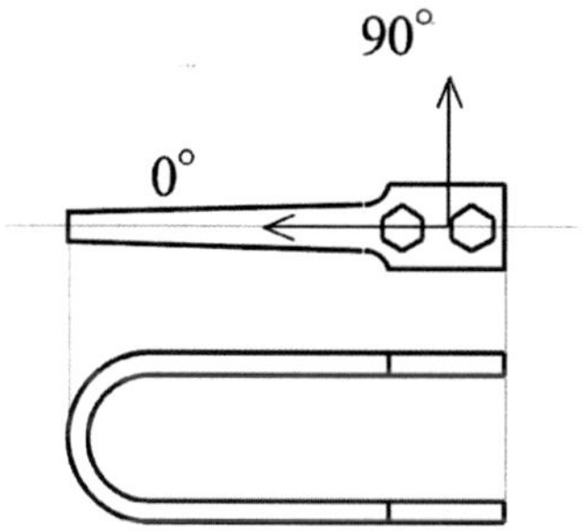


**Fig. 1.** Definition of direction in a U-shaped steel damper

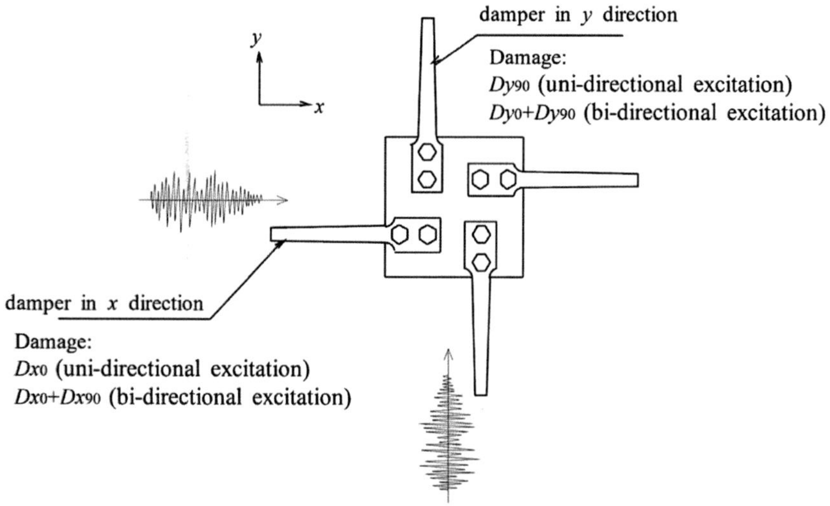


**Fig. 2.** Damage of U-shaped steel dampers

**Fig. 3.** Tortional deformation of U-shaped steel damper

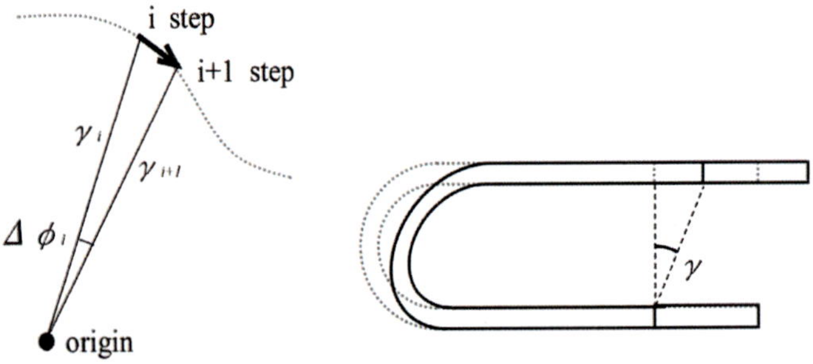


**Fig. 4.** Definition of $\gamma$ and $\Delta\phi$

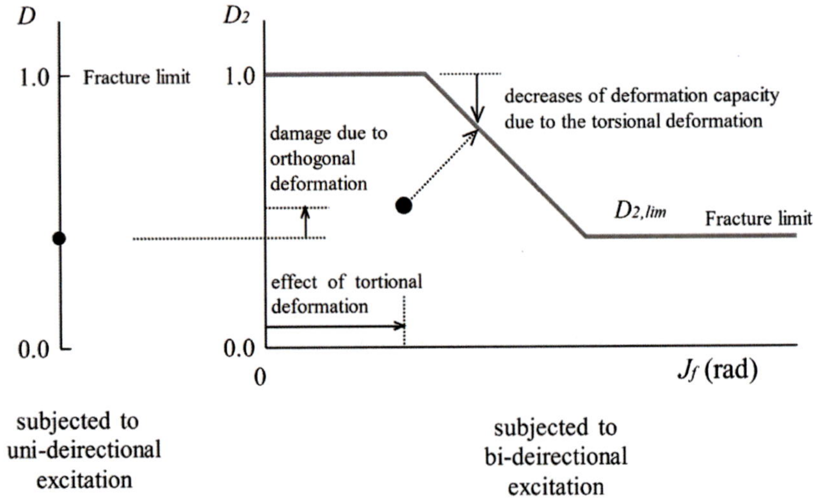


**Fig. 5.** Fracture limit of U-shaped steel damper subjected to uni-directional and bi-directional excitation

# 3 Analytical Method

## 3.1 Analytical Model

This study focuses on the response of the isolated layer, so the superstructure is simplified as single mass, and two analytical models are used. The first model is the 1D model shown in Fig. 6 (a), which is a single-degree-of-freedom system in which the isolated layer is represented by a shear spring. Another model is the 2D model shown in Fig. 6 (2), which is a two-degree-of-freedom system in which the isolated layer is represented by MSS (Multiple Shear Springs) [4].

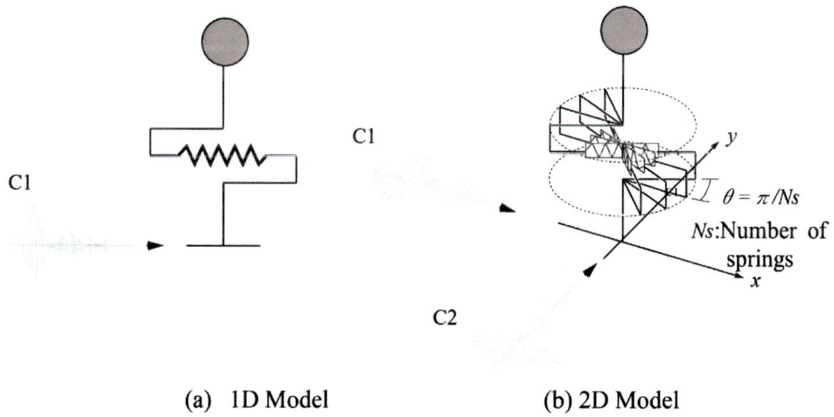


**Fig. 6.** Analytical models

In this study, a normal bilinear model is used as the hysteresis model of the U-shaped steel damper. The rubber bearing was modeled as a linear elastic spring. These were combined as parallel springs. For 1D models it is applied directly to the shear spring and for 2D models it is applied after being divided into MSS. Figure 7 shows a comparison of static analysis results and cyclic loading test results [2, 3]. It can be seen that the hysterias behavior of the U-shaped steel dampers under both 1D and 2D loading can be reproduced.

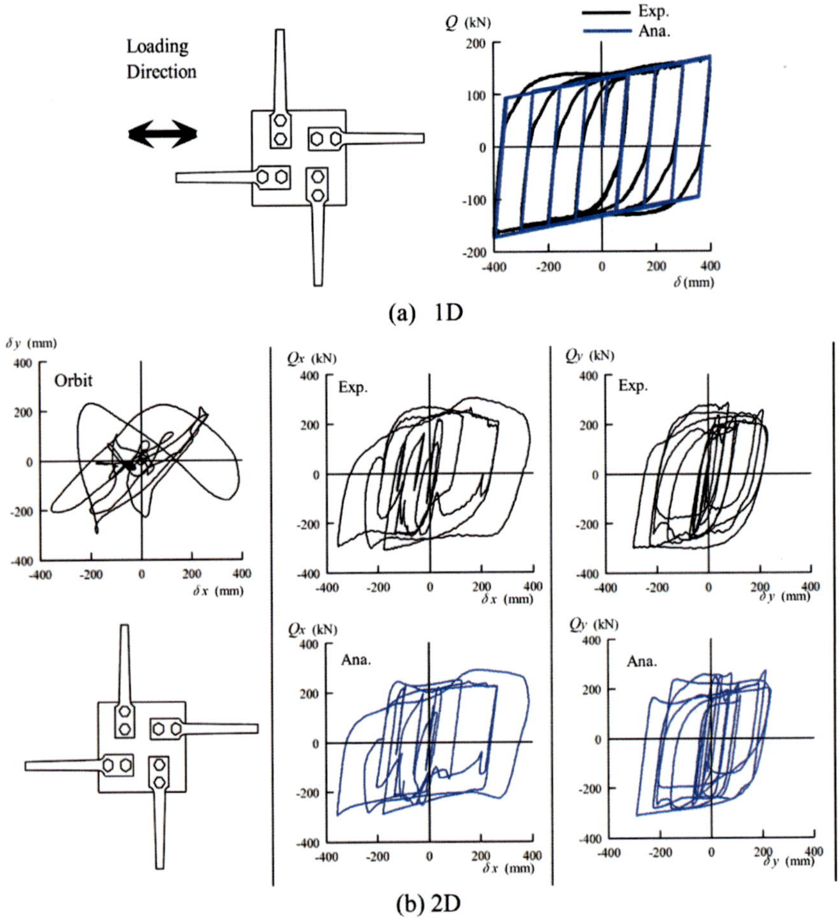


**Fig. 7.** Comparison of analytical results and experimental results

The yield deformation of the damper was set as 32.9 mm, the elastic stiffness was set as 8320 (kN/m), and the secondary stiffness was set as 144 (kN/m). Those values represent one unit (consists with 4 dampers) of generally used U-shaped steel damper. Also, the mass was set so that the restoring force of dampers would be 3% of model's own weight when the isolated layer was deformed 400 mm. The elastic stiffness of the rubber bearing was set so that the equivalent period would be 4.0 s when the isolated layer was deformed 400 mm.

### 3.2 Input Wave

10observation records of strong motion, which are considered to have various characteristics, were examined. Table 1 shows the list of input waves. The horizontal motion of each observation record consists of two orthogonal components. Among the two components, the C1 component is the larger maximum velocity (PGV), and the C2 component is the smaller one. In the analysis, the amplification factor was multiplied to the observation record to set the maximum velocity (PGV) of the C1 component 0.25 (m/s), 0.50 (m/s), 0.75 (m/s), and 1.00 (m/s). The C1 component was input to the 1D model. The C1 component and the C2 component multiplied by the same amplification factor as the C1 component were input to the 2D model.

**Table 1.** Input Waves

| Earthquake | Site | Abv. | PGV of the Original Record (m/s) | |
|---|---|---|---|---|
| | | | NS | EW |
| 1940 Imperial Valley | El Centro | EL | 0.381 | 0.488 |
| 1952 Kern County | Taft | TF | 0.182 | 0.175 |
| 1968 Tokachi-Oki | Hachinohe | HC | 0.342 | 0.370 |
| 1989 Loma Prieta | Agnews Hospital | LP | 0.176 | 0.259 |
| 1992 Cape Mendocino | Rio Dell Overpass | CM | 0.438 | 0.419 |
| 1994 Northridge | Olive View | SO | 1.292 | 0.780 |
| 1995 Kobe | Amagasaki | AM | 0.495 | 0.859 |
| 1995 Kobe | Nishi Akashi | KN | 0.373 | 0.366 |
| 1999 Chi-Chi | TCU129 | CB | 0.544 | 0.510 |
| 2011 Tohoku | JMA Sendai | JS | 0.544 | 0.545 |

## 4 Analytical Result

Figure 8 shows the damage index of the U-shaped steel damper in the 1D model $D$ in relation to the input level (PGV). Also, Fig. 9 shows the damage index of the 2D model $D_2$ in relation to the sway motion index $J_f$. The damage of a U-shaped steel damper varies greatly depending on the input wave. For uni-directional excitations, the maximum value of $D$ is about 0.055 for PGV = 0.5 (m/s), a design level ground motion in Japanese structural design. Even with very strong excitation with PGV = 1.0 (m/s), its maximum value is about 0.18. Those are not problem. On the other hand, in the case of bi-directional excitations, plots of $D_2$ for PGV = 0.5 (m/s) are far from the damage limit. But one plot of $D_2$ for PGV = 1.0 (m/s) is close to the damage limit line.

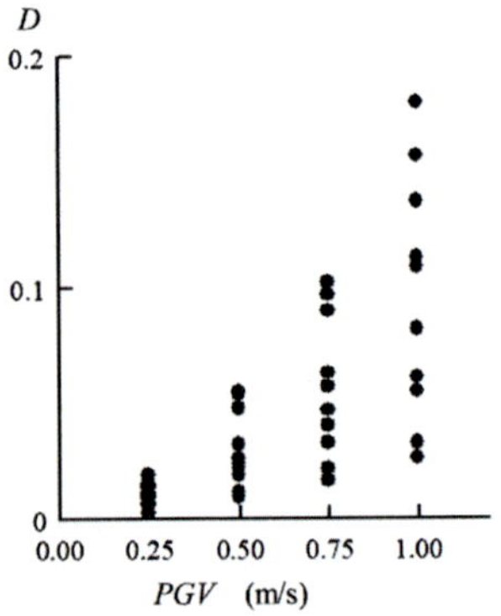


**Fig. 8.** $D$ for uni-directional excitations

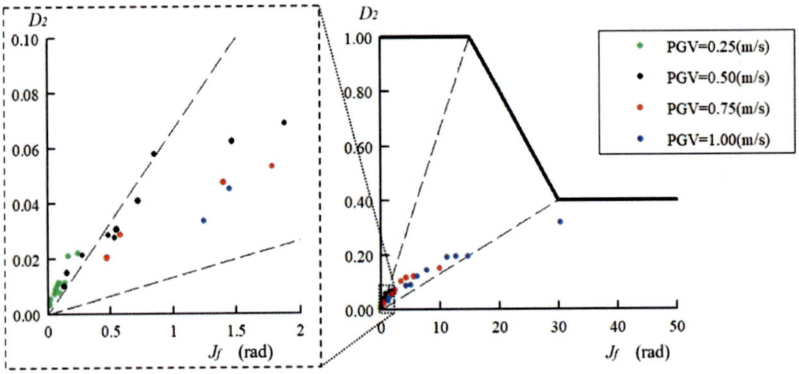


**Fig. 9.** $D_2$ for bi-directional excitations

## 5 Comparison of Damage Under Uni and Bi Directional Excitations

The plastic deformation capacity of the U-shaped steel damper, which is subjected tobi-directional external forces, is affected by torsional deformation. When the sway-motion index $J_f$ is 15 or more, it reaches the limitation of damage $D_{2,lim}$ before $D_2$ reaches 1.0. Therefore, it is not appropriate to directly compare the damage index $D$ for uni-directional excitation and the damage index $D_2$ for bi-directional excitation. Therefore, as shown in Fig. 10, the damage limit ratio $DR$, which is the ratio of $D_2$ to the limitation of damage $D_{2,lim}$ under the condition that the ratio of Jf to $D_2$ is constant is defined. Then, $DR$ is compared with $D$.

Figure 11 shows how $DR$, for bi-directional excitation increases to $D$ for uni-directional excitation with each input level (PGV). As the input level increases, the estimated damage of the U-shaped steel damper $D$ predicted by the in-plane analysis underestimates the damage generated by bi-directional excitation. $DR$, of U-shaped steel damper installed in a base isolated structure that subjected to bi-directional excitation is not only affected by the deformation in the orthogonal direction but also affected by the torsion deformation due to sway motion. As a result, it become larger than $D$ evaluated by in-plane analysis. Especially, the effect of trotional deformation due to sway motion is proportional to the cube of the response displacement of the isolated layer, so it becomes more pronounced as the input level increases.

Figure 12 shows the relationship between $D$ and $DR$ for all cases. Regardless of the input wave and input level, the relationship represented by Eq. (5) can be seen between them.

$$DR = 7 \cdot D^{1.4} \tag{5}$$

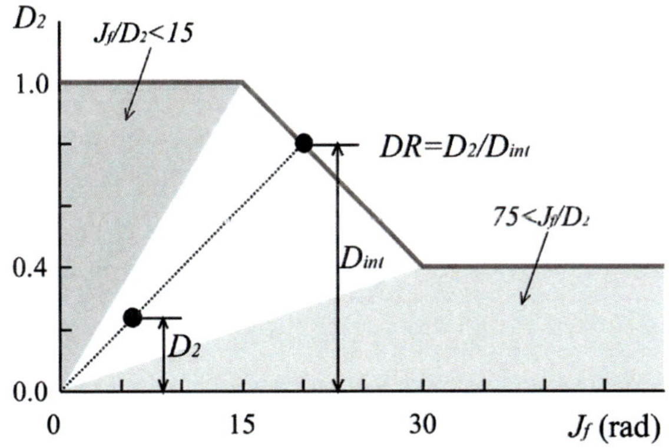


**Fig. 10.** Definition of the damage limit ratio *DR*

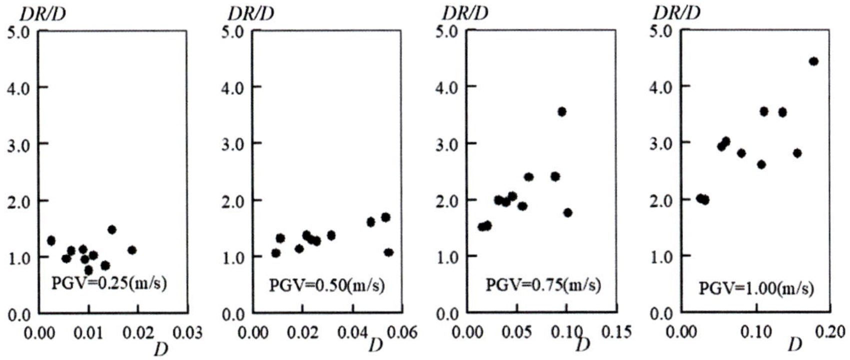


**Fig. 11.** $DR/D$ with each input level

## 6 Maximum Drift of the Isolated Layer

The maximum drift of the isolated layer is an important design value to avoid pounding to retaining wall. Figure 13 shows a comparison between the maximum drift generated by uni-directional excitation $\delta_{max}$ and the maximum drift generated by bi-directional

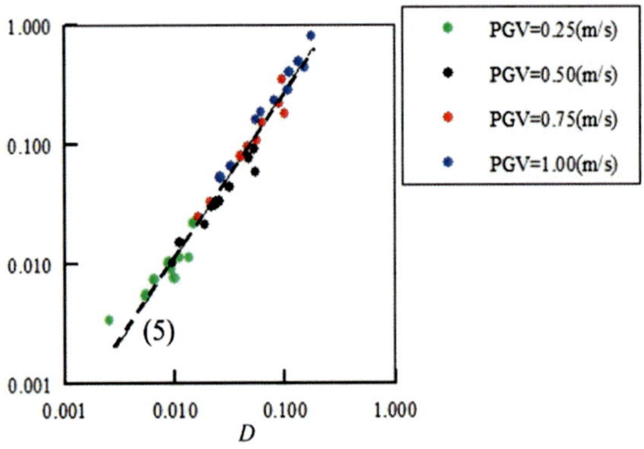


**Fig. 12.** Relationship between $D$ and $DR$

excitation $R_{max}$. It can be seen that $R_{max}$ is approximately 10% larger than $\delta_{max}$ regardless of the input wave or input level.

$$R_{max} = 1.1 \cdot \delta_{max} \tag{6}$$

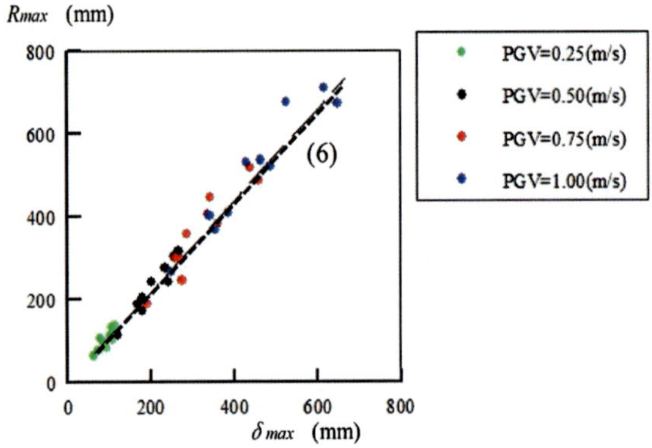


**Fig. 13.** Relationship between $Rmax$ and $\delta max$

## 7 Conclusions

The energy based seismic design is a design method based on the balance between the input energy by earthquake and the absorbed energy of the building. It has an advantage of being able to conduct plastic design conveniently. A base isolated structure absorbs energy with dampers installed in the isolated layer, so it is a structural system very suitable for the energy based seismic design. However, dampers are subjected to bi-directional excitation, so, it is necessary to estimate the influence and reflect it in design. This study focuses on U-shaped steel dampers used in base isolated structures, difference between damage evaluation based on in-plane response analysis results and damage evaluation based on bi-directional response analysis results was examined. In addition, the maximum drift of the isolated layer, which is also a important design criterion, was compared. The conclusions obtained are summarized as follows.

The plastic deformation capacity of U-shaped steel dampers subjected to bi-directional excitation is affected by torsional deformation due to sway motion. In this study, the damage limitation ratio *DR* as an index of damage for bi-directional excitation is introduced and compared it with *D* of the in-plane analysis. The ratio of *DR* to *D* becomes significant because as the input level increases, the greater the effect of sway motion.

A constant relationship was found independent of the input wave and the input level. It can be said that, from the estimated damage on steel damper by in-plane design based on energy balance, damage due to bi-directional excitation can be estimated.

Regardless of the input wave and input level, the maximum drift of the isolated layer subjected to bi-directional excitation is about 1.1 times the maximum drift of it subjected to uni-directional excitation. So, it can be estimated by in-plane structural design.

## References

1. Akiyama, H.: Earthquake-Resistant Limit-State Design for Buildings. University of Tokyo Press, Tokyo (1985)
2. Jiao, Y., et al.: Low cyclic fatigue and hysteretic behavior of U-shaped steel dampers for seismic for seismic for seismic isolated buildings under dynamic cyclic loadings. Earthq. Eng. Struct. Dyn. **44**, 1523–1538 (2015). https://doi.org/10.1002/eqe.2533
3. Ene, D., et al.: Experimental study on the bidirectional inelastic deformation capacity of U-shaped steel dampers for seismic isolated buildings. Earthq. Eng. Struct. Dyn. **45**, 173–192 (2016).https://doi.org/10.1002/eqe.2621
4. Wada, A., Hirose, K.: Building frames subjected to 2D earthquake motion. In: Seismic Engineering: Research and Practice Proceedings, Structures Congress 1989. ASCE (1989)

Check for updates

# Input Energy Spectra for Pulse-Like Ground Motions

C. Frau(✉), S. Panella, and M. Tornello

Universidad Tecnológica Nacional, Regional Mendoza, Argentina
cdfrau@frm.utn.edu.ar

**Abstract.** In energy based seismic design (EBSD) approach, effect of ground motions on structures is considered as an energy input to structures (EI). The usage of energy spectra is an effective tool in energy based seismic design (EBSD) methods, such as the use of design acceleration spectra in force-based and displacement-based methods. The obtention of input energy spectra offers an important advantage to determine the energy input to structures with the effect of ground motions. On the other side, near-fault seismic ground motions are frequently characterized by intense velocity and displacement pulses of relatively long periods that clearly distinguish them from typical far-field ground motions. Intense velocity pulse motions can affect adversely the seismic performance of structures. Based on these ideas, a correlation between the impulsivity level and input energy spectrum is presented in this study, and a new parameter is established for evaluating the input energy power. This correlation may aid to select ground motion for structural analysis in near-fault regions.

**Keywords:** input energy spectra · pulse-like ground motions · near-fault zones

## 1 Introduction

Energy concept in seismic design of structures has been widely studied over a half-century period and energy-based methods have always been considered more rational and reliable for the design and assessment of structures under seismic effects when compared to conventional force-based and displacement-based methods (Uang and Bertero 1990; Akbas and Shen 2003). In the energy-based methods earthquake effect is considered as an energy input to structures and this energy input expresses the total energy demand of the earthquake. Making a structure safe is considered as a balance of energy dissipation capacity and earthquake energy demand in these approaches. However, an important question of the energy-based seismic design is to determine the energy input to structures with earthquake motion. Energy-based earthquake resistant design was first proposed by Housner (1956), who studied the seismic energy input to structures using the velocity spectra of elastic systems. Energy-based design parameters were first defined in his research, and these formed a basis for earthquake resistant energy-based design. Some other researchers also made many previous estimations about the input energy concept, and they considered the input energy as an effective tool in earthquake-resistant design


H. Varum et al. (Eds.): IWEBSE 2023, LNCE 236, pp. 216–233, 2023.
https://doi.org/10.1007/978-3-031-36562-1_17

(Uang and Bertero 1990; Fajfar and Fischinger 1990; Manfredi 2001). Zahrah and Hall (1984), Akiyama (1985), Kuwamura and Galambos (1989), Fajfar et al. (1989) made pioneer studies like Housner about seismic energy concepts and they proposed useful analytical and empirical equations for the seismic input energy.

On the other side, it is known that near fault regions are exposed to directivity effects. When a fault ruptures toward a site, a rupture velocity slightly slower than the shear wave velocity produces an accumulation of seismic energy released during rupture (Somerville et al. 1997; Spudich and Chiou 2008); this generally results in a large pulse in the velocity-time series. Thus, near-fault seismic ground motions are frequently characterized by intense velocity and displacement pulses of relatively long periods that clearly distinguish them from typical far-field ground motions. Báez and Miranda (2000) found that the maximum ground velocity and the maximum incremental velocity are the parameters that most influence the structural response.

Intense velocity pulse motions can affect adversely the seismic performance of structures (Bertero et al. 1978; Chopra and Chintanapakdee 2001). Anderson and Bertero (1987) demonstrated that the presence of long acceleration pulses demands higher resistance of structures to stand upright. Malhotra (1999) affirms that the presence of characteristic pulses of acceleration, velocity and displacement can generate greater shear at the base of buildings and greater lateral displacements compared to records that do not have these pulses; the demand for ductility can be much higher and the additional damping added to a structure can be less effective. In structures placed into near-fault zones the damage is caused by a few cycles inelastic strain, they are in coincidence with the long velocity pulses and great amplitude (Alavi and Krawlinker 2001). In opposite, in sites placed far away faults damage is distributed during all time of the record in many cycles with minor inelastic strain (Báez and Miranda 2000).

In energy based seismic design (EBSD) approach, effect of ground motions on structures is considered as an energy input to structures. The earthquake input energy spectra are created combining the maximum input energies of single degree of freedom (SDOF) systems having a certain damping ratio for different natural vibration periods. The determination of input energy spectra is of great importance for the energy based seismic design since the total energy input to structural systems can be practically obtained via these graphs.

The usage of energy spectra is an effective tool in EBSD methods, such as the use of design acceleration spectra in force based and displacement based methods. The obtention of input energy spectra offers an important advantage to determine the energy input to structures with the effect of ground motions.

In EBSD of structures, the energy demand of an earthquake should be less than or, in limit, should be equal to, the energy dissipation capacities possessed by the structure. It is of utmost importance for structural and earthquake engineers that the seismic input energy transmitted to structures is computed exactly.

These concepts suggest assessment the relationship between input energy and impulsivity in ground motions like-pulse. With this target, in this work input energy for impulsive records set are analyzed. The set is taken from the Panella et al. (2017) and it is organized in different impulsivity levels. To evaluate the energy input a new parameter is proposed, it consider the time of the earthquake in delivering the energy. To the end a matrix to link both parameters (input energy and impulsivity) is presented. The matrix may help to a better selection of record to structural analysis.

## 2 Ranking of Pulse-Like Ground Motions

### 2.1 Pulse-Like Ground Motions Classification

Panella et al. (2017) developed a method to identifying and classifying like-pulse ground motions. Departing from a velocity time history of ground motion the "developed length of velocity" *Ldv* is defined as the length reached by the trace of velocity records as if it were "extended" like a string (Eq. 1).

$$Ld_v = \sum_{i=1}^{n} \left( \sqrt{(\Delta t)^2 + (\Delta v_i)^2} \right) \tag{1}$$

where $\Delta t$ is the time lapse of the record between two successive points $t(i+1) - t(i)$ in *s*, $\Delta vi$ are velocity increments between $t(i)$ and $t(i+1)$ in *cm/s*, and $n$ is the number of samples in the series. Given the binary character of the target classification (pulse-like or non-pulse) and enough data availability, a binary logistic regression was used to classify the records; *Ldv* and *PGV* (Peak Ground Velocity) were the parameters used (Eq. 2). The logistic regression proved the following predictive equation for the Impulsivity Index by Regression (IPR).

$$\mathrm{IP_R} = \frac{1}{1 + e^{(5 - 0.45\mathrm{PGV} + 0.01\mathrm{Ldv})}} \tag{2}$$

The Impulsivity Index by Regression takes values between 0 and 1. A ground motion qualifies as pulse type if its *IPR* is higher than 0.7 and its *PGV* is higher than 30 cm/s. Below 0.7 the record is non-pulse. It is well-known that several researchers have established as an excluding condition to classify a record as pulse-type that *PGV* should reach a minimum of 30 cm/s [18, 23]. Even though this value is not adequately justified, it imposes a minimum to the power of the pulse for a time velocity series to be classified as pulse-type. The impulsivity level is defined with an indicator based on the development length of velocity and *PGV*; the "Impulsivity Index" *IP* is as follows:

$$\mathrm{IP} = \frac{\mathrm{Ldv}}{\mathrm{PGV}} \tag{3}$$

The definition of *Ldv* captures in a simple and efficient way the impulsive aspect visually detected in a velocity time series. A relatively low *Ldv* value represents an impulsive character, whereas a high *Ldv* value represents a non-pulse or vibratory character. Pulse amplitude also plays an important role: high *PGV* reveal the presence of at least one pulse, whereas low *PGV* "dilute" pulses. In this way, a combination of low *Ldv* values and high *PGV* leads to small *IP* suggesting high impulsivity, while the opposite is the manifestation of a non-pulse character. Once a record is identified as pulse-type by *IPr*, ranks can be established using the value taken by *IP* to classify ground motions into three impulsivity levels: high, medium or moderate, and low (see Table 1); $IP = 35$ divides between pulse and non-pulse.

**Table 1.** Classification proposed for different impulsivity levels.

| Impulsivity Index IP | Impulsivity Level |
|---|---|
| IP $\leq$ 12 | High (H) |
| 12 < IP $\leq$ 20 | Medium or Moderate (M) |
| 20 < IP $\leq$ 35 | Low (L) |

## 2.2 Database and Ranking of Pulse-Like Ground Motions

For this study we selected a set of 114 impulsive records from Panella et al. database; with IP between 6.5 and 35. The seismic records used correspond to earthquakes in different parts of the world with a moment magnitude between 5.5 and 7.9. Strong motions have a Joyner-Boore distance less than 30 km. In the Appendix 1, Table A-1 shows data for the selected set. Figure 1 shows *IP* in increasing order for the 104 selected records. A continuous variation is seen (between 5 and 35), without jumps; this shows the consistency of the data for subsequent analysis.

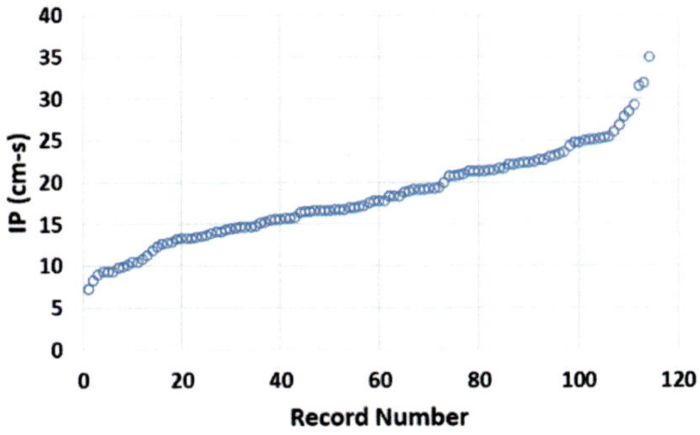


**Fig. 1.** *IP* in increasing order for the selected records.

# 3 Energy Spectra

## 3.1 Elastic Input Energy Spectrum

Starting from the fundamental definition of work, (i.e., the integral of force with respect to displacement) the so-called energy balance equation can be easily obtained by integrating the governing differential equation of motion of a SDOF system subjected to a horizontal ground motion over the relative displacement of the mass with respect to the ground:

$$\int_0^u m\ddot{u}(t)du + \int_0^u c\dot{u}(t)du + \int_0^u ku(t)du = -\int_0^u m\ddot{u}_g(t)du \tag{4}$$

The integral quantities on the left-hand side of this equation identify the different energy components of the structure named as the kinetic energy, the damping energy and the recoverable strain energy, respectively. The right-hand side of the Equation which is closely associated with the main concern of the present study, expresses the total input energy $EI(t)$.

According to the rules of mathematical analysis, the incremental displacement du can be expressed as which enables integration of the governing equation of motion with respect to duration of earthquake. Accordingly, for a specific earthquake ground motion, the relative energy input to a SDOF system, as well as the other energy components, and be theoretically obtained by integrating the equation of motion over the time (Eq. 5); Fig. 2-a shows the input energy as a time function.

$$\frac{E_I}{m} = \int_0^t \ddot{u_g}(t)\dot{u}(t)dt \tag{5}$$

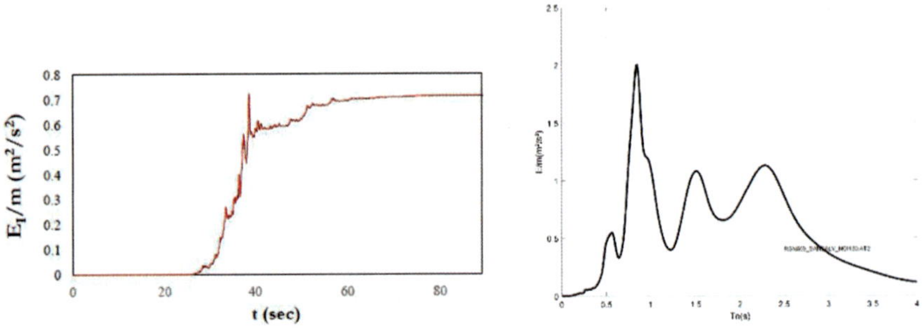


a) b)

**Fig. 2.** a) Input energy as a time function for a record with period and damping given; b) Input energy spectrum.

When this process is repeated for various SDOF systems with different periods *Tn* but the same damping ratios, a set of *EI* versus *Tn* elastic input energy spectrum are obtained. In brief definition, the seismic input energy spectra are the graphs which combine the maximum energy input values corresponds to different natural vibration periods of SDOF systems (Fig. 2-b). Figure 3 shows the input energy spectra for some records from the set selected; great variability can be seen.

### 3.2 Input Energy Power and Input Energy Power Spectrum

Considering input energy for a period given *Ti*, Input Energy Power (*IEP*) is defined as the total input energy divided by the time energy takes to enter (Eq. 6). Because the energy delivery is small at the beginning and end of the records (see Fig. 2-a), to define *IEP* the effective duration is considered. For this Arias Intensity is used; effective duration is time interval between 5% ($t_i$) and 95%($t_f$) (Arias 1970).

$$IEP = \frac{E_I/m}{(t_f - t_i)} = \frac{1}{t_f - t_i} \int_{ti}^{tf} \ddot{u_g}(t)\dot{u}(t)dt \tag{6}$$

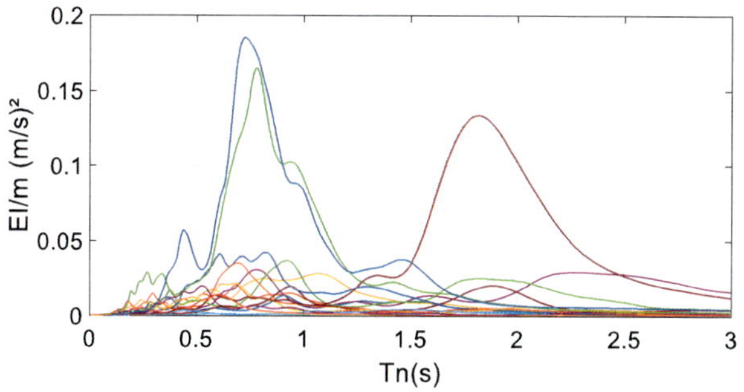


**Fig. 3.** Input energy spectra for some records from the selected set.

After calculating the *IEP* for each structural period (*Tn*), it is possible to construct the Input Energy Power Spectrum $S_{IEP}(Tn,\ \zeta)$. This spectrum takes into account, in addition to the energy that enters, the power with which it does so. It must be recognized that the total amount of energy that enters is important, but also the time in which that energy must be dissipated, that is, the power. Figure 4 shows two cases studied where one reaches higher energy but the other reaches higher power. Figure 4 shows two cases studied where once reach higher energy but the other reach higher power.

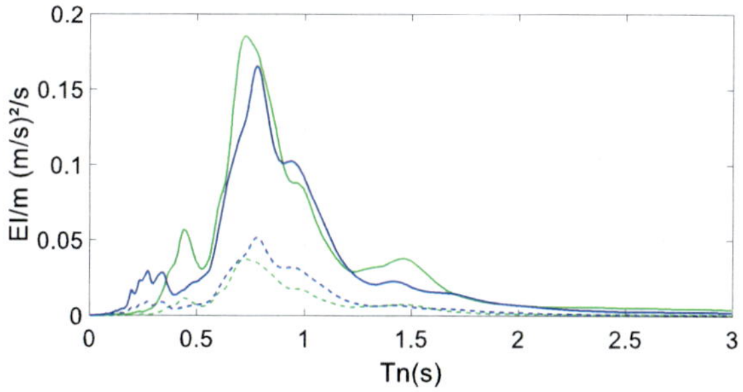


**Fig. 4.** Input Energy Spectra (continuous line) and Input Energy Power Spectrum $S_{IEP}(Tn,\ \zeta)$ (dashed line) for two records: RSN569_SANSALV_NGI270 and RSN451_ MORGAN_ CYC285 (See Appendix 1).

# 4 Energy Spectral Intensity

## 4.1 Housner Intensity

Housner (1952) defined a spectral intensity from the pseudo-velocity elastic response spectrum, which is obtained from an acceleration record. The Housner Spectral Intensity is defined as (Eq. 7):

$$SI(Housner) = \int_{0.1}^{2.5} Sv(T, \xi)dT \tag{7}$$

where *Sv* is the pseudo-velocity elastic response spectrum, *T* are the structural periods, $\zeta$ is the fraction of critical damping and 0.1–2.5 s are the periods that the integral covers. The Housner Spectral Intensity express the relative severity of earthquakes.

## 4.2 Energy Spectral Intensity

Whit the same idea of Housner Intensity, in this work a new parameter to evaluate the input energy is proposed: the Energy Spectral Intensity (*ESI*). This is defined from the Input Energy Power Spectrum according to following equation (Eq. 8).

$$ESI = \int_{0.1}^{2.5} S_{IEP}(T, \xi)dT \quad (8)$$

*ESI* quantifies the severity of seismic records by concentrating in a value the input energy and the power with which it does so in an interest periods range. Besides, it allows to compare different records each other; in special when, in many cases, the input energy spectra present great variation from a period to other (see Fig. 2-b). Figure 5 shows an example of how the parameters described above are obtained.

La Fig. 5 muestra un ejemplo de cómo se obtienen los parámetros descritos anteriormente.

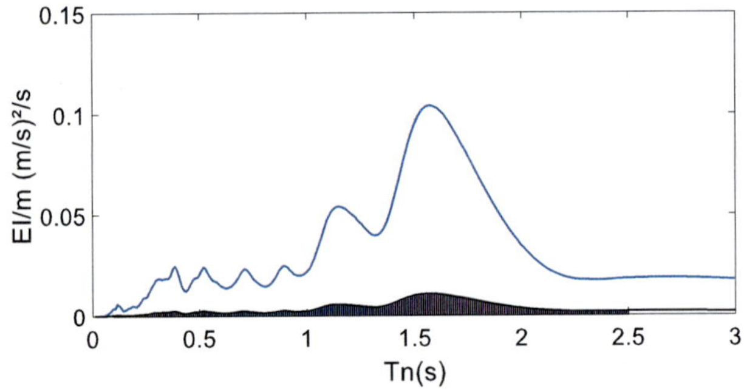


**Fig. 5.** Example of how the ESI is obtained.

For the set of selected records, the spectral intensity of energy was calculated. Figure 6 shows the values reached by *ESI* in an orderly and increasing manner. A more or less continuous variation without jumps is observed. This means that the selected record set is consistent and covers a wide range of cases.

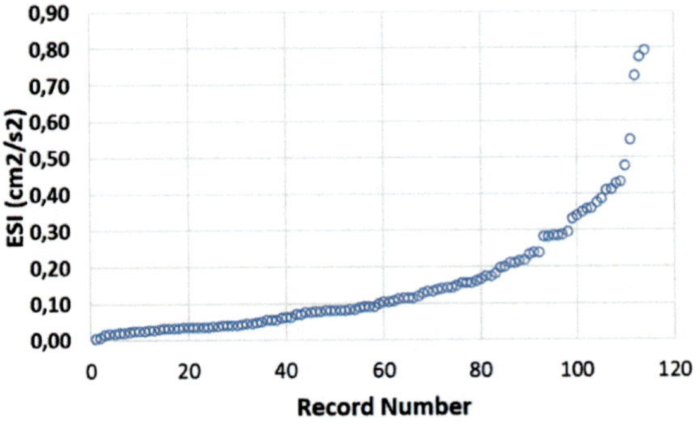


**Fig. 6.** *ESI* in increasing order for the selected records.

# 5 Energy and Impulsivity

## 5.1 Relation Between Energy and Impulsivity

This section investigates the relationship between impulsivity and energy input. For this, the impulsivity index of Panella et al. (2017) and the Energy Spectral Intensity defined in this work were considered. Figure 7 shows a scatter graph where the Impulsivity Index *IP* is represented on the abscissa, while the Energy Spectral Intensity *ESI* is represented on the ordinate.

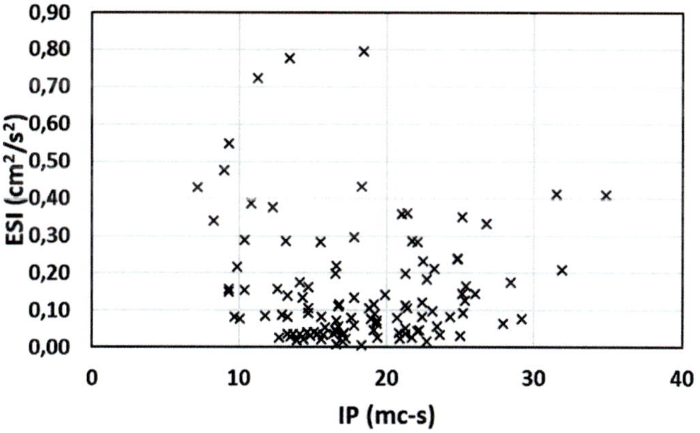


**Fig. 7.** a) *IP* vs. *ESI* for the selected records.

Figure 7 does not show a direct relationship between the impulsivity of the recording and the spectral intensity of energy. However, it can be stated that the highest energy spectral intensities correspond to records of moderate to high impulsivity ($IP < 20$); thus, records with low impulsivity do not present high *ESI*. Records with low *ESI* present impulsivity of all levels (high, moderate and low). Based on the results obtained, a classification of records according to the power of the energy input (*ESI*) is proposed; It is divided into three levels: High, Moderate and Low according to the limits shown in Table 2.

**Table 2.** Classification proposed for different Energy Spectral Intensity.

| Energy Spectral Intensity ($cm^2/s^2$) | Input Energy Level |
|---|---|
| ESI > 0.4 | High (H) |
| $0.20 < \text{ESI} \leq 0.40$ | Moderate (M) |
| $\text{ESI} \leq 0.20$ | Low (L) |

From the classification of impulsivity and energy input, a matrix is formed to select seismic records for structural analysis that involves impulsivity and the spectral intensity of energy (*ESI*). The matrix combines both parameters and is shown in Table 3.

**Table 3.** Matrix for selecting records considering impulsivity and energy input.

| SELECTION MATRIX | | IMPUSIVITY INDEX (cm,s) | | |
|---|---|---|---|---|
| | | ID ≤ 12 km | 12 < ID ≤ 20 | ID >35 |
| ENERGY SPECTRAL INTENSITY ($cm^2/s^2$) | ESI > 0,40 | $H_0$ | $H_1$ | $M_1$ |
| | 0,20 < ESI ≤ 0,40 | $H_1$ | $M_0$ | $L_1$ |
| | ESI ≤ 0,20 | $M_1$ | $L_1$ | $L_0$ |

With this classification we organized a matrix to select proper impulsive records considering energy input. This matrix combines the key parameters studied: the level impulsivity and Energy Spectral Intensity. This matrix together with the ranking of strong motions pulse-like (see Appendix 1) will allow designers to make more realistic analysis for structures that take place in near-fault regions when Energy-Based Seismic Design is used.

Both magnitudes are represented by 3 levels: (see Table 1 and 2). When Tables 2 and 3 are combined, different kind of interactions appear: a) High demand for energy and impulsivity (H); b) Moderate demand for energy and impulsivity (M); and c) Low demand for energy and impulsivity (L). The subscript zero (0) indicates strong interactions and the subscript one (1) indicates weak interactions. Thus, matrix allows to choose different levels of impulsivity and input energy for structural analysis.

## 6 Conclusions

A set of impulsive records taken from previous studies on impulsivity was selected that classifies the records according to different levels of impulsivity.

Based on the Elastic Input Energy Spectra, a new parameter was defined to evaluate the energy input of a seismic record that takes into account the power with which the energy enters: the Energy Spectral Intensity.

With the results obtained for the Energy Spectral Intensity, a ranking was proposed according to the levels of the input energy power.

Combining the classification in levels of impulsiveness and spectral energy intensity, a matrix is created to select impulsive registers according to different input energy levels. Records can be chosen from the Appendix 1.

**Acknowledgements.** The authors wish to express their gratitude to the Regional Center of Technological Development for Construction, Seismology and Earthquake Engineering (CeReDeTeC) of National Technological University (Argentina) for their support for the present study.

# Appendix 1

**Table A-1.** Parameters of the records used in this work.

| N° | ID | Nombre de Sismo | Earthquake Magnitude | J-B Dist. (km) | PGA (g) | PGV (cm/s) | IP (cm-s) | IPR | Record Duration (s) | Time IA 5–95% (s) | ESI ($cm^2/s^2$) |
|---|---|---|---|---|---|---|---|---|---|---|---|
| 1 | 174 | RSN182 IMPVALL.H H-E07230 | 6,53 | 0.6 | 0,47 | 113,1 | 7,2 | 1,0 | 36,8 | 4,8 | 0,429 |
| 2 | 159 | RSN171 IMPVALL.H H-EMO270 | 6.53 | 0.1 | 0.30 | 92.6 | 8.3 | 1,0 | 40,0 | 6,7 | 0,340 |
| 3 | 335 | RSN568 SANSALV GIC090 | 5.80 | 2,1 | 0.70 | 79.9 | 9.0 | 1.0 | 9.0 | 4,3 | 0,477 |
| 4 | 158 | RSN171 IMPVALL.H H-EMO000 | 6,53 | 0,1 | 0,32 | 73,0 | 9,3 | 1.0 | 40,0 | 8.2 | 0,158 |
| 5 | 172 | RSN181 IMPVALL.H H-E06230 | 6,53 | 0,0 | 0,45 | 113,6 | 9,3 | 1,0 | 39,1 | 8,7 | 0,148 |
| 6 | 336 | RSN568 SANSALV GIC180 | 5,80 | 2.1 | 0.42 | 62,4 | 9.3 | 1,0 | 9,0 | 3,1 | 0,549 |
| 7 | 168 | RSN179 IMPVALL.H H-E04230 | 6.53 | 4.9 | 0.37 | 80.4 | 9.7 | 1,0 | 39,1 | 10,3 | 0,082 |
| 8 | 104 | RSN150 COYOTELK G06230 | 5,74 | 0,4 | 0.42 | 44,4 | 9.9 | 1.0 | 27,1 | 3,2 | 0,216 |
| 9 | 157 | RSN170 IMPVALL.H H-ECC092 | 6,53 | 7.3 | 0.24 | 73,4 | 10,1 | 1,0 | 40,0 | 13.2 | 0,079 |
| 10 | 170 | RSN180 IMPVALL.H H-E05230 | 6,53 | 1,8 | 0.38 | 96.9 | 10,4 | 1,0 | 39.3 | 9.5 | 0,156 |
| 11 | 337 | RSN569 SANSALV NGI180 | 5.80 | 3,7 | 0,40 | 56,4 | 10,4 | 1.0 | 20,3 | 6.2 | 0.289 |
| 12 | 338 | RSN569 SANSALV NGI270 | 5.80 | 3,7 | 0,53 | 73,0 | 10.8 | 1,0 | 20,3 | 4.9 | 0,387 |

*(continued)*

**Table A-1.** (*continued*)

| N° | ID | Nombre de Sismo | Earthquake Magnitude | J-B Dist. (km) | PGA (g) | PGV (cm/s) | IP (cm-s) | IPR | Record Duration (s) | Time IA 5–95% (s) | ESI ($cm^2/s^2$) |
|---|---|---|---|---|---|---|---|---|---|---|---|
| 13 | 382 | RSN723 SUPER.B B-PTS225 | 6,54 | 1,0 | 0,43 | 134,4 | 11,3 | 1.0 | 22,3 | 10,6 | 0,723 |
| 14 | 273 | RSN412 COALINGA D-PVY045 | 5,77 | 13.2 | 0.58 | 37.5 | 11.8 | 1,0 | 21,7 | 3.8 | 0,085 |
| 15 | 329 | RSN529 PALMSPR NPS210 | 6.06 | 0.0 | 0.69 | 66.0 | 12.3 | 1,0 | 20.2 | 4.8 | 0,375 |
| 16 | 173 | RSN182 IMPVALL.H H-E07140 | 6.53 | 0.6 | 0,34 | 51,7 | 12,6 | 1,0 | 36.8 | 6.8 | 0,157 |
| 17 | 385 | RSN730 SPITAK GUK000 | 6.77 | 24,0 | 0,20 | 28,4 | 12,7 | 1,0 | 20,0 | 10,5 | 0,026 |
| 18 | 72 | RSN147 COYOTELK G02140 | 5.74 | 8,5 | 0.26 | 32,0 | 12.9 | 1.0 | 26,8 | 4,0 | 0.089 |
| 19 | 328 | RSN527 PALMSPR MVH135 | 6,06 | 3,6 | 0,22 | 40,0 | 13,2 | 1.0 | 20,1 | 6,7 | 0,285 |
| 20 | 177 | RSN184 IMPVALL.H H-EDA270 | 6,53 | 5,1 | 0,35 | 75,6 | 13,3 | 1,0 | 39.1 | 7,0 | 0,138 |
| 21 | 324 | RSN502 MTLEWIS HVR090 | 5.60 | 12,4 | 0 .15 | 19.0 | 13.3 | 0.7 | 40,0 | 5,1 | 0,036 |
| 22 | 334 | RSN564 GREECE H-KAL-NS | 6,20 | 6,5 | 0.24 | 33.5 | 13.3 | 1,0 | 29,2 | 5,0 | 0,083 |
| 23 | 294 | RSN451 MORGAN CYC285 | 6,19 | 0,2 | 1,30 | 78.5 | 13,4 | 1.0 | 30,0 | 3,2 | 0,776 |
| 24 | 154 | RSN161 IMPVALL.H H-BRA315 | 6,53 | 8,5 | 0,22 | 40.9 | 13,5 | 1,0 | 37,8 | 14,4 | 0,028 |
| 25 | 160 | RSN173 IMPVALL.H H-E10050 | 6.53 | 8.6 | 0.17 | 50,7 | 13.7 | 1,0 | 37,0 | 12,8 | 0,036 |
| 26 | 153 | RSN161 IMPVALL.H H-BRA225 | 6,53 | 8,5 | 0.16 | 36.6 | 13.9 | 1.0 | 37,8 | 14.9 | 0,019 |
| 27 | 77 | RSN148 COYOTELK G03140 | 5,74 | 6,8 | 0,26 | 29.6 | 14,1 | 1,0 | 26.8 | 8,7 | 0,035 |

(*continued*)

**Table A-1.** *(continued)*

| N° | ID | Nombre de Sismo | Earthquake Magnitude | J-B Dist. (km) | PGA (g) | PGV (cm/s) | IP (cm-s) | IPR | Record Duration (s) | Time IA 5–95% (s) | ESI ($cm^2/s^2$) |
|---|---|---|---|---|---|---|---|---|---|---|---|
| 28 | 279 | RSN415 COALINGA D-TSM360 | 5,77 | 3,7 | 1,02 | 47,8 | 14.1 | 1,0 | 21,7 | 3,7 | 0,176 |
| 29 | 345 | RSN614 WHITTIER.A A-BIR180 | 5,99 | 14,9 | 0,35 | 39.9 | 14.3 | 1.0 | 28,6 | 3,8 | 0,134 |
| 30 | 265 | RSN411 COALINGA D-PVP360 | 5,77 | 13,2 | 0.41 | 20,2 | 14,4 | 0.8 | 20,6 | 4.5 | 0,021 |
| 31 | 213 | RSN316 WESMORL PTS225 | 5,90 | 16,5 | 0,23 | 55,6 | 14,6 | 1,0 | 41,7 | 15,2 | 0,042 |
| 32 | 171 | RSN181 IMPVALL.H H-E06140 | 6,53 | 0.0 | 0,45 | 67,0 | 14,7 | 1,0 | 39,1 | 11,5 | 0,162 |
| 33 | 212 | RSN292 ITALY A-STU270 | 6,90 | 6.8 | 0.32 | 72,0 | 14,7 | 1.0 | 39.3 | 15.2 | 0,093 |
| 34 | 391 | RSN764 LOMAP GOF160 | 6,93 | 10,3 | 0.29 | 43.4 | 14.7 | 1,0 | 40,0 | 8.9 | 0.107 |
| 35 | 381 | RSN722 SUPER.B B-KRN360 | 6,54 | 18,5 | 0.14 | 29.6 | 14.8 | 1,0 | 22,0 | 12,4 | 0,032 |
| 36 | 179 | RSN185 IMPVALL.H H-HVP225 | 6,53 | 5,4 | 0.26 | 53.2 | 15.1 | 1.0 | 37,8 | 11,8 | 0,041 |
| 37 | 180 | RSN185 IMPVALL.H H-HVP315 | 6.53 | 5,4 | 0.22 | 51.5 | 15.3 | 1,0 | 37,8 | 12.8 | 0,036 |
| 38 | 278 | RSN415 COALINGA D-TSM270 | 5,77 | 3,7 | 0.78 | 47.5 | 15.5 | 1.0 | 21,7 | 4.0 | 0,283 |
| 39 | 161 | RSN173 IMPVALL.H H-E10320 | 6,53 | 8,6 | 0.23 | 46.4 | 15,6 | 1.0 | 37,0 | 12.0 | 0,082 |
| 40 | 349 | RSN668 WHITTIER.A A-NOR36 | 5.99 | 14,4 | 0.25 | 26.3 | 15,6 | 0.9 | 30,2 | 9.4 | 0,021 |
| 41 | 187 | RSN214 LIVERMOR A-KOD180 | 5,80 | 15.2 | 0,15 | 20,8 | 15,7 | 0.8 | 20,9 | 10.4 | 0,035 |
| 42 | 346 | RSN615 WHITTIER A A-DWN18 | 5,99 | 15,0 | 0,21 | 30,7 | 15,7 | 1,0 | 40,0 | 9.2 | 0,033 |

*(continued)*

**Table A-1.** (*continued*)

| N° | ID | Nombre de Sismo | Earthquake Magnitude | J-B Dist. (km) | PGA (g) | PGV (cm/s) | IP (cm-s) | IPR | Record Duration (s) | Time IA 5–95% (s) | ESI ($cm^2/s^2$) |
|---|---|---|---|---|---|---|---|---|---|---|---|
| 43 | 347 | RSN645 WHITTIER.A A-OR2010 | 5,99 | 19.8 | 0,23 | 31,5 | 15.8 | 1,0 | 32,1 | 8.0 | 0.055 |
| 44 | 217 | RSN33 PARKF TMB205 | 6,19 | 16,0 | 0,36 | 22,3 | 16,4 | 0.8 | 30,4 | 4,4 | 0.037 |
| 45 | 175 | RSN183 IMPVALL.H H-E08140 | 6,53 | 3.9 | 0,61 | 54,5 | 16,5 | 1,0 | 37,6 | 6.8 | 0,199 |
| 46 | 210 | RSN285 ITALY A-BAG270 | 6,90 | 8,1 | 0,19 | 34,7 | 16,5 | 1,0 | 36,8 | 16,1 | 0,046 |
| 47 | 105 | RSN150 COYOTELK G06320 | 5,74 | 0,4 | 0,32 | 25,4 | 16,6 | 0,9 | 27,1 | 3.5 | 0,073 |
| 48 | 182 | RSN192 IMPVALL H H-WSM180 | 6,53 | 14,8 | 0,11 | 22,6 | 16,6 | 0,8 | 40,0 | 25.6 | 0,007 |
| 49 | 327 | RSN527 PALMSPR MVH045 | 6,06 | 3,6 | 0,22 | 31,0 | 16,6 | 1,0 | 20,1 | 5.1 | 0.220 |
| 50 | 348 | RSN652 WHITTIER.A A-DEL000 | 5.99 | 22,4 | 0,30 | 32,4 | 16,6 | 1,0 | 29,7 | 11.2 | 0,041 |
| 51 | 176 | RSN183 IMPVALL. H H-E08230 | 6,53 | 3,9 | 0,47 | 52,1 | 16,7 | 1,0 | 37,6 | 5.8 | 0,112 |
| 52 | 323 | RSN496 NAHANNI S2330 | 6,76 | 0,0 | 0,36 | 32,0 | 16,7 | 1,0 | 10,0 | 7.3 | 0,056 |
| 53 | 295 | RSN459 MORGAN G06090 | 6,19 | 9,9 | 0,29 | 36,5 | 16.8 | 1,0 | 30,0 | 6.5 | 0,117 |
| 54 | 165 | RSN178 IMPVALL H H-E03230 | 6.53 | 10,8 | 0,22 | 43,3 | 17,0 | 1.0 | 39,6 | 14.1 | 0,016 |
| 55 | 322 | RSN496 NAHANNI S2240 | 6,76 | 0,0 | 0,52 | 29.6 | 17,0 | 1,0 | 10,0 | 7.2 | 0,034 |
| 56 | 91 | RSN149 COYOTELK G04360 | 5,74 | 4,8 | 0,25 | 31.9 | 17,1 | 1,0 | 27,1 | 11,0 | 0,040 |
| 57 | 188 | RSN235 MAMMOTH J J-MLS25- | 5,69 | 1,5 | 0,39 | 24,2 | 17,2 | 0.9 | 30,0 | 3.9 | 0,023 |
| 58 | 147 | RSN159 IMPVALL.H H-AGR273 | 6,53 | 0,0 | 0,19 | 41.8 | 17,6 | 1.0 | 28,4 | 12,4 | 0,081 |

(*continued*)

**Table A-1.** *(continued)*

| N° | ID | Nombre de Sismo | Earthquake Magnitude | J-B Dist. (km) | PGA (g) | PGV (cm/s) | IP (cm-s) | IPR | Record Duration (s) | Time IA 5–95% (s) | ESI ($cm^2/s^2$) |
|---|---|---|---|---|---|---|---|---|---|---|---|
| 59 | 145 | RSN158 IMPVALL.H H-AEP045 | 6,53 | 0,0 | 0,31 | 42.8 | 17.8 | 1.0 | 14,7 | 9.8 | 0,133 |
| 60 | 293 | RSN451 MORGAN CYC195 | 6.19 | 0,2 | 0,71 | 52,9 | 17,8 | 1,0 | 30,0 | 4,1 | 0.295 |
| 61 | 296 | RSN461 MORGAN HVR240 | 6.19 | 3,5 | 0,31 | 39,4 | 17,8 | 1,0 | 40,0 | 10,7 | 0,061 |
| 62 | 181 | RSN192 IMPVALL. H H-WSM090 | 6,53 | 14,8 | 0,08 | 22,2 | 18,3 | 0,7 | 40,0 | 24,7 | 0,005 |
| 63 | 342 | RSN585 BAJA CPE251 | 5.50 | 3,4 | 0,91 | 55,4 | 18,3 | 1,0 | 40,0 | 4,2 | 0,431 |
| 64 | 400 | RSN77 SFERN PUL164 | 6,61 | 0,0 | 1,22 | 114,5 | 18,4 | 1,0 | 41,7 | 7,0 | 0,794 |
| 65 | 156 | RSN170 IMPVALL.H H-ECC002 | 6,53 | 7,3 | 0,21 | 38,4 | 18,8 | 1,0 | 40,0 | 10,4 | 0,106 |
| 66 | 164 | RSN178 IMPVALL. H H-E03140 | 6,53 | 10,8 | 0,27 | 48,0 | 18,9 | 1,0 | 39,6 | 11.9 | 0,078 |
| 67 | 344 | RSN611 WHITTIER.A A-CAS000 | 5.99 | 18,3 | 0,32 | 29,5 | 19,1 | 0,9 | 31,1 | 8.0 | 0,048 |
| 68 | 383 | RSN723 SUPER.B B-PTS315 | 6,54 | 1,0 | 0,38 | 53,2 | 19,1 | 1,0 | 22,3 | 11.0 | 0,117 |
| 69 | 330 | RSN540 PALMSPR WWT180 | 6,06 | 0,0 | 0,48 | 38,5 | 19.2 | 1,0 | 20,1 | 5.5 | 0,092 |
| 70 | 183 | RSN20 NCALIF.FH H-FRN044 | 6.50 | 26,7 | 0,16 | 36,1 | 19,3 | 1,0 | 40,0 | 17,3 | 0,072 |
| 71 | 390 | RSN763 LOMAP GIL067 | 6,93 | 9,2 | 0,36 | 31,1 | 19,3 | 1,0 | 40,0 | 5.0 | 0,065 |
| 72 | 207 | RSN266 VICT CHI102 | 6,33 | 18,5 | 0,15 | 26,0 | 19,4 | 0,8 | 26,9 | 16,4 | 0,027 |
| 73 | 169 | RSN180 IMPVALL.H H-E05140 | 6,53 | 1,8 | 0,53 | 48,9 | 19,9 | 1,0 | 39,3 | 8,3 | 0,141 |
| 74 | 361 | RSN692 WHITTIER A A-EJS048 | 5.99 | 11,5 | 0,47 | 34,4 | 20,7 | 1,0 | 37.8 | 5.8 | 0.079 |

*(continued)*

**Table A-1.** (*continued*)

| N° | ID | Nombre de Sismo | Earthquake Magnitude | J-B Dist. (km) | PGA (g) | PGV (cm/s) | IP (cm-s) | IPR | Record Duration (s) | Time IA 5–95% (s) | ESI ($cm^2/s^2$) |
|---|---|---|---|---|---|---|---|---|---|---|---|
| 75 | 286 | RSN448 MORGAN AND250 | 6.19 | 3,2 | 0,42 | 25,4 | 20,8 | 0,8 | 28,4 | 6,9 | 0,038 |
| 76 | 214 | RSN316 WESMORL PTS315 | 5,90 | 16,5 | 0,15 | 32,7 | 20,9 | 1,0 | 41,7 | 18,7 | 0,024 |
| 77 | 341 | RSN585 BAJA CPE161 | 5.50 | 3,4 | 1,28 | 46,4 | 21,0 | 1,0 | 40,0 | 3,2 | 0 ,358 |
| 78 | 184 | RSN20 NCALIF.FH H-FRN314 | 6.50 | 26,7 | 0,20 | 26,2 | 21,3 | 0.8 | 40,0 | 19.4 | 0,036 |
| 79 | 167 | RSN179 IMPVALL.H H-E04140 | 6.53 | 4,9 | 0,48 | 39,7 | 21.3 | 1,0 | 39,1 | 6.7 | 0,200 |
| 80 | 196 | RSN250 MAMMOTH.L L-LUL09 | 5.94 | 9,7 | 0,41 | 34,1 | 21,3 | 1,0 | 26.0 | 7.1 | 0,115 |
| 81 | 223 | RSN359 COALINGA.H H-PV109 | 6,36 | 24,8 | 0.23 | 27,5 | 21.3 | 0,8 | 60.0 | 10.9 | 0,051 |
| 82 | 68 | RSN126 GAZLI GAZ000 | 6.80 | 3,9 | 0,70 | 66,3 | 21,4 | 1,0 | 13,5 | 6,4 | 0,360 |
| 83 | 287 | RSN448 MORGAN AND340 | 6.19 | 3,2 | 0.29 | 27.8 | 21,4 | 0.8 | 28,4 | 5.2 | 0,106 |
| 84 | 69 | RSN126 GAZLI GAZ090 | 6.80 | 3,9 | 0,86 | 67,7 | 21,7 | 1,0 | 13,5 | 7.0 | 0,286 |
| 85 | 343 | RSN595 WHITTIER A A-JAB297 | 5,99 | 10,3 | 0.22 | 28.3 | 21,7 | 0.8 | 34,3 | 10,4 | 0,025 |
| 86 | 244 | RSN407 COALINGA D-OLC270 | 5,77 | 2,0 | 0,84 | 40,0 | 22,1 | 1,0 | 21,2 | 2.8 | 0,283 |
| 87 | 362 | RSN692 WHITTIER A A-EJS318 | 5,99 | 11,5 | 0,46 | 31,6 | 22.1 | 0,9 | 37,8 | 6.0 | 0.044 |
| 88 | 90 | RSN149 COYOTELK G04270 | 5.74 | 4.8 | 0,23 | 25,8 | 22,2 | 0,7 | 27,1 | 8,2 | 0,047 |
| 89 | 67 | RSN125 FRIULLA A-TMZ270 | 6,50 | 15,0 | 0.32 | 30,5 | 22,4 | 0,9 | 36,4 | 4,9 | 0,121 |

(*continued*)

**Table A-1.** *(continued)*

| N° | ID | Nombre de Sismo | Earthquake Magnitude | J-B Dist. (km) | PGA (g) | PGV (cm/s) | IP (cm-s) | IPR | Record Duration (s) | Time IA 5–95% (s) | ESI (cm²/s²) |
|---|---|---|---|---|---|---|---|---|---|---|---|
| 90 | 162 | RSN174 IMPVALL.H H-E11230 | 6,53 | 12,6 | 0.38 | 44,6 | 22,4 | 1,0 | 39,4 | 7,9 | 0,083 |
| 91 | 387 | RSN753 LOMAP CLS000 | 6,93 | 0.2 | 0.65 | 56,0 | 22,5 | 1.0 | 40,0 | 6,9 | 0,233 |
| 92 | 333 | RSN558 CHALFANT. A A-ZAK36 | 6,19 | 6,4 | 0.40 | 44,7 | 22,7 | 1,0 | 40,0 | 8,1 | 0,183 |
| 93 | 386 | RSN737 LOMAP AGW000 | 6,93 | 24,3 | 0,17 | 33,5 | 22,7 | 0,9 | 60,0 | 21,2 | 0,016 |
| 94 | 320 | RSN495 NAHANNI S1010 | 6,76 | 2,5 | 1,11 | 43,9 | 23,1 | 1,0 | 10,3 | 7,5 | 0,099 |
| 95 | 215 | RSN319 WESMORL WSM090 | 5,90 | 6,2 | 0,38 | 44,2 | 23,2 | 1,0 | 65,0 | 6,9 | 0,210 |
| 96 | 146 | RSN159 IMPVALL.H H-AGR003 | 6,53 | 0,0 | 0.29 | 35,3 | 23,4 | 0.9 | 28,4 | 13,3 | 0,057 |
| 97 | 379 | RSN721 SUPER.B B-ICC000 | 6,54 | 18.2 | 0,36 | 48,1 | 23,6 | 1,0 | 60,0 | 28,0 | 0,033 |
| 98 | 195 | RSN250 MAMMOTH.L L-LUL00 | 5,94 | 9.7 | 0,95 | 30,3 | 24,3 | 0.8 | 26,0 | 6,5 | 0,082 |
| 99 | 178 | RSN184 IMPVALL.H H-EDA360 | 6,53 | 5,1 | 0,48 | 41,0 | 24,8 | 1,0 | 39,1 | 6,6 | 0,238 |
| 100 | 388 | RSN753 LOMAP CLS090 | 6.93 | 0.2 | 0,48 | 47,6 | 24.8 | 1.0 | 40,0 | 7,9 | 0,239 |
| 101 | 380 | RSN721 SUPER.B B-ICC090 | 6,54 | 18.2 | 0,26 | 41.8 | 25,0 | 1.0 | 60,0 | 35,7 | 0,032 |
| 102 | 71 | RSN143 TABAS TAB-T1 | 7,35 | 1,8 | 0.86 | 123,6 | 25,1 | 1,0 | 32,8 | 16,3 | 0,351 |
| 103 | 321 | RSN495 NAHANNI S1280 | 6,76 | 2.5 | 1,20 | 40,6 | 25,1 | 1,0 | 10,3 | 7,3 | 0,145 |
| 104 | 389 | RSN755 LOMAP CYC285 | 6,93 | 20,0 | 0,49 | 40,6 | 25,2 | 1.0 | 40,0 | 12,2 | 0,092 |

*(continued)*

**Table A-1.** (*continued*)

| N° | ID | Nombre de Sismo | Earthquake Magnitude | J-B Dist. (km) | PGA (g) | PGV (cm/s) | IP (cm-s) | IPR | Record Duration (s) | Time IA 5–95% (s) | ESI ($cm^2/s^2$) |
|---|---|---|---|---|---|---|---|---|---|---|---|
| 105 | 331 | RSN540 PALMSPR WWT270 | 6,06 | 0,0 | 0.63 | 30,8 | 25.3 | 0,7 | 20,1 | 3.4 | 0,127 |
| 106 | 332 | RSN558 CHALFANT.A A-ZAK27 | 6,19 | 6,4 | 0,45 | 36,9 | 25,4 | 0.9 | 40,0 | 6,2 | 0,166 |
| 107 | 224 | RSN367 COALINGA.H H-PVB04 | 6,36 | 7,7 | 0.30 | 39,4 | 26,0 | 0.9 | 58,1 | 8,3 | 0,144 |
| 108 | 225 | RSN368 COALINGA.H H-PVY04 | 6,36 | 7,7 | 0.60 | 60,5 | 26,8 | 1,0 | 58,1 | 8,2 | 0,332 |
| 109 | 211 | RSN292 ITALY A-STU000 | 6.90 | 6.8 | 0,23 | 37,0 | 27.9 | 0.8 | 39.3 | 15,0 | 0,064 |
| 110 | 216 | RSN319 WESMORL WSM180 | 5,90 | 6.2 | 0,50 | 35,8 | 28,4 | 0,7 | 65,0 | 6.1 | 0,176 |
| 111 | 384 | RSN725 SUPER.B B-POE270 | 6,54 | 11,2 | 0,48 | 41.3 | 29.2 | 0.8 | 22,3 | 13,7 | 0,078 |
| 112 | 70 | RSN143 TABAS TAB-L1 | 7,35 | 1.8 | 0,85 | 99,1 | 31,5 | 1,0 | 32.8 | 16,5 | 0,412 |
| 113 | 148 | RSN160 IMPVALL.H H-BCR140 | 6,53 | 0,4 | 0.60 | 46.8 | 31.9 | 0.8 | 37,8 | 9.6 | 0,210 |
| 114 | 401 | RSN77 SFERN PUL254 | 6,61 | 0.0 | 1,24 | 57.4 | 34.9 | 0,7 | 41.7 | 7,3 | 0,410 |

## References

Uang, C.M., Bertero, V.V.: Evaluation of seismic energy in structures. Earthq. Eng. Struct. Dyn. **19**(1), 77–90 (1990)

Akbas, B., Shen, J.: Earthquake-resistant design (EQRD) and energy concepts. Tech. J. Turk. Chamb. Civ. Eng., 2877–2901, Article no. 192 (2003). (in Turkish)

Housner, G.W.: Limit design of structures to resist earthquakes. In: Proceedings of the First World Conference on Earthquake Engineering, Oakland, California, USA, pp. 186–198 (1956)

Fajfar, P., Fischinger, M.A.: Seismic procedure including energy concept. In: Ninth European Conference on Earthquake Engineering (ECEE), Moscow, vol. 2, pp. 312–321 (1990)

Manfredi, G.: Evaluation of seismic energy demand. Earthq. Eng. Struct. Dyn. **30**(4), 485–499 (2001)

Zahrah, T.F., Hall, W.J.: Earthquake energy absorption in SDOF structures. J. Struct. Eng. **110**(8), 1757–1772 (1984)

Akiyama, H.: Earthquake-Resistant Limit-State Design for Buildings. University of Tokyo Press, Japan (1985)

Kuwamura, H., Galambos, T.V.: Earthquake load for structural reliability. J. Struct. Eng. **115**(6), 1446–1462 (1989)

Fajfar, P., Vidic, T., Fischinger, M.: Seismic demand in medium- and long-period structures. Earthq. Eng. Struct. Dyn. **18**(8), 1133–1144 (1989)

Somerville, P.G., Smith, N.F., Graves, R.W., Abrahamson, N.A.: Modification of empirical strong ground motion attenuation relations to include the amplitude and duration effects of rupture directivity. Seismol. Res. Lett. **68**,199–222 (1997)

Spudich, P., Chiou, B.S.J.: Directivity in NGA earthquake ground motions: analysis using isochrone theory. Earthq. Spectra **24**(1), 279–98 (2008)

Baez, J.I., Miranda, E.: Amplification factors to estimate inelastic displacement demands for the design of structures in the near field. In: 12th World Conference in Earthquake Engineering, New Zealand, Paper 1561 (2000)

Bertero, V.V., Mahin, S.A., Herrera, R.A.: Aseismic design implications of San Fernando earthquake records. Earthq. Eng. Struct. Dyn. **6**(1), 31–42 (1978)

Chopra, A.K., Chintanapakdee, C.: Comparing response of SDF systems to near- fault and far-fault earthquake motions in the context of spectral regions. Earthq. Eng. Struct. Dyn. **30**, 1769–1789 (2001)

Anderson, J.C., Bertero, V.V.: Uncertainties in establishing design earthquakes. J. Struct. Eng. **113**(8), 1709–1724 (1987)

Malhotra, P.K.: Response of building to near-field pulse-like ground motions. Earthq. Eng. Struct. Dyn. **28**, 1309–1326 (1999)

Alavi, B., Krawinkler, H.: Effects of near-fault ground motions on frame structures. Technical report Blume Center Report 138, Stanford, California (2001)

Panella, D.S., Tornello, M.E., Frau, C.D.: A simple and intuitive procedure to identify pulse-like ground motions. Soil Dyn. Earthq. Eng. **94**, 234–243 (2017)

Arias, A.: A measure of earthquake intensity. In: Hansen, R. (ed.) Seismic Design for Nuclear Power Plants, pp. 438–483. MIT Press, Cambridge (1970)

Housner, G.W.: Intensity of ground motion during strong earthquakes. Earthquake Research Laboratory. California Institute of Technology (1952)

# Correlation Between Seismic Energy Demand and Damage Potential Under Pulse-Like Ground Motions

G. Angelucci[1](✉), G. Quaranta[2], F. Mollaioli[1], and S. K. Kunnath[3]

[1] Sapienza University of Rome, 00197 Rome, Italy
giulia.angelucci@uniroma1.it
[2] Sapienza University of Rome, 00184 Rome, Italy
[3] University of California, Davis, CA 95616, USA

**Abstract.** An energy-based perspective can provide a more rational way for quantifying the damage potential in structures subjected to strong ground shakings as compared to traditional displacement-based methods. This is basically due to the possibility of accounting for maximum deformation and cumulative damage simultaneously. Unlike typical far-field ground motions, where the input energy is progressively accumulated over time, near-fault records often impart sudden and intense energy demands to the structural systems by means of distinctive velocity pulses. These abrupt energy spikes force the structures to dissipate the input energy through few large plastic cycles, thereby imposing high values of the deformation demand within a short time window. Since most of the structural damage in this case is related to the seismic demands during this short duration, viscous and hysteretic energies accumulated at the end of the ground motion mainly depend on the contribution of a few significative peaks occurring in the early phase of the response. This paper thus introduces new energy-based intensity measures for pulse-like seismic ground motions. Consistent correlations are presented between the proposed intensity measures, the ground motion characteristics, and a representative engineering demand parameter (i.e., the maximum inter-storey drift). Towards this objective, an extensive database of nonlinear dynamic analyses is compiled for a set of sixty near-fault records to simulate the seismic response of reinforced concrete structures representative of the Italian residential buildings portfolio ranging from two to ten storeys in bare frame configurations. Machine learning techniques are finally implemented to find general, yet accurate and robust, correlations between a suite of seismic intensity measures and demand potential.

**Keywords:** Energy-based seismic engineering · Hysteretic energy · Machine learning · Near-Fault seismic ground motion · Seismic Demand

## 1 Introduction

Within the framework of Performance-Based Earthquake Engineering (PBEE), a crucial aspect in the risk analysis of structures at a specified site is the evaluation of the probabilities of a structure to exceed a given limit state, usually expressed in terms of an


H. Varum et al. (Eds.): IWEBSE 2023, LNCE 236, pp. 234–249, 2023.
https://doi.org/10.1007/978-3-031-36562-1_18

Engineering Demand Parameter (EDP), conditioned to different values of a designated ground motion Intensity Measure (IM). As the accuracy of the assessment analysis is closely affected by the correlations between the selected IM and the predicted EDP, the selection of suitable parameters to quantify the severity of a seismic event is evidently a key issue to be addressed.

Previous research efforts have been devoted to evaluate the predictive capacity of IMs available in the literature [1–4]. It was observed that the efficiency of these intensity measures generally depends on the particular structural system as well as the specific response parameter of interest. For instance, spectral acceleration and/or spectral displacement at the fundamental period of the structure may be preferred in the case of regular buildings where most of the mass participates in the first mode. However, their predictive capabilities for irregular or tall buildings might be arguable as they cannot adequately account for period elongation due to the excursion of the structure into the inelastic range or the influence of higher modes.

Accordingly, alternative scalar intensity measures [5–7], spectrum intensity measures [8, 9] and vector intensity measures [10, 11] have been proposed to explicitly consider such aspects and improve the correlations with respect to the predicted EDPs. Among them, energy-based IMs are found to be better predictors both for ordinary and pulse-like ground motions given the well-correlation with the main characteristics of the ground motion (amplitude, frequency content and duration) as well as the properties of the structure [12, 13]. Moreover, energy-based methodologies provide more rational indicators for quantifying the expected damage caused by severe ground shakings as compared to traditional displacement-based methods, given the possibility of accounting for maximum deformation and cumulative effects of repeated cycles of inelastic response simultaneously.

It is now well recognized that, in typical far-field ground motions, input energy is progressively accumulated over time and the damage potential depends not only on the maximum deformation but also the contribution of low-cycle fatigue effects [14, 15]. In contrast, near-fault records often impart sudden and intense energy demands to the structural systems by means of distinctive velocity pulses whereby most of the damage is a result of instantaneous energy demand associated with intense pulse effects and few plastic cycles [16, 17]. Most of the structural damage in this latter case is directly related to the seismic demands conveyed in a short time window; thus, dissipated viscous and hysteretic energy accumulated at the end of the ground motion record mainly depends on the contribution of a few significative peaks occurring in this early phase of the response.

By recognizing that the zero-crossing points in the velocity time response of an inelastic SDOF oscillator directly refer to the absolute peak displacements of each hysteresis cycle, Kalkan and Kunnath [18] proposed the Effective Cyclic Energy (ECE) parameter, which is defined as the peak-to-peak dissipated energy demand imposed to the structure over the ground motion duration.

With the aim of estimating the expected damage related to ground motion features, this paper proposes new energy-based parameters based on the summation of effective peaks of the seismic energy demand imparted by pulse-like seismic records. The effectiveness of the new parameters in predicting the seismic demand of reinforced concrete multi-story buildings is evaluated using an extensive set of actual impulsive near-fault records and their predictive capability is compared to that of other intensity measures already available in the literature.

Consistent correlations are presented between the proposed intensity measures, the ground motion characteristics, and the peak deformation of the selected structures considering different reinforced concrete structures ranging from two to ten storeys.

## 2 Methodology

### 2.1 Near-Fault Pulse-Like Database

It is nowadays well aware that rupture directivity effects occurring in sites close to causative faults typically trigger sudden, large amounts of input energy in a relatively short time. The destructive potential associated with such seismic events is mainly due to intense embedded long period velocity pulses.

In order to estimate the expected structural damage attributable to near-fault ground motion characteristics, a set of sixty records was suitably extracted from a larger database. The pulse features are identified and processed by using the identification method based on the Variational Mode Decomposition (VMD) technique proposed by Quaranta and Mollaioli [19].

The selected ground motions were recorded from earthquakes having a magnitude range from 5.0 to 7.5, soil preferred shear-velocities between 186 m/s and 1000 m/s, epicentral distances varying from 4.4 to 98.2 km and closest site-to-source distances between 0.07 and 35.7 km. The full set of values of the most relevant seismological parameters and related variability within the considered dataset is provided in Fig. 1.

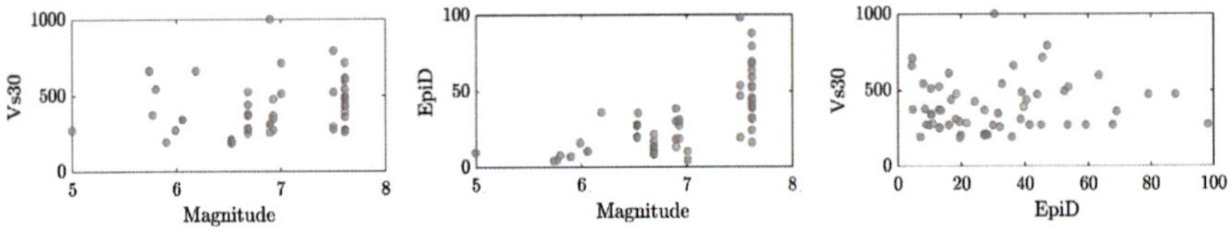


**Fig. 1.** Variability of magnitude, soil preferred shear-velocities ($V_{S30}$) and epicentral distances (*EpiD*) within the considered dataset.

### 2.2 Numerical Models

In order to simulate the seismic response of a variety of reinforced concrete structural typologies representative of the Italian buildings portfolio, multi-storey reinforced concrete frames of different heights are considered (namely, 2, 4, 6, 8 and 10 stories).

The structural systems are modelled as two-dimensional, two-bay generic frames with constant storey height of 3.2 m and beam span of 5.0 m. The fundamental periods of the first two vibration modes are provided in Table 1.

**Table 1.** Periods of vibration of the analysed reinforced concrete frames.

| RC Frame | $T_1$ (s) | $T_2$ (s) |
|---|---|---|
| 2-storey | 0.575 | 0.234 |
| 4-storey | 0.954 | 0.358 |
| 6-storey | 1.316 | 0.477 |
| 8-storey | 1.613 | 0.580 |
| 10-storey | 1.951 | 0.696 |

For the sake of generalizing the findings of this study, the bare structural frames are not intended to reproduce actual specific structures. Instead, simplified Equivalent Shear-Type Models (ESTMs) are employed to estimate the dynamic responses over the near-fault ground motions.

Each bare frame was designed in such a way to have three different resistance capacities expressed through seismic coefficients $C_y$ (i.e., the ratio between the maximum base shear and the weight of the building) equal to 0.08, 0.15 and 0.25, respectively. The lowest $C_y$ value is representative of weaker buildings designed for gravity loads only and with materials having poor mechanical properties. Higher values of $C_y$ refer to code-compliant structures designed for seismic loads. This is in line with the scope of this work to simulate a wide range of possible structural types, characterized by different levels of stiffness and strength degradation, and different pinching behaviours. Accordingly, different hysteretic models are employed, so that the sensitivity of results with respect to different mechanical properties can be promptly assessed.

A generalized hysteretic behaviour is illustrated in Fig. 2, whose cyclic response is governed by four control parameters: $p$, $\alpha$, $\beta$, $\gamma$ and $\rho$. The strain hardening ratio $p$ controls the post-yielding stiffness (i.e., when $p$ is equal to 0 the model describes the behaviour of a perfectly-plastic system). The parameter $\alpha$ represents the unloading stiffness ratio and it is larger than 20 if unloading degradation is negligible. The strength degradation is ruled by $\beta$: if it is equal to 0, then no strength degradation occurs. The pinching parameter $\gamma$ ranges between 0 and 1 (where 1 is the value that corresponds to a cyclic behaviour with no pinching). For further details, including description of the set of cyclic rules adopted to define the cyclic behaviour, the reader is referred to Mollaioli and Bruno [20].

Table 2 reports the values of the model parameters that adjust the inelastic loading reversals of the three analysed systems. Specifically, each multi-story structure is modelled considering three different cyclic behaviours, thus generating fifteen reinforced concrete building models subjected to sixty impulsive near-fault records.

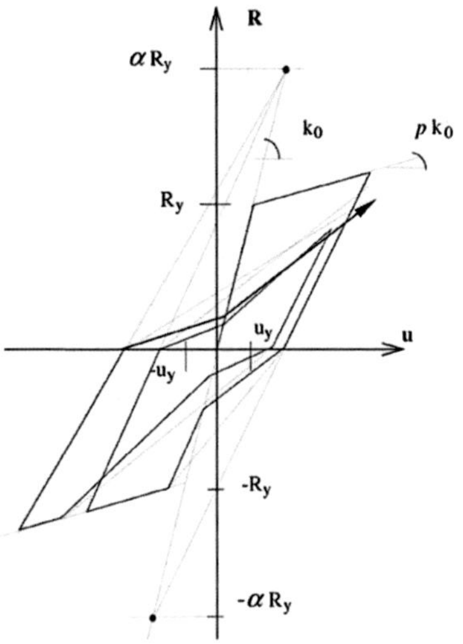


**Fig. 2.** Typical cyclic behaviour of the four-parameter hysteretic model.

The inter-storey drift index IDI (i.e., relative displacement between two consecutive stories normalized by the story height) is selected as a representative response parameter for the equivalent shear-type models since it is well-correlated with the local deformation demand at each story. Owing to the fact that the expected structural damage is associated with the maximum value of the peak inter-storey drift over all stories, the $IDI_{max}$ is here accepted as relevant engineering demand parameter.

**Table 2.** Model parameters controlling the cyclic behaviour of the three analysed systems.

| $C_y$ | $p$ | $\alpha$ | $\beta$ | $\gamma$ | $\rho$ |
|---|---|---|---|---|---|
| 0.08 | 0.01 | 2.0 | 0.1 | 0.6 | 0.8 |
| 0.15 | 0.05 | 2.0 | 0.0 | 0.8 | 0.8 |
| 0.25 | 0.05 | 2.0 | 0.0 | 0.8 | 0.8 |

## 2.3 Intensity Measures

Among the intensity measures currently available in the literature, those under investigation in this study are listed in Table 3. IMs for which Ground Motion Prediction Equations (GMPEs) exist are marked with an asterisk. The listed parameters include non-structure-specific IMs (i.e., calculated directly from the ground motion record) and structure-specific IMs (i.e., obtained from the response spectra at the fundamental period of the structural system). They can be further classified into acceleration-, velocity- and displacement-related IMs.

Most common non-structure-specific intensity measures are the time-domain parameters of the strong ground motion, such as PGA (peak ground acceleration), PGV (peak ground velocity), PGD (peak ground displacement), and $I_A$ (Arias intensity).

Structure-specific intensity measures include the spectral quantities $S_{pa}$ (5% damped pseudo acceleration spectral value at specified period), $S_{pv}$ (5% damped velocity spectral value at specified period), $C_y$ (5% damped seismic coefficient spectral value at specified period), $u_{max}$ (5% damped maximum displacement spectral value at specified period), $E_{Ir}$ (5% damped relative input energy spectral value at specified period). The average spectral acceleration, $S_{a,avg}$ is a complex scalar IM defined as the geometric mean of the log-spectral accelerations in the range of periods of interest 0.2 T–2 T, where T is the fundamental period of the structure [5, 6].

In addition, integral intensity measures are evaluated. These include $V_{EIr}SI$ (relative input equivalent velocity spectrum intensity) and $V_{ED}SI$ (damping equivalent velocity spectrum intensity) calculated through integration of the energy response spectra in the period range 0.1–3.0 s [21].

**Table 3.** Considered non-structure and structure-specific intensity measures available in the current literature.

| Notation | Definition |
|---|---|
| *Acceleration-related* | |
| PGA* | Peak ground acceleration |
| $I_A$* | Arias intensity |
| $C_y$ ($T_1$)* | Spectral seismic coefficient at vibration period |
| $S_{a,avg}$ | Average spectral acceleration |
| $S_{pa}$ ($T_1$) | Spectral acceleration at vibration period |
| *Velocity-related* | |
| PGV* | Peak ground velocity |
| $I_H$* | Housner intensity |
| $S_{pv}$ ($T_1$) | Spectral velocity at vibration period |
| $E_{Ir}$ ($T_1$) | Relative input energy at vibration period |
| $V_{EIr}SI$* | Relative input equivalent velocity spectrum intensity |
| $V_{ED}SI$ | Damping equivalent velocity spectrum intensity |
| $MV_{EIr}SI$ | Modified $V_{EIr}SI$ |
| $MV_{ED}SI$* | Modified $V_{ED}SI$ |
| $ECE_{MOD}$ | Modified Effective Cyclic Energy |
| *Displacement-related* | |
| PGD* | Peak ground displacement |
| $u_{max}$ ($T_1$) | Spectral maximum displacement at vibration period |

* Available GMPEs.

Modified versions of these IMs (i.e., $MV_{EIr}SI$ and $MV_{ED}SI$) are also considered by changing the period range of integration into 0.2 T–2 T. This array of periods might cover higher mode response and also the structural period elongation caused by the nonlinear behaviour due to the accumulation of damage [3, 4]. Integration of the pseudo-velocity spectrum in the period range of 0.1–2.5 s is used for the Housner Intensity ($I_H$).

Finally, a modified version of the effective cyclic energy ($ECE_{MOD}$) proposed by Kalkan and Kunnath [18] is considered in this work to obtain an intensity measure based on ground motion attributes only. Specifically, the peak value between two zero-crossing points in the velocity time-history is assumed instead of those in the velocity time-response. Therefore, the intensity measure is related solely to reversals of the velocigram while the effective system response is not explicitly involved.

## 2.4 Proposed Energy-Based IMs

This work aims to identify new intensity measures by expressing the energy demand imparted to the structure as a function of the ground motion characteristics only, while being independent of the specific features of the earthquake resisting system. Towards this objective, methods based on the assessment of the dynamic response of equivalent single-degree-of-freedom (ESDOF) systems are deemed to be sufficiently accurate to evaluate global energy demands. This conveniently expedites the complex computation of hysteretic and damping energies for MDOF systems.

In agreement with past research studies, in fact, the hysteretic energy component of multi-story frames can be easily predicted using ESDOF oscillators derived from the modal properties of MDOF models [22, 23]. However, it should be emphasized here that, since ESDOF systems neglect significant effects of the seismic response of multi-storey frames due to the contribution of higher modes, the energy demands between the two numerical models tend to drastically differ as the number of stories increases [24].

A systematic evaluation of inelastic SDOF systems equivalent to the multi-story frames presented in Sect. 2.2 is performed.

Since damage is directly related to the energy dissipated by the structural components, the combined effects of damping energy $E_D$ and hysteretic energy $E_H$ is a good indicator of the damage level attained by the structure through plastic excursions.

As illustrated in Fig. 3 for a general near-fault pulse-like earthquake, the accumulation of dissipated energy (i.e., $E_H + E_D$) over the ground motion time window clearly shows an abrupt variation of the seismic demand. This trend is typical for impulsive seismic records where distinctive velocity pulses impart sudden and intense energy demands to the structural systems, thereby imposing high values of the deformation demand within a relatively short time window.

It is now well recognized that, for ordinary far-field ground motions, the input energy is progressively accumulated over time so that the structural damage depends not only on the maximum deformation but also on the contribution of low-cycle fatigue effects. Conversely, for near-fault records, most of the structural damage is related to the seismic demands imparted in a short duration. Viscous and hysteretic energies accumulated at the end of the ground motion mainly depend on the contribution of a few significative peaks occurring in this phase of the response.

However, regardless of whether the seismic ground motion is far-field or near-field, an energy-based standpoint can provide a more rational way for quantifying the damage potential in structures as compared to traditional displacement-based methods. Owing to this fact, a new energy-based measure is introduced which is defined as the summation of effective steps of the seismic energy demand $\Delta E_{H+D}$ (i.e., the coloured bar on the right of Fig. 3).

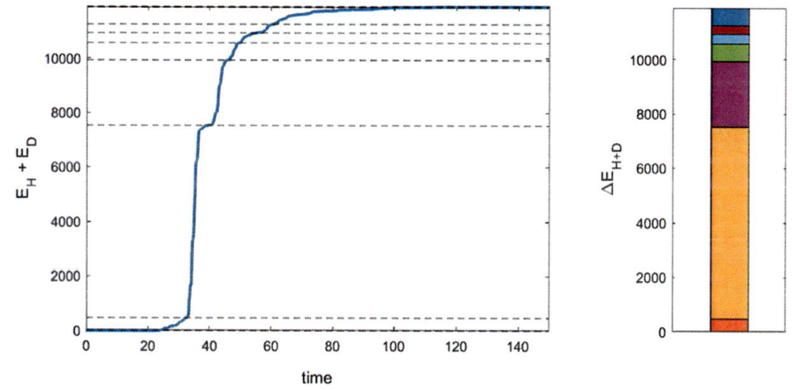


**Fig. 3.** Time variation of dissipated energy (sum of damping energy and hysteretic energy) and identification of seismic energy demand peaks.

It can be inferred that the peak structural response is associated with a few effective hysteretic cycles, at least only one, which dictate a sudden excursion of the structure in the inelastic range. Thus, the maximum deformation (e.g., maximum inter-storey drift) of the MDOF system is likely to be well-correlated with the contribution of a reduced number of energy peaks. This intuition is in line with the findings provided by Decanini et al. [25], who first recognized the correlation between the spectral shapes of input energy and inter-story drift index, as IDI spectral values are more influenced by the energy content of the earthquake than the displacement spectra. Consistently with the definition of dissipated energy jumps, additional seismic energy demand measures are identified considering hysteretic energy only $\Delta E_H$ or relative input energy $\Delta E_{Ir}$.

For a comprehensive characterization of the structural damage, it is useful to also take into account the rate at which energy is supplied and dissipated in the structure, in addition to the energy dissipation parameters [26]. Therefore, the spikes of the power time history are calculated through derivation of the energy measures of interest. In this study, the following power peaks are considered: peaks of input power ($PPE_{Ir}$), peaks of power dissipated by the structure through damping and inelastic deformations ($PPE_{H+D}$) and peaks of power dissipated by the structure through inelastic deformations only ($PPE_H$). For the sake of conciseness, a complete description of the energy- related and power-related intensity measures proposed in this study is given in Table 4.

**Table 4.** Energy- and power-related intensity measures proposed in this study.

| Notation | Definition |
|---|---|
| *Energy-related* | |
| $\Delta E_{Ir}$ | Effective peaks of input energy |
| $\Delta E_{H+D}$ | Effective peaks of dissipated energy |
| $\Delta E_{H}$ | Effective peaks of hysteretic energy |
| *Power-related* | |
| $PPE_{Ir}$ | Peaks of input power |
| $PPE_{H+D}$ | Peaks of power dissipated |
| $PPE_{H}$ | Peaks of power dissipated via inelastic deformations |

## 3 IMs Correlation

The predictive capacity of existing IMs listed in Table 3 and those proposed in Table 4 are investigated by analysing the response of multi-storey RC frame structures characterized by different fundamental periods of vibration and three values of seismic coefficient $C_y$.

For the purpose of applying machine learning techniques, a training database of nonlinear dynamic analyses is compiled using the set of near-fault pulse-like records. The investigated IMs are correlated with the peak IDI of the equivalent shear-type models. For the sake of brevity, only a representative selection of intensity parameters is provided from Fig. 4 to Fig. 12.

Overall, the predictive capacity of all considered IMs is shown to be design-dependent and a comprehensive trend cannot be established. The largest dispersion is observed when the seismic coefficient is equal to 0.08 while the predictions provide the minimum scattering as the value of $C_y$ increases. This is practically due to the poor seismic performance of existing buildings designed according to very low base shear values.

The results in Fig. 4, Fig. 5 and Fig. 6 confirm, on average, that conventional IMs, such as PGA, PGV and $S_{pa}$ ($T_1$), do not appear to exhibit satisfactory correlation with the selected demand parameter. Instead, the predictive capability increases in case of the Housner intensity ($I_H$), $S_{pv}$ ($T_1$) and relative input equivalent velocity spectrum intensity ($V_{EIr}$SI), as shown in Fig. 7, Fig. 8 and Fig. 9, respectively. The improved correlation of $I_H$ and $V_{EIr}$SI compared to traditional spectral parameters is to be found in the opportunity to implicitly capture the effects of inelasticity as well as ground motion duration. As previously observed by Mollaioli et al. [12], $V_{EIr}$SI is an energy-based ground motion parameter directly related to the number and amplitudes of the response cycles, thus reflecting cumulative effects by virtue of the integration over time involved in their computation.

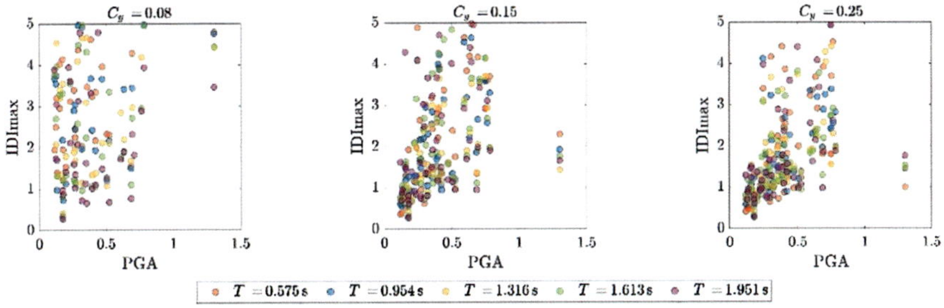


**Fig. 4.** Correlation of peak ground acceleration with the peak IDI.

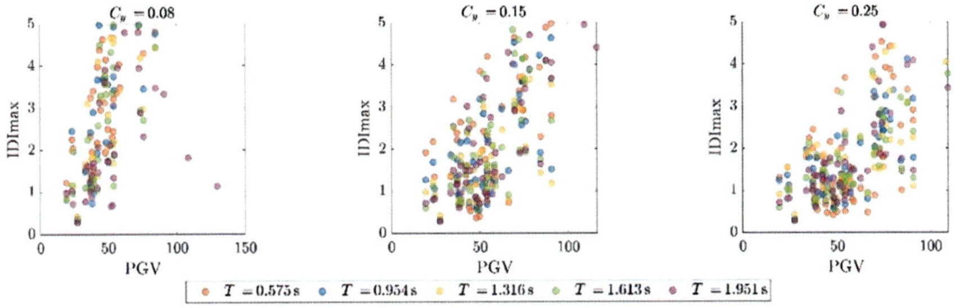


**Fig. 5.** Correlation of peak ground velocity with the peak IDI.

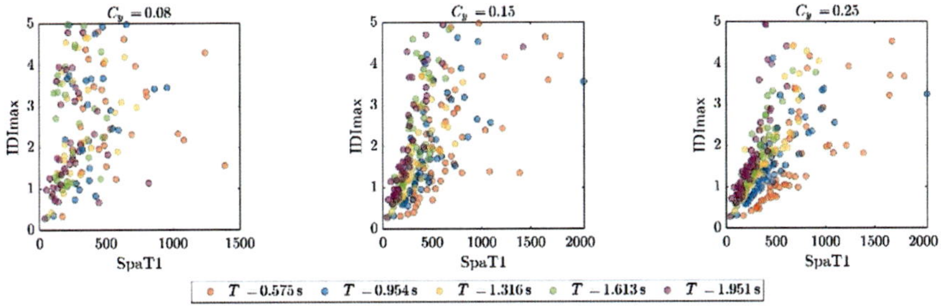


**Fig. 6.** Correlation of spectral acceleration at vibration period with the peak IDI.

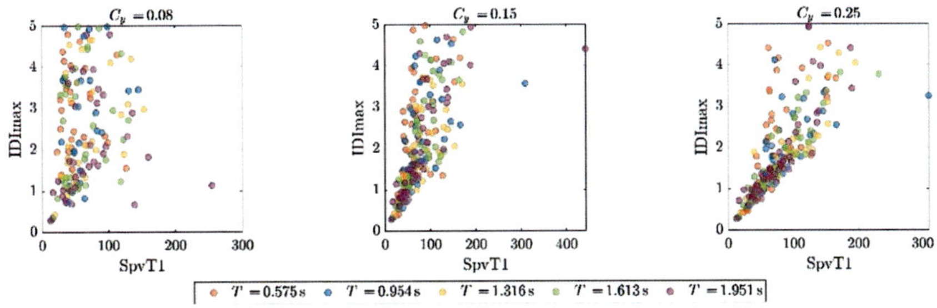


**Fig. 7.** Correlation of spectral velocity at vibration period with the peak IDI.

On the other hand, the modified version of the ECE intensity measure evaluated in the velocity time-history provides superior correlation to the peak displacement demand as shown in Fig. 10. This result is certainly attributable to the fact the dominant pulse-like and, often high-amplitude, component of near-fault records is especially evident in the fault-normal velocity waveform.

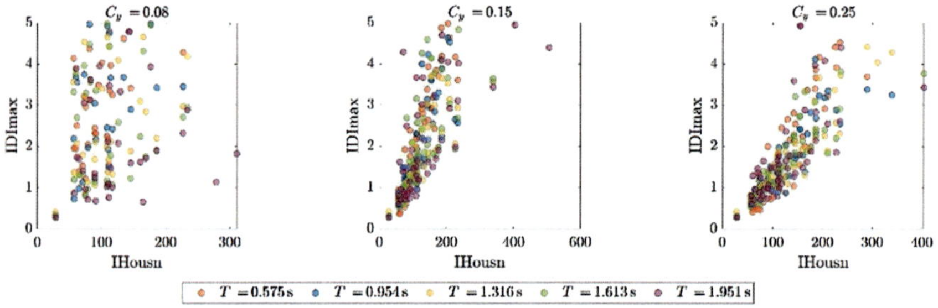


**Fig. 8.** Correlation of Housner intensity with the peak IDI.

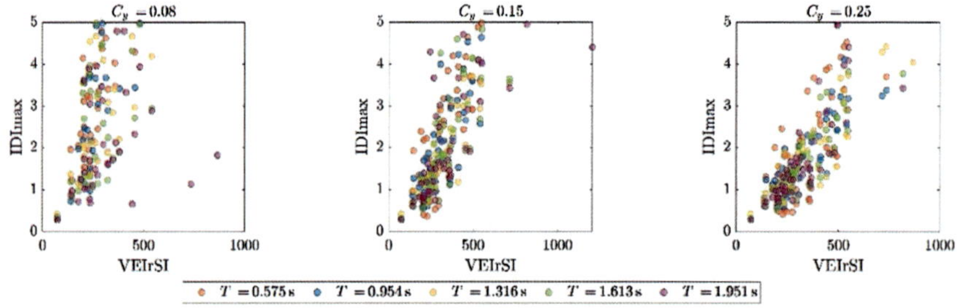


**Fig. 9.** Correlation of relative input equivalent velocity spectrum intensity with the peak IDI.

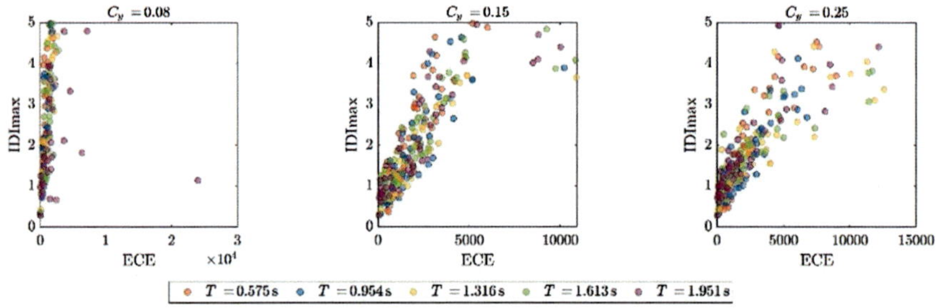


**Fig. 10.** Correlation of modified effective cyclic energy with the peak IDI.

Among the proposed measures of seismic energy demand, the effective peaks of hysteretic energy $\Delta E_H$ is found to be the best predictor for the damage potential of RC multi-story buildings. As the building database is very extensive, this observation does not apply directly to all the observed data, instead the good prediction of the measure is estimated on average. Good correlations and reasonable scattering, in fact, is obtained for IDI values below 3% as illustrated in Fig. 11. The proposed power-based parameters show enhanced predictive capabilities overall when high values of base shear coefficient are employed for the seismic design of frame structures (Fig. 12). However, an increase in data dispersion is evident when Cy reduces, reaching the largest scattering in case of structures with seismic coefficient equal to 0.08.

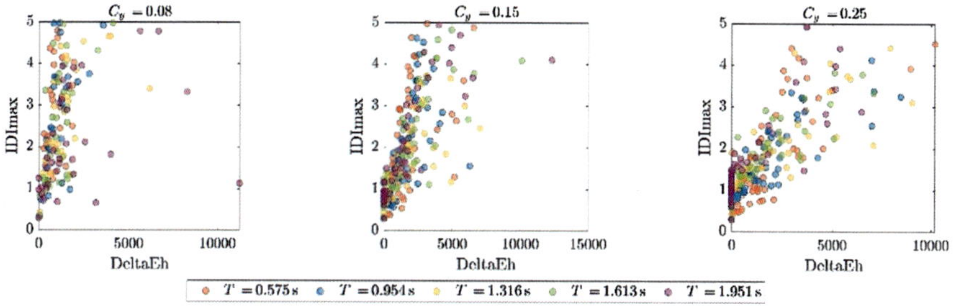


**Fig. 11.** Correlation of effective peaks of hysteretic energy with the peak IDI.

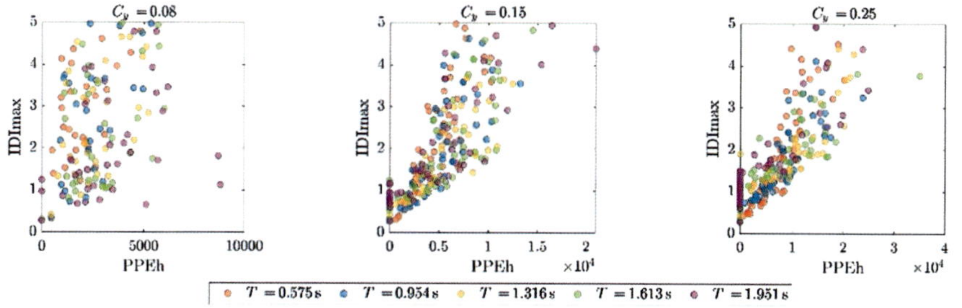


**Fig. 12.** Correlation of peaks of power dissipated via inelastic deformations with the peak IDI.

## 4 Predictive Models

A large training database has been finally prepared to derive general, yet accurate, correlations between considered intensity measures and damage potential via machine learning techniques. Furthermore, a validation dataset is compiled. This latter serves at testing the efficiency and robustness of the obtained correlations.

Towards this objective, two refined case studies are considered consisting of three-bay reinforced concrete frames of 4-storey and 6-storey. The structures are representative of existing buildings located in a high seismic zone. The related seismic coefficient values are 0.301 for the 4-storey building and 0.235 for the 6-storey building. Fibre elements with concentrated plasticity are used for modelling beams and columns of the two-dimensional frames. Finally, the response of the selected case studies is evaluated via nonlinear dynamic analyses through OpenSEES using the same set of near-fault pulse-like ground motions considered to collect the training data. All the details on the design and modelling of the structural members can be found in the appendix of Mollaioli et al.

[1]. The associated periods of the first vibration mode are 0.967 s and 1.165 s for the 4-storey and the 6-storey structure, respectively.

The absolute energy formulation for the MDOF models of the multi-storey buildings subjected to seismic ground motions can be derived in agreement with Uang and Bertero [27] as follows:

$$E_K + E_D + E_H = E_I \tag{1}$$

$$\frac{1}{2}\dot{\mathbf{u}}_t^{\mathrm{T}}\mathbf{m}\dot{\mathbf{u}}_t + \int \dot{\mathbf{u}}^{\mathrm{T}}\mathbf{c}d\mathbf{u} + \int \mathbf{f}_s^{\mathrm{T}}d\mathbf{u} = \int \left(\sum_{i-1}^{N} m_i \ddot{u}_{t,i}\right)du_g \tag{2}$$

where **m** and **c** are the mass matrix and viscous damping matrix, respectively; **u** and $\mathbf{u}_t$ are the relative and absolute displacement vectors, respectively; $m_i$ is the lumped mass associated with the $i$-th floor, $\ddot{u}_{t,i}$ is the absolute acceleration recorded at the $i$-th floor and $N$ is the number of stories. Alternatively, the kinetic energy $E_K$ can be calculated as:

$$E_K = \frac{1}{2}\sum_{i=1}^{N} m_i \dot{u}_{t,i}^2 \tag{3}$$

where $\dot{u}_{t,i}$ is the absolute velocity recorded at the $i$-th floor.

In a similar manner, Kalkan and Kunnath [18] expressed the relative energy imparted to a MDOF system as:

$$E_K' + E_D + E_H = E_I' \tag{4}$$

$$\frac{1}{2}\dot{\mathbf{u}}^{\mathrm{T}}\mathbf{m}\dot{\mathbf{u}} + \int \dot{\mathbf{u}}^{\mathrm{T}}\mathbf{c}\,d\mathbf{u} + \int \mathbf{f}_s^{\mathrm{T}}d\mathbf{u} = -\int \left(\sum_{i-1}^{N} m_i \ddot{u}_i\right)du_g \tag{5}$$

where $\ddot{u}_i$ is the relative acceleration recorded at the $i$-th story.

Consistently with Eq. (2) and Eq. (5), the relevant quantities needed for the computation of the proposed energy-based measures are evaluated for the refined 4-storey and 6-storey buildings. A machine learning technique is employed to validate the efficiency of the predictive models.

Figure 13 compares actual values of the maximum IDI with corresponding predictions obtained by means of two machine learning techniques, namely least-square boosting (LSBoost) and random forest (RM). This figure demonstrates that the accuracy of the obtained predictive formulations for the maximum IDI is satisfactory. The robust correlation is further confirmed by the coefficient $R^2$ for the training and the validation database, which are very similar to each other with values close to 80%. Overall, results are in agreement with those reported by Wen et al. [28] for non-pulse-like seismic ground motions.

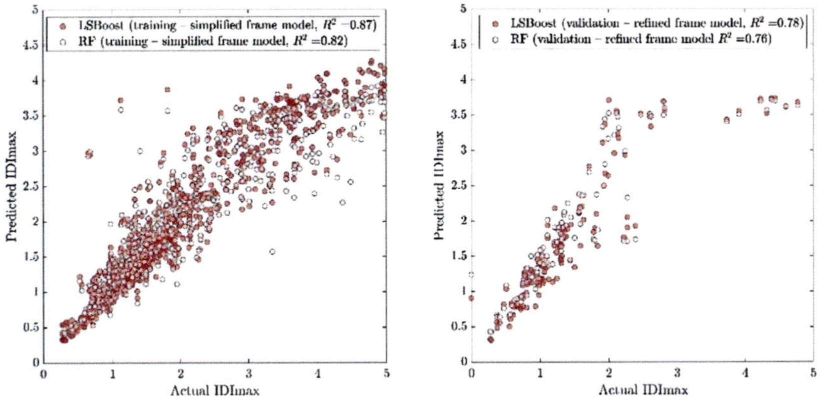


**Fig. 13.** Comparison between actual and predicted maximum inter-story drift ratio for the training database (on the left) and the validation database (on the right).

## 5 Conclusions

The work dealt with here aims to identify the intensity measures that better correlate with the seismic response of typical reinforced concrete structures subjected to severe near-fault pulse-like ground motions. In order to fulfil this objective, the prediction capability of existing, widely adopted IMs were investigated by assessing the response of several bare frames structures ranging from two to ten storeys characterized by different seismic coefficient ratios. The IM that among others suitably predicts the seismic performance of the analysed structures is the parameter proposed by Kalkan and Kunnath, which was conveniently modified in this work to account for the earthquake record time history only.

In addition to IMs already available in the literature, new energy-based parameters were proposed for quantifying the damage potential in RC structures subjected to impulsive ground shakings. For near-fault records, in fact, earthquake damage is mainly associated to sudden and intense seismic input energy imparted to structures by distinctive velocity pulses, peak deformation is likely to be well correlated with the contribution of these few, large peaks of energy demand. With this premise, new energy-based and power-based intensity measures were introduced and their predictive capabilities were evaluated. The results confirmed that the effective peaks of hysteretic energy $\Delta E_H$ provides the best prediction of damage potential in RC multi-story buildings with bare frame configuration. Robust and accurate correlations were obtained via machine learning techniques to predict damage potential in terms of maximum inter-storey drift.

**Acknowledgements.** This work has been supported by the Italian Ministry of Education, University and Research (MIUR), Sapienza University of Rome (Grant n. RG11916B4C241B32) and the Italian Network of University Laboratories of Seismic Engineering (ReLUIS).

## References

1. Mollaioli, F., Lucchini, A., Cheng, Y., Monti, G.: Intensity measures for the seismic response prediction of base-isolated buildings. Bull. Earthq. Eng. **11**(5), 1841–1866 (2013). https://doi.org/10.1007/s10518-013-9431-x
2. Donaire-Ávila, J., Mollaioli, F., Lucchini, A., Benavent-Climent, A.: Intensity measures for the seismic response prediction of mid-rise buildings with hysteretic dampers. Eng. Struct. **102**, 278–295 (2015)
3. Kohrangi, M., Bazzurro, P., Vamvatsikos, D.: Vector and scalar IMs in structural response estimation, Part I: hazard analysis. Earthq. Spectra **32**(3), 1507–1524 (2016)
4. Kohrangi, M., Bazzurro, P., Vamvatsikos, D.: Vector and scalar IMs in structural response estimation, part II: building demand assessment. Earthq. Spectra **32**(3), 1525–1543 (2016)
5. Cordova, P.P., Deierlein, G.G., Mehanny, S.S., Cornell, C.A.: Development of a two-parameter seismic intensity measure and probabilistic assessment procedure. In: The Second US-Japan Workshop on Performance-Based Earthquake Engineering Methodology for Reinforced Concrete Building Structures, vol. 20 (2000)
6. Bianchini, M., Diotallevi, P., Baker, J.W.: Prediction of inelastic structural response using an average of spectral accelerations. In: 10th International Conference on Structural Safety and Reliability (ICOSSAR09), vol. 1317, pp. 2164–2171 (2009)
7. Fiore, A., Mollaioli, F., Quaranta, G., Marano, G.C.: Seismic response prediction of reinforced concrete buildings through nonlinear combinations of intensity measures. Bull. Earthq. Eng. **16**(12), 6047–6076 (2018). https://doi.org/10.1007/s10518-018-0430-9
8. Yakut, A., Yılmaz, H.: Correlation of deformation demands with ground motion intensity. J. Struct. Eng. **134**(12), 1818–1828 (2008)
9. Jayaram, N., Mollaioli, F., Bazzurro, P., De Sortis, A., Bruno, S.: Prediction of structural response of reinforced concrete frames subjected to earthquake ground motions. In: 9th US National and 10th Canadian Conference on Earthquake Engineering, pp. 428–437 (2010)
10. Baker, J.W., Allin Cornell, C.: A vector-valued ground motion intensity measure consisting of spectral acceleration and epsilon. Earthq. Eng. Struct. Dyn. **34**(10), 1193–1217 (2005)
11. Luco, N., Manuel, L., Baldava, S., Bazzurro, P.: Correlation of damage of steel moment-resisting frames to a vector-valued set of ground motion parameters. In: Proceedings of the 9th International Conference on Structural Safety and Reliability (ICOSSAR05), pp. 2719–2726 (2005)
12. Mollaioli, F., Bruno, S., Decanini, L., Saragoni, R.: Correlations between energy and displacement demands for performance-based seismic engineering. Pure Appl. Geophys. **168**, 237–259 (2011)
13. Mollaioli, F., Lucchini, A., Donaire-Ávila, J., Benavent-Climent, A.: On the importance of energy-based parameters. In: COMPDYN 2019 7th ECCOMAS Thematic Conference on Computational Methods in Structural Dynamics and Earthquake Engineering, pp. 24–26 (2019)
14. Kunnath, S.K., Chai, Y.H.: Cumulative damage-based inelastic cyclic demand spectrum. Earthq. Eng. Struct. Dyn. **33**(4), 499–520 (2004)
15. Teran-Gilmore, A., Jirsa, J.O.: A damage model for practical seismic design that accounts for low cycle fatigue. Earthq. Spectra **21**(3), 803–832 (2005)
16. Kalkan, E., Kunnath, S.K.: Effects of fling step and forward directivity on seismic response of buildings. Earthq. Spectra **22**(2), 367–390 (2006)
17. AlShawa, O., Angelucci, G., Mollaioli, F., Quaranta, G.: Quantification of energy-related parameters for near-fault pulse-like seismic ground motions. Appl. Sci. **10**(21), 7578 (2020)
18. Kalkan, E., Kunnath, S.K.: Effective cyclic energy as a measure of seismic demand. J. Earthq. Eng. **11**(5), 725–751 (2007)

19. Quaranta, G., Mollaioli, F.: Analysis of near-fault pulse-like seismic signals through variational mode decomposition technique. Eng. Struct. **193**, 121–135 (2019)
20. Mollaioli, F., Bruno, S.: Influence of site effects on inelastic displacement ratios for SDOF and MDOF systems. Comput. Math. Appl. **55**(2), 184–207 (2008)
21. Decanini, L.D., Mollaioli, F.: An energy-based methodology for the assessment of seismic demand. Soil Dyn. Earthq. Eng. **21**(2), 113–137 (2001)
22. Chou, C.C., Uang, C.M.: A procedure for evaluating seismic energy demand of framed structures. Earthq. Eng. Struct. Dyn. **32**(2), 229–244 (2003)
23. Tso, W.K., Zhu, T.J., Heidebrecht, A.C.: Seismic energy demands on reinforced concrete moment-resisting frames. Earthq. Eng. Struct. Dyn. **22**(6), 533–545 (1993)
24. Decanini, L., Mollaioli, F., Mura, A.: Equivalent SDOF systems for the estimation of seismic response of multistory frame structures. WIT Trans. Built Environ. **57**(1993) (2001)
25. Decanini, L., Mollaioli, F., Saragoni, R.: Energy and displacement demands imposed by near-source ground motions. In: Proceedings of the 12th World Conference on Earthquake Engineering (2000)
26. Pathak, S., Khennane, A., Al-Deen, S.: Power based seismic collapse criterion for ductile and non-ductile framed structures. Bull. Earthq. Eng. **18**(13), 5983–6014 (2020). https://doi.org/10.1007/s10518-020-00925-w
27. Uang, C.M., Bertero, V.V.: Evaluation of seismic energy in structures. Earthq. Eng. Struct. Dyn. **19**(1), 77–90 (1990)
28. Wen, W., Zhang, C., Zhai, C.: Rapid seismic response prediction of RC frames based on deep learning and limited building information. Eng. Struct. **267**, 114638 (2022)

# Study on Spandrel-Pier Interaction in Masonry Walls

Kaushal P. Patel and R. N. Dubey(✉)

Indian Institute of Technology Roorkee, Roorkee, India
{kpatel,rn.dubey}@eq.iitr.ac.in

**Abstract.** Most of the time, masonry walls are tested in the laboratory or simulated in finite element software by considering them as piers. The spandrels connect the piers in actual masonry buildings, which may considerably affect their behavior. It has been observed during past earthquakes that spandrels are critical elements in masonry buildings, which must be carefully examined while analyzing the masonry walls. Past studies have shown that spandrel geometry has a noticeable impact on masonry wall behavior. However, not enough attention has been paid to it. The present study investigates the influence of spandrel depths on the in-plane behavior of masonry walls. Further, the effect of spacing between openings, which in turn affects the spandrel length, has also been studied. Masonry walls have been modeled using the concrete damage plasticity (CDP) model in the finite element software Abaqus. The results are obtained in terms of pushover curves. Further, the influence of spandrel lengths and depths on the damage pattern of the walls has also been examined.

**Keywords:** Spandrel · Pier · Coupling · URM walls · CDP model

## 1 Introduction

Masonry buildings are an assemblage of piers and spandrels. Piers are the primary elements in masonry buildings, which resist the gravity and lateral load acting on the buildings. At the same time, the spandrels are considered secondary elements, which usually lie above the openings. Spandrels connect the piers adjacent to openings with each other. In traditional design practices, the piers are modeled without considering the effect of spandrels (Tomazevic 1998). However, past earthquakes have shown that spandrels may suffer severe damage, affecting the seismic performance of piers. During the earthquake, masonry piers generally show rocking/toe-crushing, sliding shear, or diagonal tension failure modes, while spandrels are assumed to fail under shear. Many researchers have tried to understand the spandrel-pier interaction in masonry buildings in past studies by experimental or numerical approaches (Milani et al. 2009; Beyer and Mangalathu 2013; Hejazi and Soltani 2020).

Beyer and Mangalathu (2013) studied the various strength models of masonry spandrels available in past research and recommended the equations to predict the strength of spandrels. Beyer and Mangalathu (2014) studied the force-deformation relationship of


H. Varum et al. (Eds.): IWEBSE 2023, LNCE 236, pp. 250–257, 2023.
https://doi.org/10.1007/978-3-031-36562-1_19

masonry spandrel with different types of arches, material properties, and vertical compression. The authors found that the strength of the spandrel is sensitive to its cohesion and aspect ratio. Parisi et al. (2014) studied the influence of varying types of spandrels on the lateral behavior of masonry walls. It was observed that different type of spandrels considerably affects the behavior of masonry walls in terms of strength and damage patterns. Foraboschi (2009) studied the coupling effect between masonry spandrels and piers. It was noticed that the flexure and shear strength of spandrels are meaningful parameters affecting masonry walls' behavior. Aldemir et al. (2013) proposed the hybrid equivalent frame modeling approach to capture the more accurate behavior of masonry walls. In this method, the properties of the masonry walls obtained using the micro-modeling approach are further used in the equivalent frame model to obtain the global performance of masonry buildings. It covers the drawbacks of both macro and micro-modeling approaches. Lagomarsino et al. (2013) developed the TREMURI program for the nonlinear seismic analysis of masonry structures using an equivalent frame modeling approach. This approach considers piers and spandrels' flexure and shear failure modes. The proposed modeling approach was able to predict the global behavior of masonry structures with reasonable accuracy. Bracchi et al. (2015) studied the effect of various uncertainty parameters on the seismic response of masonry walls.

The authors observed that uncertainty related to modeling and analysis could considerably affect the behavior of masonry walls. Knox et al. (2017) performed a full-scale experimental test on pier-spandrel substructures. It was observed that the behavior of spandrel has a significant impact on the behavior of the piers. Sandoli et al. (2020) proposed the hybrid equivalent frame model to model the spandrel-pier interaction more realistically. The spandrels were modeled as a diagonal strut element. The authors found that both the flexural and shear failure modes affect the behavior of spandrels. Singh et al. (2013) used the equivalent frame modeling approach to identify the seismic vulnerability of masonry buildings in India. The masonry walls were modeled as a pier-spandrel assembly. The hinges were assigned to the top and bottom of the pier to capture the in-plane failure modes, while in the case of the spandrel, the shear hinge was assigned at the center to capture its brittle behavior under shear. A similar approach has been used by various researchers in past studies (Marino et al. 2019; Cattari et al. 2021).

Most of the past research has focused on the behavior of spandrels of various configurations. However, few studies consider the spandrel-pier interaction in unreinforced masonry walls. Therefore, the present study focused on the effect of varying spandrel depths and lengths on the in-plane behavior of masonry walls. In this study, masonry walls are modeled using a homogenized model (concrete damage plasticity model) because of its higher accuracy in the analysis. The results are obtained in the form of pushover curves, and the effect of various spandrel configurations on the damage patterns of the masonry walls has been investigated.

## 2 Modeling Approach

The masonry walls are modeled using the concrete damage plasticity (CDP) model. This model was basically developed for quasi-brittle materials like concrete, but the behavior of masonry is quite similar to concrete. Therefore, it has been used in past studies to analyze masonry walls. It has been discussed in detail by Patel and Dubey (2018), Dubey and Patel (2020), and Patel and Dubey (2022). It follows the non-associated flow rule assumption. The yield surface of the CDP model in plane stress conditions is shown in Fig. 1.

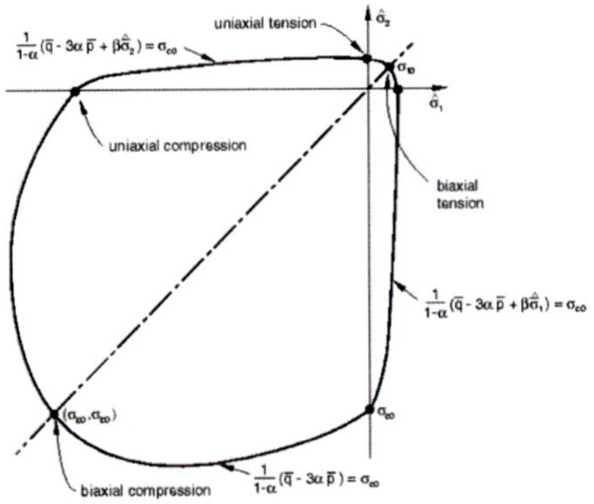


**Fig. 1.** Yield surface of CDP model in plane stress (Abaqus theory manual)

Figure 1 shows that the failure surface is symmetric to the line at 45°. $p$ and $q$ represent the hydrostatic stress and Von-Mises equivalent stress. $\alpha$ incorporates the ratio of biaxial to uniaixial stress. The value of $\beta$ can be obtained using $\alpha$, compression, and tension stresses. $\sigma_{c0}$ and $\sigma_{t0}$ represent the initial compression and tension yield stresses, respectively. Material properties of the masonry in compression and tension have been considered from Kaushik et al. (2007) and Chen et al. (2008), respectively. The damage in the tension after the peak strength has been modeled using the tension stiffening criteria. In which the degradation in the strength has been defined using strain-softening criteria. A similar approach has also been used in the case of compression. The stiffness of the CDP model in compression and tension after the damage initiation has been defined using the damage parameters. The damage parameters ($d_t$ for tension and $d_c$ for compression) have zero value in undamaged conditions, and they reach the value of 1 in the case of complete damage. The CDP model also requires some other parameters for the material model, which have been considered as default values. In which eccentricity is considered as 0.1, stress ratio as 1.16, and $K$ as 0.667. The value of the dilation angle is considered 30 degrees based on the experimental evidence of masonry walls. The model has been validated for masonry walls by Patel and Dubey (2018) and Dubey and Patel (2020).

## 3 Geometry, Load, and Boundary Conditions of the Walls

Masonry walls consisting of piers and spandrels have been analyzed under in-plane loading, as shown in Fig. 2. Hw denotes the height of the wall, while $H_p$ represents the height of the piers. $L_1$ and $L_3$ represent the width of the pier on the left and right sides of the opening, which is considered the same (1.5 m) in all the analyses. $L_2$ shows the length of the opening. The walls are fixed at the bottom, and the in-plane displacement is applied at the top of the walls. The walls are subjected to 1 MPa pre-compression initially at their top. The thickness of the wall is considered 230 mm. The walls are analyzed for the varying height of the spandrel ($H_s = H_w - H_p$) to examine the spandrel-pier interaction incorporating varying opening lengths ($L_2$).

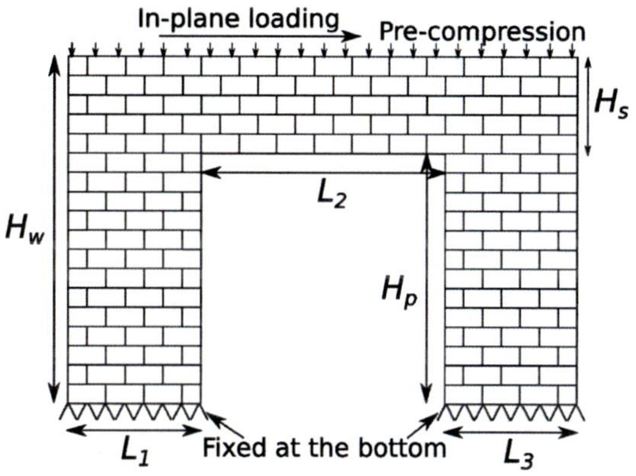


**Fig. 2.** The geometry of the wall

The properties of masonry have been assumed based on Indian standard bricks (Kaushik et al. 2007). The compressive and tensile strength of brick has been considered as 20 MPa and 1 MPa, respectively. The compressive strength of mortar is 15 MPa. The prism strength of masonry has been obtained based on the strength of brick and mortar as 6.5 MPa with Young's modulus as 4047 MPa. The masonry's tension and compression stress-strain curves have been obtained based on these properties, as shown in Fig. 3.

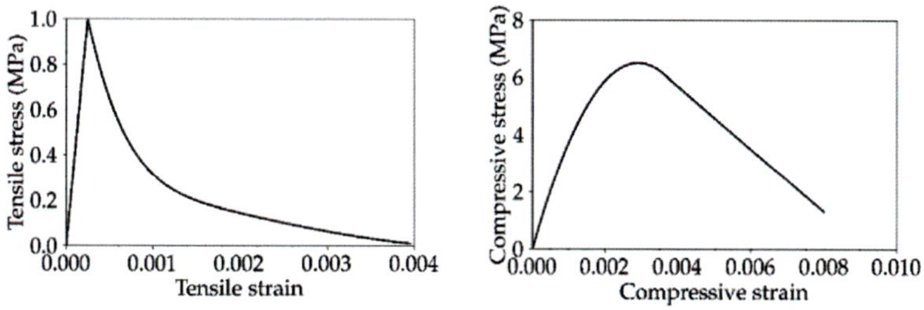


**Fig. 3.** Tension and compression stress-strain curves of masonry

## 4 Results and Discussion

The masonry walls have been analyzed for varying spandrel depths ($H_s$) and opening lengths ($L_2$), and the results are presented in this section. The piers' length (L1 and L3) and height (Hp) are considered 1.5 m and 3 m, respectively. The force-displacement curves of masonry walls with varying spandrel depths are shown in Fig. 4.

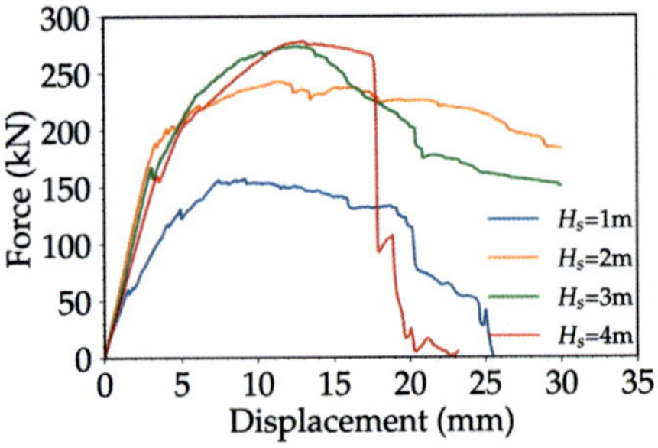


**Fig. 4.** Influence of spandrel depths on the behavior of walls ($L_2 = 3$ m)

It has been observed from Fig. 4 that the walls' strength increases with the increase in spandrel depth from 1 m to 4 m. The opening length ($L_2$) for all the walls has been considered 3 m. The stiffness of the walls is also increasing with an increase in the spandrel depth. It is interesting to note that walls with a spandrel depth of 4 m show sudden failure after reaching the ultimate strength, while in other cases, walls show ductility before failure. It can be explained using the damage pattern of the walls, as shown in Fig. 5, with the help of tensile damage contours. Wall with $H_s$ of 1 m shows lower strength due to failure of the spandrel and joint failure between piers and spandrel. With the increase in the spandrel depth, the damage to the walls is more concentrated towards the top and bottom end of the piers and above the opening in terms of vertical tensile cracks.

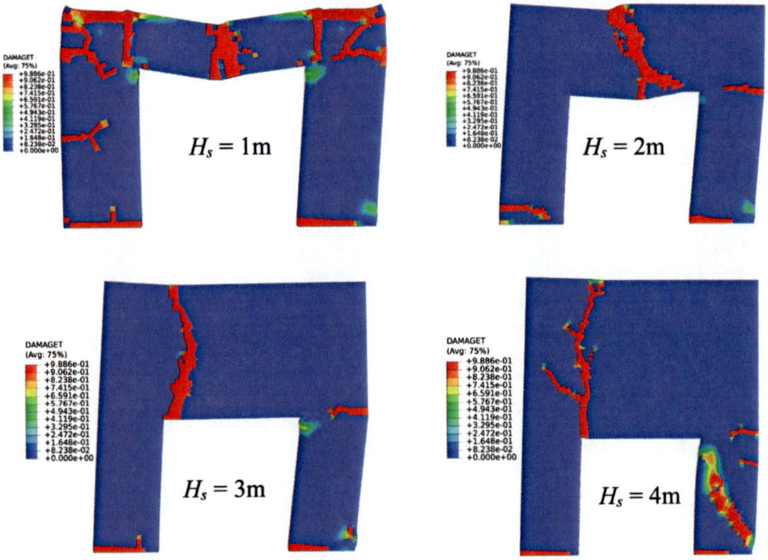


**Fig. 5.** Tensile damage contours of walls with varying spandrel depths ($L_2 = 3$ m)

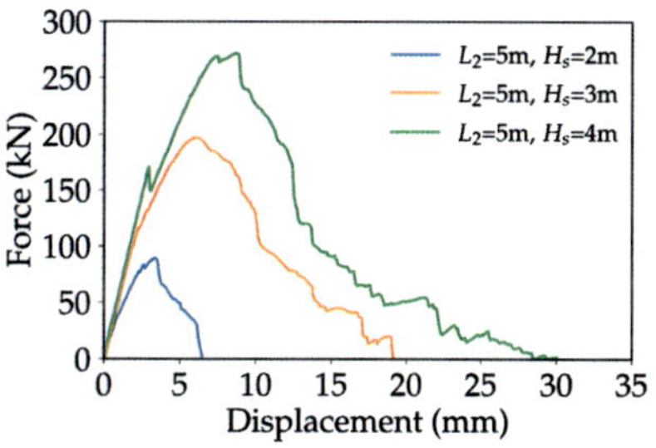


**Fig. 6.** Influence of wide openings on the spandrel-pier interaction

The wall with $H_s$ of 4 m shows sudden failure due to diagonal tensile cracking in the pier on the right side of the opening. The vertical compression acting on the walls increases with an increase in the spandrel depth, and the strength of masonry walls also increases along with it. Masonry walls generally fail in diagonal tension in case of high pre-compression acting on them, and it is considered a brittle failure mode of the masonry walls. Therefore, the wall with a 4 m spandrel depth shows brittle behavior.

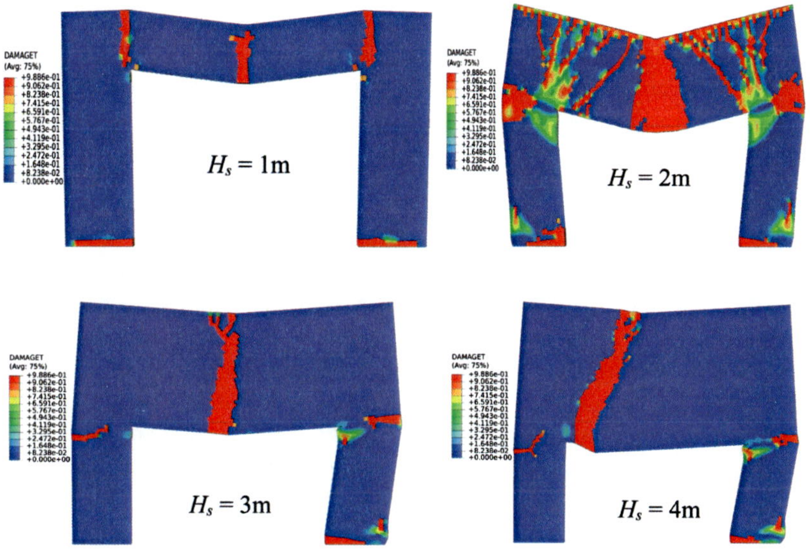


**Fig. 7.** Tensile damage contours of walls with opening length, $L_2$ of 5 m

The effect of large-size openings on the spandrel-pier interaction has been investigated for an opening length ($L_2$) of 5 m. The results are plotted as pushover curves, as shown in Fig. 6. The results indicate that the strength of the walls increases with an increase in the spandrel depth, as observed earlier. However, the strength of the walls reduces with an increase in the opening length for the same value of $H_s$. The strength of the walls with $H_s$ of 2 m with opening lengths of 3 m and 5 m have been obtained as 242.7 kN and 89.58 kN, respectively. So, a 63% reduction in strength has been observed due to an increase in the opening length. Similarly, for walls with $H_s$ of 3 m and 4 m, reductions in the strengths have been obtained as 28% and 2%, respectively. It indicates that spandrel does not have a noticeable influence on piers for a significant value of $H_s$. Further, the walls with $L_2$ of 5 m show brittle behavior compared to those with $L_2$ of 3 m. In Fig. 6, the results are not shown for the walls with $H_s$ of 1 m due to their failure under the gravity loading caused by large openings.

The tension damage pattern of the walls, with $H_s$ equal to 1, 2, 3, and 4 m having $L_2$ as 5 m, are shown in Fig. 7. It depicts that walls with $H_s$ of 1 m and 2 m are damaged severely in the spandrel as well as the near intersection of piers and spandrel. It reduces their strength considerably, while walls with $H_s$ of 3 m and 4 m have shown damage at the top and bottom of the pier with vertical tensile cracking in the spandrel. Hence, they show a more stable failure mechanism with comparatively higher strengths.

## 5 Conclusions

The spandrels are an integral part of the masonry walls. The geometry of the spandrel considerably influences the behavior of the piers. The present study investigates the influence of varying spandrel depths and lengths on the spandrel-pier interaction. The conclusions from the present study are:

- The strength of the walls increases with an increase in the spandrel depth. The axial compression acting on the walls increases with an increase in the spandrel depth, which enhances the load-carrying capacity of the walls. However, very high spandrel depth may also cause piers to fail in diagonal tension failure, which is a brittle failure mode.
- Walls with small spandrel depths are prone to fail by developing tensile cracks in the spandrel and near the corner.
- Walls with large opening lengths and small spandrel depths have much less vertical load-carrying capacity. It may fail suddenly under rocking action during the earthquake.
- The ductility of the walls is reduced with an increase in the length of the opening.
- Walls with large spandrel depths have less influence on the performance of the masonry walls, irrespective of opening lengths. Therefore, spandrel-pier interaction does not have much importance under this condition. However, it needs further investigation by considering walls with varying aspect ratios and pre-compressions.

## References

ABAQUS analysis user's manual 6.16-EF: Dassault Systems Simulia Corp., Providence (2016)

Aldemir, A., Altuğ Erberik, M., Demirel, I.O., Sucuoğlu, H.: Seismic performance assessment of unreinforced masonry buildings with a hybrid modeling approach. Earthq. Spectra **29**(1), 33–57 (2013)

Beyer, K., Mangalathu, S.: Review of strength models for masonry spandrels. Bull. Earthq. Eng. **11**(2), 521–542 (2013)

Beyer, K., Mangalathu, S.: Numerical study on the peak strength of masonry spandrels with arches. J. Earthq. Eng. **18**(2), 169–186 (2014)

Cattari, S., Camilletti, D., D'Altri, A.M., Lagomarsino, S.: On the use of continuum finite element and equivalent frame models for the seismic assessment of masonry walls. J. Build. Eng. **43**, 102519 (2021)

Chen, Y., Ashour, A.F., Garrity, S.W.: Moment/thrust interaction diagrams for reinforced masonry sections. Constr. Build. Mater. **22**(5), 763–770 (2008)

Dubey, R., Patel, K.P.: Homogenized model for in-plane analysis of unreinforced masonry walls. In: 17th World Conference on Earthquake Engineering, Sendai, Japan (2020)

Foraboschi, P.: Coupling effect between masonry spandrels and piers. Mater. Struct. **42**(3), 279–300 (2009)

Hejazi, M., Soltani, Y.: Parametric study of the effect of hollow spandrel (Konu) on structural behaviour of Persian brick masonry barrel vaults. Eng. Fail. Anal. **118**, 104838 (2020)

Kaushik, H.B., Rai, D.C., Jain, S.K.: Stress-strain characteristics of clay brick masonry under uniaxial compression. J. Mater. Civ. Eng. **19**(9), 728–739 (2007)

Knox, C.L., Dizhur, D., Ingham, J.M.: Experimental cyclic testing of URM pier-spandrel substructures. J. Struct. Eng. **143**(2), 04016177 (2017)

Lagomarsino, S., Penna, A., Galasco, A., Cattari, S.: TREMURI program: an equivalent frame model for the nonlinear seismic analysis of masonry buildings. Eng. Struct. **56**, 1787–1799 (2013)

Marino, S., Cattari, S., Lagomarsino, S., Dizhur, D., Ingham, J.M.: Post-earthquake damage simulation of two colonial unreinforced clay brick masonry buildings using the equivalent frame approach. Structures **19**, 212–226 (2019)

Milani, G., Beyer, K., Dazio, A.: Upper bound limit analysis of meso-mechanical spandrel models for the pushover analysis of 2D masonry frames. Eng. Struct. **31**(11), 2696–2710 (2009)

Parisi, F., Augenti, N., Prota, A.: Implications of the spandrel type on the lateral behavior of unreinforced masonry walls. Earthq. Eng. Struct. Dyn. **43**(12), 1867–1887 (2014)

Patel, K., Dubey, R.N.: Effect of opening on the strength of masonry wall. In: 16th Symposium on Earthquake Engineering, IIT Roorkee, Roorkee, India (2018)

Patel, K., Dubey, R.N.: Effect of flanges on the in-plane behavior of the masonry walls. Eng. Struct. **273**, 115059 (2022)

Sandoli, A., Musella, C., Lignola, G.P., Calderoni, B., Prota, A.: Spandrel panels in masonry buildings: effectiveness of the diagonal strut model within the equivalent frame model. Structures **27**, 879–893 (2020)

Singh, Y., Lang, D.H., Prasad, J.S.R., Deoliya, R.: An analytical study on the seismic vulnerability of masonry buildings in India. J. Earthq. Eng. **17**(3), 399–422 (2013)

Tomazevic, M.: Earthquake-Resistant Design of Masonry Buildings. Imperial College Press, London (1998)

# Constitutive Model Characterization of the RC Columns Using Energy Terms

Bilal Güngör[1], Serkan Hasanoğlu[2], Furkan Çalım[3], Ziya Muderrisoglu[4], Ahmet Anıl Dindar[1](✉), Ali Bozer[5], Hasan Özkaynak[4], and Ahmet Güllü[6]

[1] Department of Civil Engineering, Gebze Technical University, Kocaeli, Turkey
bgungor@gtu.edu.tr
[2] University School for Advanced Studies IUSS, Pavia, Italy
[3] Faculty of Civil Engineering, Istanbul Technical University, Istanbul, Turkey
[4] Department of Civil Engineering, Beykent University, Istanbul, Turkey
[5] Department of Civil Engineering, Eskişehir Technical University, Eskişehir, Turkey
[6] Ingram School of Engineering, Texas State University, Texas, USA

**Abstract.** The accuracy level of predicting the hysteretic response of structural members is crucial in energy-based design procedures. Probabilistic methodologies are generally implemented to consider the uncertainties related to demand and capacity-based parameters in structural analyses. Accordingly, implementing the large variability on critical parameters that affect the accuracy level of predicted response is still a challenging issue. This proceeding focuses on an improved calibration methodology proposed to select the optimal analytical modeling parameters that takes into account the widely-used modeling techniques. The input data used for the calibration procedure is compiled from quasi-static experiments conducted on reinforced concrete column members exist in the literature. The finite element models of 62 test units are established to predict the energy dissipation characteristics by utilizing a set of modeling assumptions. Here, the optimal ranges of critical analytical model parameters are evaluated to increase the accuracy level in predicting the target energy dissipation capacity of a member. A comparison between the predicted responses for default and the calibrated model parameters is revealed. Results provide a basis for an efficient calibration methodology to get more accurate capacity predictions. Since the predictions for the damage level of structural members is directly related to the accuracy level of an established analytical model, this attempt is expected to have a considerable impact on the energy-based design of structural members.

**Keywords:** Energy-Based Design Procedures · Modeling Uncertainties · Energy Dissipation Capacity · Calibration of Model Parameters

## 1 Introduction

The performance of a structural system is generally predicted via two major approaches: experience, and modeling capability. Due to the advances in software technology, complex structural models are easily constructed to consider the requirements available in


H. Varum et al. (Eds.): IWEBSE 2023, LNCE 236, pp. 258–268, 2023.
https://doi.org/10.1007/978-3-031-36562-1_20

seismic codes. However, the accuracy of predicted results is still a challenging issue in the case of implementing the large variability on key concepts such as demand and capacity. In general, the strong ground motion characteristics and the energy dissipation capacity of a structure are utilized to identify the demand and capacity terms, respectively. In addition to the conventional seismic design procedures (i.e., strength-based and displacement-based), the accuracy of constructed finite element model to simulate the hysteretic behavior of structural components under cyclic loads is also crucial in energy-based design. The accuracy of simulated hysteretic models is very often tested by comparing the responses of existing datasets of test units available in the literature [1–3]. Modeling uncertainties that arise depending on the assumptions, simplifications, and idealizations considered in a finite element model are explicitly taken into account during the analysis. Moreover, analytical model parameters are also calibrated by implementing the widely used calibration techniques and modeling approaches. The calibration of numerical models has been an active interest for predicting the actual behavior in a more accurate way. For this purpose, the reinforced concrete (RC) column's responses have been calibrated using the finite element modeling assumptions [4, 5] and the hysteretic behavior models [6, 7]. The force-based or displacement-based beam-column elements are generally considered in modeling techniques to simulate the column members' responses in finite element analysis. The hysteretic behavior columns are implemented via widely-used numerical modeling technique such as lumped-plasticity [8–10], and distributed-plasticity [11–13]. However, the accuracy of predicted responses is significantly affected due to the type, and sensitivity level of the calibrated hysteretic model-based parameters [14–16]. The majority of recently proposed hysteretic behavior models consist of a hysteretic rule, a backbone curve, and a damage rule representing the target damage state conditions [17]. Moreover, based on the numbers and uncertainties in model-based parameters, excessive computational effort or validation procedures [18, 19] are generally required to calibrate these models. To achieve this, optimization-based calibration techniques are being rapidly implemented. Chisari et al. [20] proposed a calibration procedure for cyclic models of steel members to minimize the differences between actual and predicted responses via a multi-objective optimization framework. Manikantan et al. [21] investigated the efficiency of different optimization methods (i.e., genetic algorithms, particle swarm techniques, and the homotophy technique) in predicting the Bouc-Wen model parameters. They concluded that the accuracy of the predicted response is acceptable with a considerably low error in case of using the homotopy method. Noori et al. [22] employed a multi-objective optimization algorithm to optimize the parameters of the extended Bouc-Wen model (i.e., as the Bouc-Wen-Baber-Noori -BWBN- hysteresis model). Specifically, the hysteretic restoring force was considered as an objective function for an optimization problem. Ling et al. [23] recommended formulations for unloading stiffness and pinching parameters to be implemented in the Pivot hysteresis model. The energy dissipation results of 113 RC column specimens were optimized by using the Simulated Annealing approach to derive the equations for failure mode-based parameters. Here, the fundamental purpose of the studies mentioned above is to calibrate the hysteretic model parameters for commonly used models in applications via optimization techniques.

In this study, an improved calibration methodology that aims to select the optimal hysteretic modeling parameters for a set of alternative modeling techniques, is proposed. For this purpose, a Python [24] based framework is developed to calibrate the model parameters by using the quoFEM application [25]. Here, a comparison between the predicted responses of 62 RC column test units evaluated using the default and calibrated hysteretic model-based parameters, is presented. The effect of using alternative finite element modeling techniques is also discussed by taking into account the dissipated energy capacities of test specimens. Statistical parameters of calibrated model parameters are provided for selected alternatives. Results of this study provide a basis for an efficient calibration methodology to evaluate the optimal hysteretic model parameters those are commonly used to predict the hysteretic response and the dissipated energy capacity of RC column type members.

## 2 Proposed Framework for the Calibration Strategy

The proposed calibration-based framework contributes to the existing improvement attempts in terms of reducing the modeling uncertainties through hysteretic behavior model parameters. In particular, the large variability in hysteretic rule-based parameters considered in the scope of widely-used software frameworks significantly affects the predicted responses of structural members. This case is illustrated in Fig. 1 [26]. Here, the phase transition (i.e., from elastic to plastic regions) controlling parameter, $R_0$ implemented in isotropic hardening behavior incorporated Giuffré-Menegotto-Pinto Model (Steel02 material model) [27] of OpenSeesPy software framework [28], is selected as an indicator for a cyclic response of a cantilever RC column member. Moreover, as the higher values of $R_0$ parameter (e.g., $R_0 = 20$) is assumed to increase the level of sharp change in stress-strain curve slope at yield plateau [29], the predicted dissipated energy level during cyclic loading is also affected. Other phase controlling parameters (i.e., curvature variation parameters of $c_{R1}$ and $c_{R2}$) are considered as constant. It is obviously seen from Fig. 1 that, adjusting the $R_0$ parameter that fundamentally governs the initial curvature between the elastic and post-yield slope of the curve affects the predicted response. Very often in practice, an increment in the accuracy level of predicted responses is ensured by adjusting these rule-based parameters manually for recommended values in literature.

The calibration-based framework proposed in this proceeding is aimed to improve the accuracy of the predicted hysteretic response of RC column type structural members. For this purpose, analytical models constructed via OpenSeesPy software are implemented to quoFEM application for calibrating the commonly-used material-model parameters.

Procedure mainly depends on minimizing the differences between predicted and experimental responses of RC column test units. The adaptive nonlinear least square minimization algorithm (i.e., NL2SOL) proposed by Dennis et al. [30] is implemented to calibrate model-based parameters in a deterministic way as follows:

$$\text{minf}(x) = 0.5 * \Sigma_{i=1}^{n}\left[f_{m,i}(x) - f_{e,i}\right]^2 \tag{1}$$

where $f_{m,i}(x)$ is the predicted base shear force at loading step-i for the considered material model having vector $x$ of hysteretic-rule-based parameters, and $f_{e,i}$ represents the base

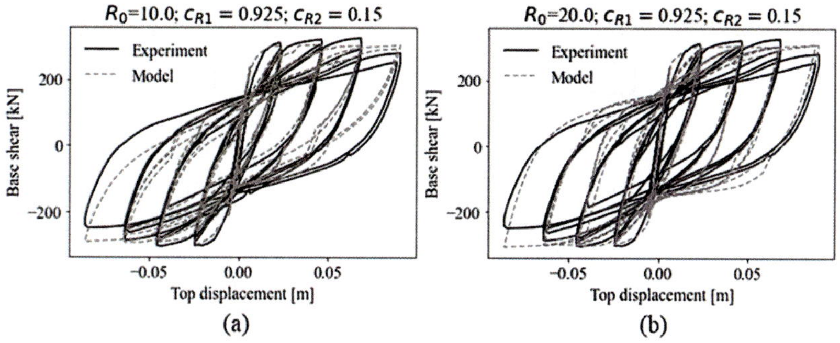


**Fig. 1.** The sensitivity of predicted hysteretic responses on the phase transition controlling parameter, $R_O$: (a) $R_O = 10$, (b) $R_O = 20$.

shear force captured for the test unit at $i^{th}$ loading step. The hysteretic-rule-based parameters are calibrated during the iteration process between the upper and lower bounds selected based on the intervals recommended in literature. A flowchart representing the fundamental steps of proposed framework is depicted in Fig. 2.

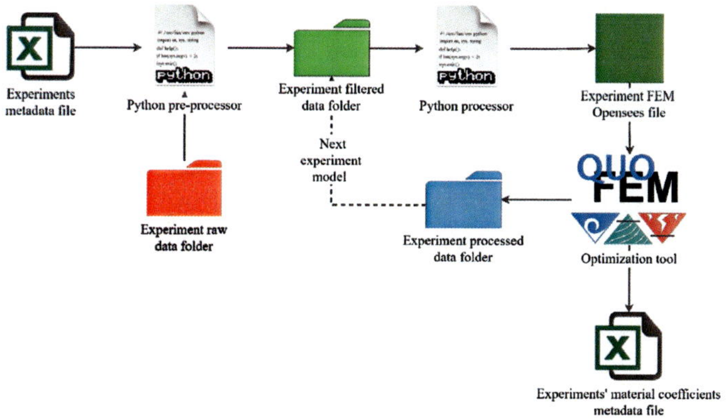


**Fig. 2.** A general overview of the proposed framework.

## 3 Predicting the Dissipated Energy Capacity

### 3.1 Dataset Properties

In this study, a set of 62 RC column type test units are analyzed to calibrate and to generalize the hysteretic model parameters that fundamentally affect the accuracy of predicted responses. For this purpose, a dataset is assembled by arranging the input data of existing databases in the literature [2]. Specifically, the compiled dataset consists of flexure-dominated RC column units, only. Thus, the effects of shear deformations on the hysteretic behavior are not considered within the scope of this study as the calibration process is entirely focused on flexure-based hysteretic model parameters. Accordingly, general characteristics of selected RC column test units are: (1) compressive strength of concrete, $f_c \leq 80$ MPa, (2) yield strengths of longitudinal, $f_{yl}$ & transverse rebars, $f_{yt} \leq 500$ MPa, (3) test units having square & rectangular type cross-sections are preferred, (4) quite slender components having a flexural response, only (i.e., shear span-to-depth ratio, $a/d \geq 3$, (5) axial load ratios $\rho_{axial}$ between 0.15 and 0.40, (6) cantilever, double-ended, and double-curvature type test configurations. Statistical parameters of fundamental properties of a compiled dataset are provided in Table 1.

**Table 1.** Statistical parameters of various properties for selected dataset.

| Property/Parameter | $f_c$ [MPa] | $f_{yl}$ [MPa] | $f_{yt}$ [MPa] | $a/d$ [−] | $\rho_{axial}$ [−] |
|---|---|---|---|---|---|
| Minimum | 17.59 | 330.91 | 254.93 | 3.28 | 0.15 |
| Maximum | 79.98 | 496.87 | 475.88 | 7.50 | 0.40 |
| Mean | 49.48 | 434.13 | 410.70 | 4.88 | 0.24 |
| Standard dev | 25.64 | 40.56 | 49.42 | 1.03 | 0.06 |

### 3.2 Finite Element Modeling Approaches

The OpenSeesPy software framework [27] is utilized to establish the finite element models of considered test units. *Force-based* beam-column element formulation is implemented to calibrate the hysteretic model parameters of two commonly-used section hysteresis models: (1) *Takeda model*, and (2) *Fiber-section model*. The effect of selected modeling strategies on predicted response is also investigated via considering alternative test configurations and hysteresis models (Fig. 3).

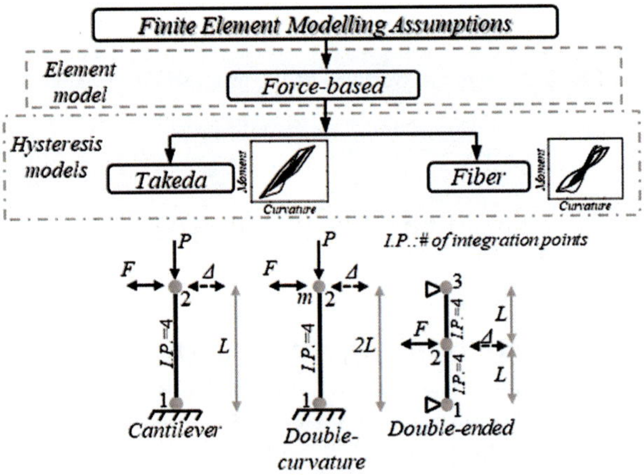


**Fig. 3.** Finite element model representation of the experiments considered in this study.

The curvature exhibited along the member mainly depends on sectional forces in force-based elements. Several approaches are considered (e.g., discretization of members, adjusting the number of integration points) to capture the accurate plastic curvature accumulations along the plastic hinging region. In this study, analytical models of force-based elements are constructed as single members between the main nodes (Fig. 3). The numerical integration method of 2-point Gauss-Radau is implemented in force-based models. Moreover, both considered section hysteresis models in scope of this study, represent the stiffness degradation and unloading responses under cyclic loads. The *hysteretic* type model implemented in OpenSeesPy framework is taken into account to simulate the hysteretic response of a component. Model mainly consists of six stress-strain based quantities and four additional parameters that identify the reloading and ductility functions (i.e., pinching factors as $p_X$ and $p_Y$; damage factors conditioned on ductility as $d_1$, $d_2$). Additionally, *fiber-section* type hysteresis model is also utilized to obtain the sectional hysteresis by taking into account the strain-stress characteristics of each fiber. Here, fibers represent the small pieces constructed by dividing the cross-sections of members. The strain-stress models of *Steel02* and *Concrete07* available in OpenSeesPy framework are utilized to capture the cyclic responses of rebar steel and concrete materials, respectively. In order to increase the accuracy level of predicted responses due to hysteresis rule-based parameters, the *hysteretic* material model pinching factors ($p_X$ and $p_Y$) and damage factors conditioned on ductility ($d_1$ and $d_2$); and *Steel02* material model phase transition controlling parameter of $R_0$ are calibrated. As a result, the finite element models are established by considering the finite element formulations and section hysteresis described above (i.e., $62 \times 2 = 124$ analytical models). The nonlinear behavior is provided by section definitions those are assigned to integration points where nonlinear behavior is expected to occur.

### 3.3 Dissipated Energy Capacity

The main purpose of this study is to investigate the affects of hysteretic rule-based parameters on the prediction of cyclic response of flexure-dominated RC column members. In general, an increment in the accuracy of predicted cumulative dissipated energy level during the cyclic loading as well as the total dissipated energy capacity, is also expected as a similar tendency between the predicted and actual responses is observed. Here, the energy dissipation capacity, $E_{cd}$ of the RC type column members is calculated via the area enclosed by the force-displacement hysteresis captured during the cyclic loading. The effectiveness of calibrated parameters is assessed by using a normalized mean error parameter, $\mu_{\varepsilon,norm}$ that is explicitly based on the differences between the actual and predicted base shear values detected at each cyclic loading step (Eq. 2).

$$\mu_{\varepsilon,norm} = \left(1/n \times \Sigma_{i=1}^{n} \left| \left[ V_{act}^{i} - V_{pred}^{i} \right] \right| \right) \Big/ \max|V_{act}| \quad (2)$$

where $n$ is the number of test units, $V_{pred}$ and $V_{act}$ represent the model-base predicted and actual base shear forces, respectively.

### 3.4 Calibration Results and Sensitivity of the Calibrated Parameters

Following the analyses based on a proposed framework in Fig. 2, the predicted responses are evaluated using the alternative modeling approaches and the calibrated model Parameters (Fig. 4).

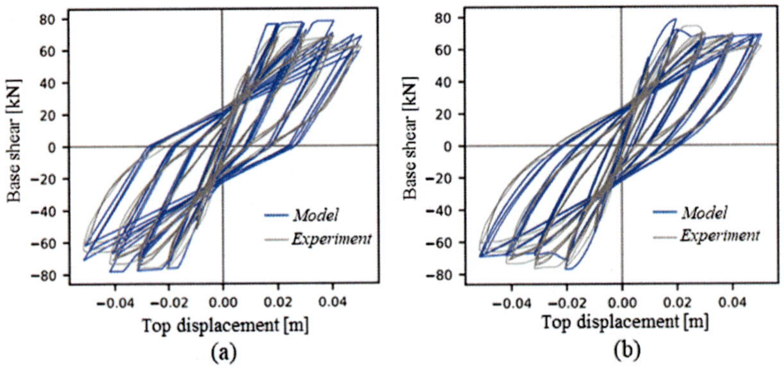


**Fig. 4.** Predicted responses obtained using the calibrated parameters for a test unit [31] in selected dataset: (a) Takeda model, $\mu_{\varepsilon,norm} = 0.07$, (b) Fiber-section based model, $\mu_{\varepsilon,norm} = 0.06$.

The statistical parameters of optimal hysteresis rule-based parameters ($p_X$, $p_Y$, $d_1$, $d_2$, $R_0$) are provided in Table 2.

**Table 2.** Statistical parameters of optimal hysteresis rule-based parameters.

| Property/Parameter | Takeda | | | | Fiber-section |
|---|---|---|---|---|---|
| | $p_X$ | $p_Y$ | $d_1$ | $d_2$ | $R_0$ |
| Minimum | 0.10 | 0.10 | 0.0 | 0.0 | 5.00 |
| Maximum | 1.00 | 1.00 | 1.00 | 1.00 | 30.00 |
| Mean | 0.55 | 0.74 | 0.08 | 0.16 | 18.20 |
| Standard dev | 0.30 | 0.28 | 0.22 | 0.35 | 8.00 |
| Recommended intervals [32] | 0–1.0 | 0–1.0 | 0–1.0 | 0–1.0 | 10–20 |

The sensitivity of calibrated parameters is investigated by considering the mean error parameter for a selected data set. To achieve this, variability of mean error parameter is compared for the responses evaluated via optimal values of hysteresis rule-based parameters and recommended values in literature (i.e., $p_X = 1.0$, $p_Y = 1.0$, $d_1 = 0.0$, $d_2 = 0.0$, $R_0 = 20.0$) for each individual test unit (Fig. 5). Figure 5 shows that, in general, the normalized mean error parameters representing the differences between the predicted and the actual cyclic responses is lower for calibrated hysteresis rule-based (i.e., for both Takeda and fiber-based behavior models) results those obtained for the commonly-used parameters without implementing a calibration process. This case shows the effectiveness of proposed calibration process in terms of increasing the accuracy levels in predicting the cyclic response of considered structural members.

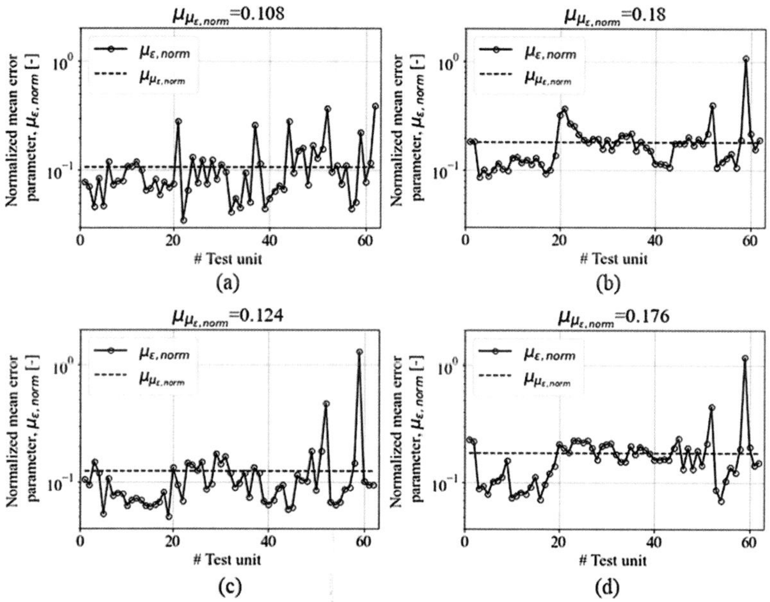


**Fig. 5.** Variations in mean error parameter for a selected dataset: (a) Using the calibrated parameters (Takeda hysteresis), (b) $p_X = 1.0$, $p_Y = 1.0$, $d_1 = 0.0$, $d_2 = 0.0$ (Takeda hysteresis), (c) Using the calibrated parameters (Fiber section), (d) $R_0 = 20.0$ (Fiber section).

Moreover, to compare the accuracy level of predicting the dissipated energy capacities, the correlation plots between the actual dissipated energy levels, $E_{cd,act}$ and the predicted capacities using the Takeda and Fiber section behavior models with calibrated (represented as Takeda w/clb and Fiber w/clb with green color, respectively) and non-calibrated parameters (represented as Takeda w/o clb and Fiber w/o clb with orange color, respectively), $E_{cd,pred}$ are provided in Figs. 6a and 6b, respectively. Preliminary results show that implementing a calibration framework shows good agreement with the actual dissipated energy levels captured during the experiments.

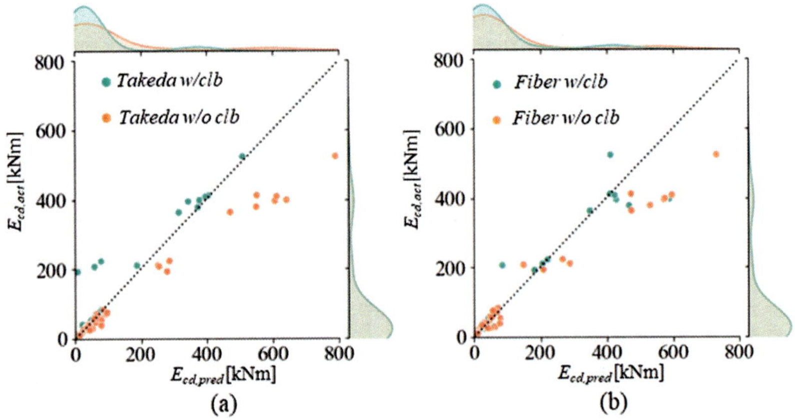


**Fig. 6.** Correlation plots between the actual (i.e., experimental) and predicted dissipated energy capacities evaluated using the: (a) Takeda hysteresis-based model, (b) Fiber section hysteresis-based model.

## 4 Conclusions

An improved calibration methodology that is based on selecting the optimal analytical modeling parameters is presented in this paper. For this purpose, the analytical models of 62 RC column type test units are constructed via commonly-used hysteresis models to calibrate the hysteretic rule-based parameters that fundamentally affect the accuracy of predicted responses. The results of the investigation are summarized below:

- A calibration-based framework is proposed to evaluate the optimal hysteresis rule-based parameters for predicting the dissipated energy capacities of flexure-dominated RC column type members subjected to cyclic loading.
- The comparisons in cyclic behavior prediction obtained via calibrated and non-calibrated hysteresis rule-based parameters indicate that, the accuracy level in prediction of cyclic behavior is higher for calibrated models (i.e., 40% and 30% lower relative differences in $\mu_{\mu\varepsilon,norm}$ parameters for Takeda and Fiber-section based calibrated models compared to that parameters evaluated via models using commonly-used hysteresis rule-based parameters).
- The effectiveness of calibration-based framework is also investigated in terms of reducing the level of hysteresis-rule based modeling uncertainty for the dissipated energy level during the cyclic loading. Preliminary results show that, the correlation between the actual and predicted dissipated energy capacities show good agreement for both modeling assumption (i.e., Takeda and Fiber section-based models).

## References


1. Browning, J.A., Pujol S., Eigenmann R., Ramirez J.A.: NEEShub Databases. Concrete International (2013). https://datacenterhub.org/resources/databases. Accessed 01 Oct 2021
2. Rathje, E., et al.: DesignSafe: A new cyberinfrastructure for natural hazards engineering. ASCE Nat. Haz. Rev. **18**(3) (2017)
3. Berry, M., Parrish, M., Eberhard, M.: PEER structural performance database user's manual (version 1.0). University of California, Berkeley (2004)
4. Bathe, K.: Finite Element Procedures. Prentice-Hall, Upper Saddle River (1996)
5. Taucer, F., Spacone, E., Filippou, F: A fiber beam-column element for seismic response analysis of reinforced concrete structures. University of California. Berkeley, California: Earthquake Engineering Research Center (1991)
6. Sezen, H., Chowdhury, T.: Hysteretic model for reinforced concrete columns including the effect of shear and axial load failure. J. Struct. Eng. **135**(2), 139–146 (2009)
7. Ghannoum, W.M., Moehle, J.P.: Dynamic collapse analysis of a concrete frame sustaining column axial failures. ACI Struct. J. **109**(3), 403–412 (2012)
8. Ibarra, L.F., Medina, R.A., Krawinkler, H.: Hysteretic models that incorporate strength and stifness deterioration. Earthquake Eng. Struct. Dynam. **34**(12), 1489–1511 (2005)
9. Goulet, C.A., et al.: Evaluation of the seismic performance of a code-conforming reinforced-concrete frame building from seismic hazard to collapse safety and economic losses. Earthquake Eng. Struct. Dynam. **36**(13), 1973–1997 (2007)
10. Haselton, C.B., Liel, A.B., Deierlein, G.G., Dean, B.S., Chou, J.H.: Seismic collapse safety of reinforced concrete buildings. I: assessment of ductile moment frames. J. Struct. Eng. ASCE **137**(4), 481–491 (2011)
11. CEB (1996b): RC frames under earthquake loading: State of the art report. No.231. European Committee for Concrete, Lausanne (1996)
12. ATC20: Procedures for postearthquake safety evaluation of buildings. Applied Technology Council, California (1989)
13. NIST: Nonlinear structural analysis for seismic design. NIST GCR 10–917–5, National Institute of Standards and Technology, Gaithersburg (2010)
14. Yazgan, U., Dazio, A.: Simulating maximum and residual displacements of RC structures: I. Accur. Earthq. Spectra **27**(4), 1087–1202 (2011)
15. Ghaemian, S., Muderrisoglu, Z., Yazgan, U.: The effect of finite element modeling assumptions on collapse capacity of an RC frame building. Earthq. Struct. **18**(5), 555–565 (2020)
16. Di Domenico, M., Ricci, P., Verderame, G.M.: Empirical calibration of hysteretic parameters for modelling the seismic response of reinforced concrete columns with plain bars. Eng. Struct. **237**, 112120 (2021)
17. Lee, C.S., Han, S.W.: An accurate numerical model simulating hysteretic behavior of reinforced concrete columns irrespective of types of loading protocols. Int. J. Concr. Struct. Mater. **15**(1), 1–16 (2021)
18. Skalomenos, K.A., Hatzigeorgiou, G.D., Beskos, D.E.: Parameter identification of three hysteretic models for the simulation of the response of CFT columns to cyclic loading. Eng. Struct. **61**, 44–60 (2014)
19. Lignos, D.G., Krawinkler, H., Whittaker, A.S.: Prediction and validation of sidesway collapse of two scale models of a 4-story steel moment frame. Earthq. Eng. Struct. Dynam. **40**(7), 807–825 (2011)
20. Chisari, C., Francavilla, A.B., Latour, M., Piluso, V., Rizzano, G., Amadio, C.: Critical issues in parameter calibration of cyclic models for steel members. Eng. Struct. **132**, 123–138 (2017)

21. Manikantan, R., Ghosh Mondal, T., Suriya Prakash, S., Vyasarayani, C.P.: Parameter identification of Bouc-Wen type hysteresis models using homotopy optimization. Mech. Based Des. Struct. Mach. **50**(1), 26–47 (2020)
22. Li, Z., Noori, M., Zhao, Y., Wan, C., Feng, D., Altabey, W.A.: A multi-objective optimization algorithm for Bouc-Wen-Baber-Noori model to identify reinforced concrete columns failing in different modes. Proc. Inst. Mech. Eng. Part L: J. Mater. Des. Appl. **235**(9), 2165–2182 (2021)
23. Ling, Y.C., Mogili, S., Hwang, S.J.: Parameter optimization for Pivot hysteresis model for reinforced concrete columns with different failure modes. Earthq. Eng. Struct. Dynam. **51**(10), 2167–2187 (2022)
24. Van Rossum, G., Drake, F. L.: Python 3 reference manual. CreateSpace (2009)
25. McKenna, F., Yi, S., Aakash, B.S., Zsarnoczay, A., Gardner, M., Elhaddad, W.: NHERI-SimCenter/quoFEM: Version 3.2.0 (v3.2.0). Zenodo (2022)
26. Saatcioglu, M., Ozcebe, G.: Response of reinforced concrete columns to simulated seismic loading. Struct. J. **86**(1), 3–12 (1989)
27. Filippou, F.C., Bertero, V.V., Popov, E.P.: Effects of Bond Deterioration on Hysteretic Behavior of Reinforced Concrete Joints. University of California Press, Berkeley (1983)
28. Zhu, M., McKenna, F., Scott, M.H.: OpenSeesPy: python library for the OpenSees finite element framework. SoftwareX **7**, 6–11 (2018)
29. Carreño, R., Lotfizadeh, K.H., Conte, J.P., Restrepo, J.I.: Material model parameters for the Giuffrè-Menegotto-Pinto uniaxial steel stress-strain model. J. Struct. Eng. **146**(2), 04019205 (2020)
30. Dennis, J.E., Jr., Gay, D.M., Walsh, R.E.: An adaptive nonlinear least-squares algorithm. ACM Trans. Math. Softw. (TOMS) **7**(3), 348–368 (1981)
31. Atalay, M.B., Penzien, J.: The Seismic Behavior of Critical Regions of Reinforced Concrete Components as Influenced by Moment, Shear and Axial Force. Earthquake Engineering Research Center, University of California, Berkeley (1975)
32. Mazzoni, S., McKenna, F., Scott, M.H., Fenves, G.L.: OpenSees command language manual. Pac. Earthq. Eng. Res. (PEER) Center **264**(1), 137–158 (2006)

# Damage Index Model for Masonry Infill Walls Subjected to Out-of-Plane Loadings

André Furtado[1](✉), Hugo Rodrigues[2], and António Arêde[3]

[1] CERIS, Instituto Superior Técnico, Lisboa, Portugal
andre.furtado@tecnico.ulisboa.pt
[2] RISCO, Universidade de Aveiro, Aveiro, Portugal
[3] CONSTRUCT, Faculdade de Engenharia da Universidade do Porto, Porto, Portugal

**Abstract.** The definition of damage states for structural and non-structural elements is necessary to support the seismic vulnerability analyses and identify the most vulnerable elements that need to be retrofitted. The studies on masonry infill walls focused on multiple experimental studies to characterize their in-plane and/or out-of-plane (OOP) behaviour. Some codes have already proposed damage states for masonry infill walls only under in-plane loadings. Also, the study of the normalized energy dissipation and the hysteretic viscous damping when subjected to OOP loadings were not explored. It is recognized that the OOP behaviour of masonry infill walls is still nowadays a topic needing further investigations. This work's main objective is to propose a damage index for infill walls under pure OOP seismic loadings. A detailed analysis of the damage observed in masonry infill walls tested under pure OOP loadings was performed. The damage observed in masonry infill walls made with different masonry units, with and without openings, is carefully detailed. After that, a pilot damage index model is proposed for masonry infill walls made of hollow clay horizontal brick units.

**Keywords:** masonry infill walls · out-of-plane behaviour · experimental testing; Damage state · damage index

## 1 Introduction

Several post-earthquake survey damage assessment reports recognize that the masonry infill walls' Out-Of-Plane (OOP) seismic performance was responsible for multiple collapses, casualties, and economic losses [1–3]. Calvi and Bolognini [4] concluded that the damage state of the infill walls plays an essential role in the definition of damage states of Reinforced Concrete (RC). The authors justified this sentence by saying that the infill walls cannot accommodate large deformations without damaging or collapsing. For example, in well-designed RC frames, according to recent seismic standards (e.g. Eurocode 8), the walls can be severely damaged or even collapse before the RC frame is damaged.

Based on this motivation, several research works have been developed to assess the masonry infill walls seismic behaviour through experimental testing. The experimental


H. Varum et al. (Eds.): IWEBSE 2023, LNCE 236, pp. 269–281, 2023.
https://doi.org/10.1007/978-3-031-36562-1_21

works available in the literature are divided into two approaches: i) the IP testing of infilled RC specimens [5]; and the OOP testing of infill walls surrounded by RC elements (i.e. RC frame) [6, 7]. Concerning the IP testing of the infilled RC frames, the authors investigated the infill walls' impact on the lateral response of RC frames designed with low or high seismic demands. The effect of the openings, masonry units and strengthening was also the focus of multiple tests [4, 8]. From those studies, it was concluded that the presence of the infill walls increases the lateral stiffness, strength and energy dissipation of the frame. However, shear failure of RC elements can also occur due to the infill walls and their openings (i.e. development of short-column mechanism) [9–11].

Based on this motivation, the primary purpose of this manuscript was to propose a pilot damage index for masonry infill walls made of hollow clay horizontal brick units. A database containing thirteen OOP tests performed in three different research works was used to achieve these goals.

## 2 Experimental Campaign

### 2.1 Description of the Specimens

The database used for this research work comprises a total of 13 masonry infill walls that were subjected to OOP tests, using a distributed loading approach (i.e., airbags or pneumatic actuators), and were subjected to a charge-discharge (half-cyclic) loading sequence. The database is subdivided into two groups, namely: Group 1 - four tests carried out by Furtado, Rodrigues [12]; Group 2 – four specimens tested by Furtado, Rodrigues [13].

Table 1 presents a summary of information concerning the specimens selected for this research study. Nomenclature of the specimens, geometric dimensions, type of OOP loading application, masonry unit and possible variables under investigation are presented. Please note that the loading shape applied by the airbags is parabolic with high amplitude at the panel mid-height. On the other hand, the loading shape applied by the pneumatic actuators is purely uniform along the entire wall height.

The monotonic OOP tests and the OOP tests of infill walls with prior IP damage available in the literature were excluded from this study since the focus of this study are the pure OOP cyclic or semi-cyclic tests.

### 2.2 Description of the Specimens

The specimens from Group 1 were constructed using the same workmanship and the same construction methodology. No mechanical connection was included in the panel-frame boundaries. The connection between the upper beam and the panel, which determines a three or four-sided constraint, is performed by applying the mortar. Furtado, Rodrigues [12] studied the effect of plaster (Inf_06), panel support conditions (Inf_05) and axial load (Inf_04) applied on the adjacent columns on the OOP behaviour of full-scale masonry infill walls. The walls' geometric dimensions were $4.20 \times 2.30\text{m}^2$ (length × height), as shown in Fig. 1a. Hollow clay horizontal brick units 150 mm thick were used. Specimen Inf_02 was the reference test since it was built without plaster, fully supported in the

**Table 1.** Summary of the database specimens.

| Group | Specimen | Wall geometric dimensions (mm) Length × Height | OOP loading application | Masonry unit | Plaster | Openings | Axial load ratio on each column | Panel support conditions | WorkmanshipD |
|---|---|---|---|---|---|---|---|---|---|
| 1 | Inf_02 | 4200 × 2300 | Airbags | HCHB 150 mm | No | No | No | Full | No |
| | Inf_04 | | | | No | No | 0.16 (axial load: 270kN) | Full | No |
| | Inf_05 | | | | No | No | No | 2/3 thickness supported | No |
| | Inf_06 | | | | 10 mm | No | No | Full | No |
| 2 | Inf_08 | 4200 × 2300 | Pneumatic actuators | HCHB 150 mm | 10 mm | No | No | Full | No |
| | Inf_09 | | | | 10 mm | No | No | Full | Yes |
| | Inf_15 | | | | 10 mm | Window (15%) | No | Full | No |
| | Inf_16 | | | | 10 mm | Door (22% $^{F}$) | No | Full | No |

bottom RC beam and without axial load on the adjacent RC columns. The specimen Inf_04 was constructed with the same geometric and material characteristics as the reference specimen Inf_02. In each column, an axial load of 270 kN (axial load ratio equal to 0.16) was applied to simulate the dead loads' effect. Inf_05 was constructed with the same characteristics and tested without axial load on the columns. The difference was that the wall was built with 2/3 of its thickness supported in the bottom beam. Finally, the specimen Inf_06 was constructed with 1 cm plaster. All the walls were built without reinforcement and openings. More details concerning the material and mechanical properties can be found in [12] and in Subsect. 2.3. The OOP loading was applied using airbags.

Group 2 is comprised of five specimens, four tested by Furtado, Rodrigues [13] and one tested by De Risi, Furtado [14]. Furtado, Rodrigues [13] tested four full-scale with the same geometry of the specimens presented in Group 1. The walls were built using the hollow clay horizontal bricks 150 mm thick and 10 mm plaster. The reference specimen in Group 2 was the wall Inf_08. Specimens Inf_08 and Inf_09 were built with different qualified operators but using the same construction methodology and materials. Moreover, the authors investigated the effect of the opening with the specimens Inf_15 and Inf_16. Specimen Inf_15 is a wall with a central window with the geometric dimensions of 1.25 × 1.15 $m^2$ (opening area ratio is 15%.), length and height, respectively, as shown in Fig. 1b. Finally, Inf_16 was an infill wall with a central door with the geometric dimensions of 1.25 × 1.70 $m^2$, respectively, length and height, as shown in Fig. 1c. The opening area ratio is 22%. The dimensions and detailing of the surrounding RC frame are the same as the one used by Furtado, Rodrigues [12]. No gap or reinforcement was used between the wall and the frame. In both specimens, an RC lintel was built at the top of the opening with 1.65 m length and 0.10 m height. The longitudinal reinforcement used in the lintel was 3ø6 mm diameter. Detailed information concerning the material and mechanical properties can be found in [13]. The specimens of Group 2 were subjected to distributed OOP loads applied by pneumatic actuators.

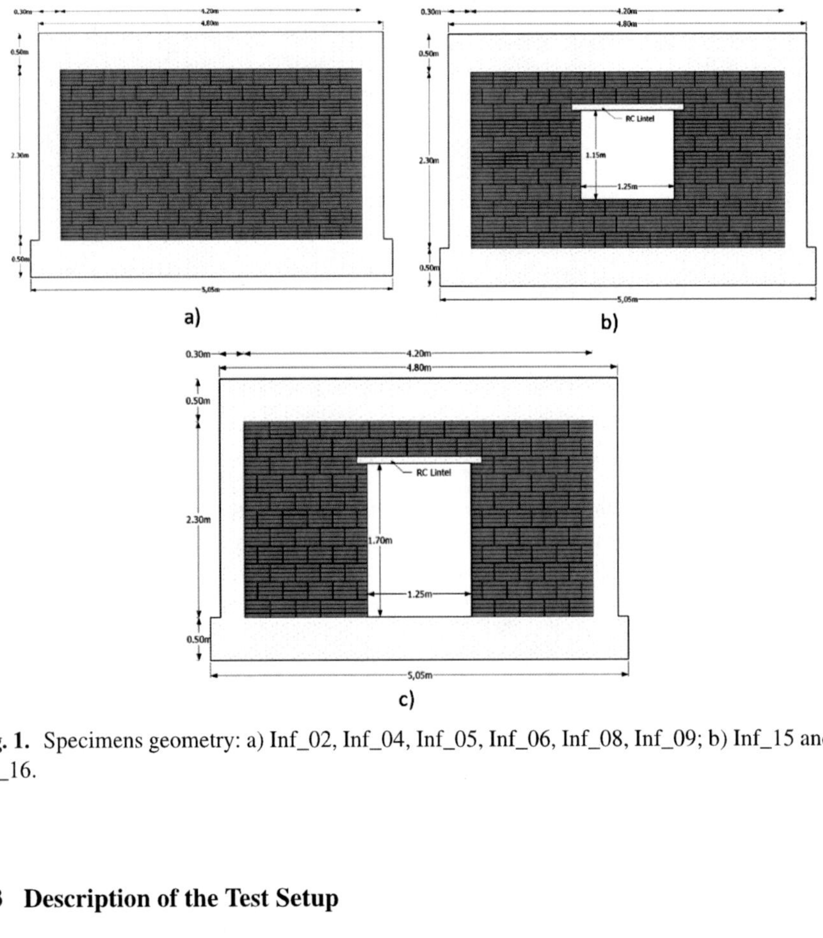


**Fig. 1.** Specimens geometry: a) Inf_02, Inf_04, Inf_05, Inf_06, Inf_08, Inf_09; b) Inf_15 and; c) Inf_16.

## 2.3 Description of the Test Setup

The OOP tests performed in group 1 consisted of applying a distributed OOP load through nylon airbags. A self-equilibrated reaction steel structure was attached to the RC frame in twelve different points, and the airbags were placed between the panel and the reaction structure. The steel reaction structure is composed of five vertical steel I-shape profiles and four tube profiles. A wood plate with 10 mm thickness was placed in the front of the steel structure to support the airbags. In this group, the wall Inf_04 was tested with axial load applied on the adjacent columns.

The axial load was applied by means of a hydraulic jack inserted between a steel cap placed on the top of each column and an upper steel profile, which, in turn, was connected to the foundation steel shape resorting to a pair of high-strength rods per column. Hinged connections were adopted between these rods and the top and foundation steel shapes. The axial load applied to the columns was continuously measured by load cells inserted between the jacks and the top of each column.

The test setup used to perform the OOP tests of specimens from Group 2 is similar to the one used in Group 1. Again, the distributed OOP loading was applied through several pneumatic actuators that mobilized the entire infill panel surface with wood plates (one per actuator) placed between the actuators and the panel. All the pneumatic actuators are linked to four horizontal alignments, which react against five vertical alignments. The horizontal alignments are coupled with hinged devices that allow lateral sliding. The steel reaction structure was also attached to the RC frame in the same twelve points used in the test setup from group 1, with steel bars coupled with load cells to monitor the OOP loadings. This self-equilibrated system balances the transmission of the OOP loadings to the reaction frame. The test setup configuration used to test Inf_08 and Inf_09 can be found in [12, 14, 15].

The disposition of the pneumatic actuators was re-arranged to test the walls with the openings. For this, some pneumatic actuators were removed in the opening region. This test setup allowed the authors to perform the OOP tests until the wall collapse without damaging any equipment. No axial load was applied at the top of the adjacent RC columns in the tests of this group. More details regarding this test setup can be found in [16].

Concerning the instrumentation, several displacement sensors were used to monitor the infill wall and frame OOP displacements. The reference measurement used to compute the OOP drift used in Sect. 3 was the sensor placed in the geometric center of the wall.

# 3 Experimental Results

## 3.1 Damage Limit States and Observed Damages

Each test was stopped at the end of each target displacement for each specimen to observe and record the damage evolution. This procedure allowed registering the development of new cracks and the expansion of the existing ones. Based on those observations, the damage observed can be grouped into four different damage states, namely: First cracking (damage state 1); Moderate damage (damage state 2); Extensive damage (damage state 3); and Collapse (damage state 4). Table 2 summarizes the damage states herein proposed for the masonry infill walls subjected to OOP loadings, based on the observation of the typical damage evolution.

Following the definition of the different damage states, the OOP drift corresponding to the distinct damage states was identified to understand the effect of each variable on the damages observed in the masonry infill walls. Table 3 provides a data summary of the drift corresponding to each damage state. Table 4 provides the Δdrift corresponding to each damage state by each specimen. This parameter represents the ratio between the OOP drift reached by the specimen and the OOP drift in which the maximum peak load was achieved. The main intention is to understand how far from the peak load each damage state occurs. Based on these results, it is possible to compare the damage observed in each wall and the corresponding OOP drift demand. In addition, it is plotted

**Table 2.** Proposal of damage states and the corresponding description.

| Damage state | Description of the damages observed |
|---|---|
| First Cracking (Damage state 1) | – corresponds to the first visible crack, which usually corresponds to the reduction of the panel stiffness (other authors considered the development of the first cracking as the first damage state of an infill wall under pure OOP loads [17]) |
| Moderate Damage (Damage state 2) | – extension of the first crack observed in damage state 1<br>– development of cracks with a corresponding extension higher than 30 cm with thickness not larger than 1 mm<br>– slight plaster detachment in the wall-frame interfaces<br>– slight OOP detachment of the wall from the surrounding frame |
| Extensive Damage (Damage state 3) | – The cracking pattern of the wall is evident (e.g. trilinear cracking; linear cracking; 5-linear cracking)<br>– Cracks with different lengths and with thicknesses larger than 1 mm<br>– Severe plaster detachment in the wall-frame interfaces<br>– Evident OOP detachment of the wall from the surrounding frame |
| Collapse (Damage state 4) | – Partial collapse of parts of the walls in the case of panels with openings<br>– Total collapse of the wall in the case of panels without openings |

in Fig. 2 the OOP drift associated with each damage state. Based on these results, the following observations can be drawn, namely:

The first cracking occurred in Group 1 between 0.11% and 0.47% (4.27 times higher), as shown in Fig. 2a. Reducing the panel support width caused the increment of the OOP drift demand necessary to develop the first crack. The reference specimen Inf_02 stands out significantly from the values recorded in the remaining specimens since it is 60%, 77% and 70% lower than the specimens Inf_04, Inf_05 and Inf_06, respectively. In Group 2, the OOP drift varied from 0.11% to 0.27% (2.36 times), justified by the openings as shown in Fig. 2b. For example, Inf_08 and Inf_09 (different workmanship) reached the first crack for almost the same OOP drift demand (0.11% and 0.12%, respectively). The openings lead to the delay of the first crack development. The panel with the largest opening area (i.e., Inf_16) reached the first crack for an OOP drift 2 times higher than Inf_15.

The moderate damage was achieved in Group 1 for OOP drift values between 0.63% (Inf_04) and 1.39% (Inf_06). Considering all the data results, the moderate damage observed in Inf_04 occurred for a low OOP drift. Again, if we look at the conventional failure, it happened for a similar drift (0.69%). It is important to remember that during the OOP test of the wall Inf_04, a pure vertical crack suddenly occurred with an OOP detachment in the top and bottom interfaces of the wall. This behaviour may have had

**Table 3.** Data summary concerning the drift corresponding to each damage state.

| Specimen | Drift corresponding to DS1 (%) | Drift corresponding to DS2 (%) | Drift corresponding to DS3 (%) | Drift corresponding to DS4 (%) | Conventional Failure (%) |
|---|---|---|---|---|---|
| Inf_02 | 0.11 | 1.13 | 2.49 | N/A[2] | 3.91 |
| Inf_04 | 0.28 | 0.63 | 3.56 | N/A[2] | 0.69 |
| Inf_05 | 0.47 | 1.30 | 1.91 | N/A[2] | N/A[1] |
| Inf_06 | 0.37 | 1.39 | 2.62 | N/A[2] | N/A[1] |
| Inf_08 | 0.11 | 1.09 | 2.02 | 3.20 | N/A[3] |
| Inf_09 | 0.12 | 1.21 | 2.34 | 2.58 | N/A[3] |
| Inf_15 | 0.26 | 1.19 | 2.10 | N/A[2] | 3.92 |
| Inf_16 | 0.13 | 1.15 | 1.95 | 3.24 | 3.24 |
| Range Values | [0.11–0.47] | [1.09–1.69] | [1.91–2.78] | [2.58–3.83] | [3.24–3.92] |

N/A[1] – Not applicable since it did not occur a 20% reduction of the panel OOP maximum strength (post-peak stage)
N/A[2] – Not applicable since it did not occur panel OOP collapse
N/A[3] – Not applicable since the collapse happened before a 20% reduction of the panel OOP maximum strength

some relationship with applying the axial load in the envelope RC columns that could have compressed the infill wall and modified the OOP response. Since the behaviour of Inf_04 was significantly different from the remaining walls, it will be analyzed separately from the remaining.

Regarding Group 2, moderate damage occurred for approximated OOP drift values (1.09%-1.69%). The extensive damage was reached in Group 1 for OOP drift demands between 1.91% (Inf_05) and 3.56% (Inf_04). In Group 2, it was observed that the extensive damage state was observed for OOP drift demands around 1.95% (Inf_16). The $\Delta$drift ranges between 0.44 and 1.17, with a mean value of 0.78 and a CoV equal to 29.99%;

The conventional failure was only reached for specimens Inf_02, Inf_04 and Inf_15. On the one hand, the lowest result can be justified by the axial load on the columns. The red circle observed in Fig. 2a is the result reached by Inf_04. Finally, the OOP collapse occurred in the testing of walls Inf_08, Inf_09 and Inf_16. The collapse happened for the specimen Inf_09 for a drift of 2.58%.

Moreover, the information herein collected is essential to define OOP drift intervals. The different damage state levels can be observed in masonry infill walls subjected to pure OOP loadings. This novel information can help support future investigations. The drift variation (i.e. minimum and maximum value) OOP drift results for each damage state level was observed and aggregated in Fig. 3. This Figure shows the range values considering the contribution of all specimens. It can be seen that: the first cracking occurred for OOP drift values ranging from 0.11% to 1.09%; moderate damage between 1.09% and 1.91%; extensive damage between 1.91% and 2.78%; conventional failure

**Table 4.** Data summary concerning the Δdrift corresponding to each damage state.

| Specimen | Δdrift corresponding to DS1 | Δdrift corresponding to DS2 | Δdrift corresponding to DS3 | Δdrift corresponding to DS4 | Δdrift corresponding to conventional failure |
|---|---|---|---|---|---|
| Inf_02 | 0.08 | 0.78 | 1.73 | N/A[2] | 2.72 |
| Inf_04 | 0.19 | 0.44 | 2.47 | N/A[2] | 1.10 |
| Inf_05 | 0.33 | 0.90 | 1.33 | N/A[2] | N/A[1] |
| Inf_06 | 0.26 | 0.97 | 1.82 | N/A[2] | N/A[1] |
| Inf_08 | 0.08 | 0.76 | 1.40 | 2.22 | N/A[3] |
| Inf_09 | 0.08 | 0.84 | 1.63 | 1.79 | N/A[3] |
| Inf_15 | 0.18 | 0.83 | 1.46 | N/A[2] | 1.43 |
| Inf_16 | 0.09 | 0.80 | 1.35 | 2.25 | 2.35 |
| Range Values* | [0.08–0.33] | [0.44–1.17] | [1.33–2.47] | [1.79–2.66] | [1.10–2.72] |

N/A[1] – Not applicable since it did not occur a 20% reduction of the panel OOP maximum strength (post-peak stage)
N/A[2] – Not applicable since it did not occur panel OOP collapse
N/A[3] – Not applicable since the collapse happened before a 20% reduction of the panel OOP maximum strength

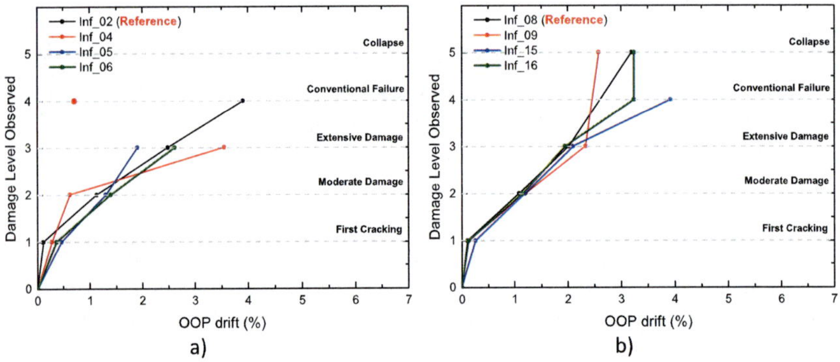


**Fig. 2.** Drift associated with each damage state: a) Group 1; and b) Group 2.

between 3.24% and 3.92% and collapse between 2.58% and 3.83%. Please note that damage level 1 was observed between 0.11% and 0.47%, but damage level 2 was only observed for drifts higher than 1.09%. Thus, it was extended the drift range related to damage level 1 until 1.09%. The same assumption was performed to the moderate damage (the upper drift limit was shifted to 1.91%).

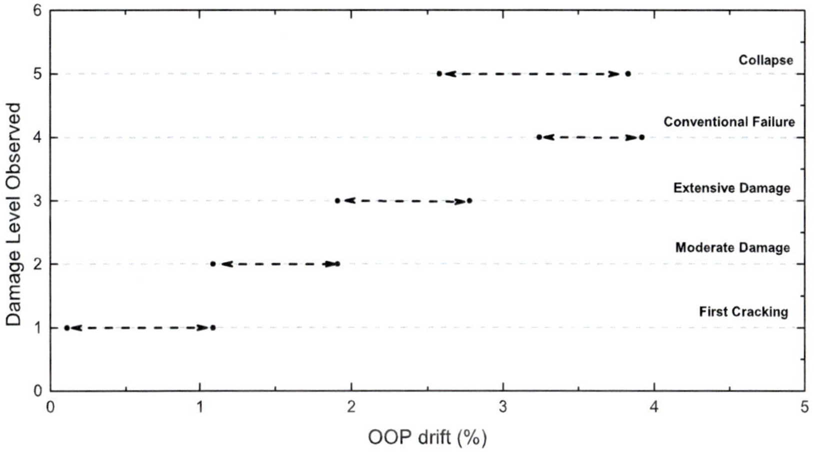


**Fig. 3.** Drift range obtained from the experimental tests.

## 3.2 Proposal of a Damage Index Quantification

The goal for an optimum damage assessment method should be its ability for general applicability, i.e., it must be able and valid for a set of structural systems based on a simple formulation to generate easily interpretable results consequently. The development of damage quantification methods for infill walls under pure OOP loadings considering the hysteretic energy dissipation was not yet approached in the literature. Nevertheless, in general, there are methods to quantify OOP (or IP/OOP) damage states based on displacement domains [18].

The practical usefulness of a damage index (DI) model is to provide an insight concerning the damage state of walls based on the OOP displacement, energy dissipation and maximum strength. The proposed DI model is based on the damage evolution of infill walls tested under OOP loadings and can be used to assess the damage condition of walls simulated using detailed micro-modelling or simplified macro-modelling approaches. It can also be very useful for calibrating the hysteretic OOP behaviour of the numerical models.

As it can be learned from the literature, damage indexes can be classified as cumulative or noncumulative. Noncumulative indexes relate the state of damage to peak response quantities and do not account for cyclic loading effects. Cumulative indices include part or all of the loading history to predict the capacity reduction due to cyclic repetitive loading. The following reviews of damage indexes can only be found in the literature dedicated to reinforced concrete structures [19, 20].

The methodology to compute the DI is adapted from the method proposed by Park and Ang [21]. The DI is defined by a linear combination of the normalized maximum deformation and the normalized dissipated hysteretic energy resulting from cyclic loading. The DI is therefore expressed by Eq. 1:

$$DI = \frac{d_{max}}{d_u} + \beta \frac{\int dE}{F_{max} d_u} \tag{1}$$

where $d_{max}$ is the maximum OOP displacement reached in each loading cycle and $d_u$ is the maximum OOP displacement reached by the wall. $\int dE$ is the area of the dissipated energy and $F_{max}$ is the maximum OOP strength predicted for the infill wall, and can be estimated using Eq. 2. Finally, β is the degradation parameter representing the influence of cyclic response on the damage. The DI range values vary from 0 to greater than 1 (being at the end of the test $d_{max}/d_u = 1$ and therefore DI = 1 + something). The total damage (collapse) is assumed when DI ≥ 1. The proposed DI is based on the original Park and Ang [21] proposal. The authors suggested to use an analytical equation to estimate $F_{max}$. Thus the calibration of the DI based on analytical expressions may allow the use of this approach in the absence of experimental data. Concerning Eq. 2, the authors understand the reviewer's point of view. They agree that this expression is mainly useful for vertical arch mechanisms. Nevertheless, this equation also seems to predict the maximum strength of the tested walls well. Other expressions can be used to estimate the maximum strength more accurately for other walls with other expected failure mechanisms (e.g. double-arching mechanism). Future calibrations must be performed using other prediction equations to estimate the Fmax of infill walls with different failure modes.

The methodology adopted to compute the DI model comprises five steps, as described below:

i) $d_{max}$ is determined in each cycle;
ii) $d_u$ is defined as the maximum OOP displacement reached by the wall;
iii) $\int dE$ is computed;
iv) $F_{max}$ is computed based on the analytical model Eurocode 6 [22] proposed to estimate the OOP strength of masonry walls or infills, respectively, without strengthening strategies. This model suggests that the maximum OOP distributed load ($q_{max}$) can be estimated based on a one-way (vertical) arch resisting mechanism. Thus the maximum horizontal uniformly distributed load ($q_{max}$) can be defined as in Eq. (2). Then the $F_{max}$ was estimated, taking into account the maximum distributed load $q_{max}$ and the panel area;

$$q_{max} = f_{mv}\left(\frac{t_w}{H_w}\right)^2 \tag{2}$$

v) β is adjusted to fit with damage observations (e.g. the collapse only occur for DI ≥ 1).

Finally, it must be underlined that the wall hysteretic behaviour causes damage. Part of this damage is directly proportional to the wall displacement, i.e. $\frac{d_{max}}{d_u}$, , and the other part is proportional to the energy dissipation (i.e. $\frac{\int dE}{F_{max} d_u}$). The β is used to weigh the energy dissipation effect on the damage quantification.

For each masonry infill wall, the β estimated is presented in Table 5.

The results are plotted in Fig. 4 in terms of DI evolution during the cyclic tests. The corresponding drift for each damage state observed during the tests (reported in the previous section) is presented in the DI curves. The DI values were obtained with the specific β values that were calibrated based on the experimental force-displacement results and damage observations.

**Table 5.** β parameter assumed to calculate the damage index.

| Specimens | β |
|---|---|
| Inf_02 | 0.046 |
| Inf_05 | 0.068 |
| Inf_06 | 0.039 |
| Inf_08 | 0.061 |
| Inf_09 | 0.045 |
| Inf_15 | 0.026 |
| Inf_16 | 0.055 |

From these results, it is possible to observe that the first cracking of the walls occurred for DI between 0.03 and 0.21. The moderate damage state was reached for DI values between 0.15 and 0.56, the extensive damage ranging from 0.26 to 1.04. The collapse occurred for DI values upper than 1. Based on this, the following damage index range values for each specific damage state are herein proposed: first cracking $0 < DI \leq 0.25$; moderate damage $0.25 < DI \leq 0.60$; extensive damage $0.60 < DI \leq 1$ and collapse DI $\leq 1$.

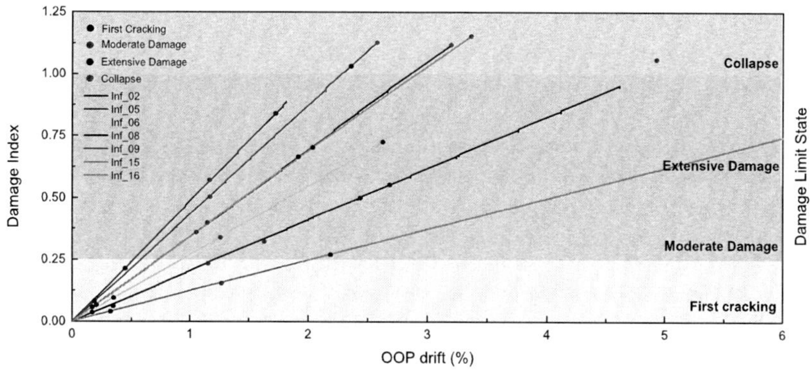


**Fig. 4.** Damage index evolution of the tested masonry infill walls.

The average value obtained for the β is 0.046 with a CoV of 29.56%. Nevertheless, the determination of the β parameter would benefit from more tests to reduce the coefficient of variation and effect of a high number of variables. However, this is the first proposal for a DI model for masonry infill walls. The β parameter found can be used in future simulations of masonry infill walls made of hollow clay horizontal bricks subjected to pure OOP loadings. Future works are needed to validate the β values herein proposed and compute β values for masonry infill walls with different characteristics (i.e. masonry units, boundary conditions, among others).

## 4 Conclusions

The study of the OOP behaviour of masonry infill walls has been the focus of numerous studies during the last years. Most of the studies available in the literature concerning the masonry infill walls' OOP behaviour cover their experimental response in terms of damage evolution, force-displacement, maximum strength and deformation capacity of the walls. However, none of them proposes a specific damage index for walls subjected to OOP loadings or discusses their hysteretic viscous damping, which is justified by the fact that many of those tests were carried out fundamentally applying a monotonic OOP load, not provoking the walls to collapse. However, it is fundamental to increase the knowledge concerning their non-linear behaviour when subjected to cyclic loadings.

Based on this motivation, the primary purpose of this manuscript was to provide the first proposal of damage states for masonry infill walls subjected to OOP loadings. After that, a pilot damage index model was proposed for masonry infill walls made of hollow clay horizontal brick units.

Regarding the damage states proposed for masonry infill walls subjected to OOP loadings, four damage states are proposed, namely: first cracking (damage state 1); moderate damage (damage state 2); extensive damage (damage state 3) and collapse (damage state 4). These damage states are proposed based on observing the damage evolution of the OOP response of the infill walls.

Moreover, a pilot application of a damage index (DI) model for masonry infill walls made of hollow clay bricks was carried out based on the proposal of Park and Ang [21]. Based on this proposal, it was observed that the first cracking ranged for a DI between 0 and 0.25, the moderate damage between 0.25 and 0.60, the extensive damage between 0.60 and 1 and the collapse for DI higher than 1. It was found an average value of 0.046 for β.

**Acknowledgments.** The first author is grateful for the Foundation for Science and Technology's support through funding UIDB/04625/2020 from the research unit CERIS. This work was also financially supported by: Project POCI-01*0145-FEDER-007457 - CONSTRUCT - Institute of R&D In Structures and Construction funded by FEDER funds through COMPETE2020 - Programa Operacional Competitividade e Internacionalização and by national funds through FCT - Fundação para a Ciência e a Tecnologia. This work was also supported by the Foundation for Science and Technology (FCT) - Aveiro Research Centre for Risks and Sustainability in Construction (RISCO), Universidade de Aveiro, Portugal [FCT/UIDB/ECI/04450/2020].

## References

1. Luca, F., et al.: The structural role played by masonry infills on RC building performances after the 2011 Lorca, Spain, earthquake. Bull. Earthq. Eng. **12**, 1999–2026 (2014)
2. Hermanns, L., Fraile, A., Alarcón, E., Álvarez, R.: Performance of buildings with masonry infill walls during the 2011 Lorca earthquake. Bull. Earthq. Eng. **12**(5), 1977–1997 (2013). https://doi.org/10.1007/s10518-013-9499-3
3. Masi, A., et al.: Seismic response of RC buildings during the Mw 6.0 August 24, 2016 Central Italy earthquake: the Amatrice case study. Bull. Earthq. Eng. **17**(10), 5631–5654 (2019)

4. Calvi, G.M., Bolognini, D.: Seismic response of reinforced concrete frames infilled with weakly reinforced masonry panels. J. Earthq. Eng. **5**(2), 153–185 (2001)
5. De Risi, M.T., et al.: In-plane behaviour and damage assessment of masonry infills with hollow clay bricks in RC frames. Eng. Struct. **168**, 257–275 (2018)
6. Furtado, A., et al.: Out-of-plane behavior of masonry infilled RC frames based on the experimental tests available: a systematic review. Constr. Build. Mater. **168**, 831–848 (2018)
7. Anić, F., Penava, D., Abrahamczyk, L., Sarhosis, V.: A review of experimental and analytical studies on the out-of-plane behaviour of masonry infilled frames. Bull. Earthq. Eng. **18**(5), 2191–2246 (2019). https://doi.org/10.1007/s10518-019-00771-5
8. Mansouri, A., Marefat, M.S., Khanmohammadi, M.: Experimental evaluation of seismic performance of low-shear strength masonry infills with openings in reinforced concrete frames with deficient seismic details. Struct. Design Tall Spec. Build. **23**(15), 1190–1210 (2014)
9. Chiou, T.-C., Hwang, S.-J.: Tests on cyclic behavior of reinforced concrete frames with brick infill. Earthq. Eng. Struct. Dynam. **44**(12), 1939–1958 (2015)
10. Blasi, G., De Luca, F., Aiello, M.A.: Brittle failure in RC masonry infilled frames: the role of infill overstrength. Eng. Struct. **177**, 506–518 (2018)
11. Asteris, P.G.: Lateral stiffness of brick masonry infilled plane frames. J. Struct. Eng. **129**(8), 1071–1079 (2003)
12. Furtado, A., et al.: Experimental evaluation of out-of-plane capacity of masonry infill walls. Eng. Struct. **111**, 48–63 (2016)
13. Furtado, A., et al.: The use of textile-reinforced mortar as a strengthening technique for the infill walls out-of-plane behaviour. Compos. Struct. **255**, 113029 (2021)
14. De Risi, M.T., et al.: Experimental analysis of strengthening solutions for the out-of-plane collapse of masonry infills in RC structures through textile reinforced mortars. Eng. Struct. **207**, 110203 (2020)
15. Furtado, A., et al.: Experimental investigation on the possible effect of previous damage, workmanship and test setup on the out-of-plane behaviour of masonry infill walls. J. Earthq. Eng. **26**, 1–32 (2021)
16. Furtado, A., et al.: Effect of the panel width support and columns axial load on the infill masonry walls out-of-plane behavior. J. Earthq. Eng. **24**(4), 653–681 (2020)
17. Ricci, P., Di Domenico, M., Verderame, G.M.: Experimental investigation of the influence of slenderness ratio and of the in-plane/out-of-plane interaction on the out-of-plane strength of URM infill walls. Constr. Build. Mater. **191**, 507–522 (2018)
18. Donà, M., et al.: A new macro-model to analyse the combined in-plane/out-of-plane behaviour of unreinforced and strengthened infill walls. Eng. Struct. **250**, 113487 (2022)
19. Williams, M.S., Sexsmith, R.G.: Seismic damage indices for concrete structures: a state-of-the-art review. Earthq. Spectra **11**(2), 319–349 (1995)
20. Chung, Y., Meyerand, C., Shinozuka, M.: Seismic damage assessment of reinforced concrete members, NCEER-87-0022. National Center for Earthquake Engineering Research (1987)
21. Park, Y.J., Ang, A.H.S.: Mechanistic seismic damage model for reinforced concrete. J. Struct. Eng. **111**(4), 722–739 (1985)
22. CEN, Eurocode 6: Part 1-1 – General Rules for buildings – Rules for reinforced and unreinforced masonry, European Committee for Standardisation, Brussels (2005)

# Numerical Evaluation of the Energy Dissipation Capacity of Steel Members Subjected to Biaxial Bending

E. Cerqueira(✉), C. Eshaghi, R. Peres, and J. M. Castro

Faculdade de Engenharia Civil, Universidade do Porto, Porto, Portugal
up202100753@edu.fe.up.pt

**Abstract.** The behaviour of steel structures under seismic actions has been widely studied in recent decades. The monotonic behaviour of steel elements subjected to uniaxial bending is well understood. However, the characterization of the cyclic behaviour and energy dissipation capacity still requires investigation, namely to what concerns steel elements subjected to biaxial bending, an issue of paramount interest for the seismic analysis and performance assessment of steel buildings. The consideration of modelling and damage assessment criteria derived for members subjected to uniaxial bending may not be fully adequate for the seismic analysis of buildings composed by structural steel members subjected to biaxial bending.

This research aims to investigate the behaviour of steel members subjected to biaxial cyclic bending and compare it with the behaviour under uniaxial loading conditions, with a focus on the energy dissipation capacity of the members. Detailed 3D finite element analyses are conducted to analyse the effects of bidirectional lateral cyclic loading on the hysteretic response of steel members under different uniaxial and biaxial lateral displacement-controlled load paths. The energy dissipation capacity of the members is evaluated in terms of cumulative dissipated energy and individual cycle dissipated energy. The numerical results indicate that biaxial cyclic loading paths can significantly influence the rotation capacity of steel members, however, in terms of ductility and energy dissipation capacity, this study shows that loads with bidirectional trajectories can be negligible, since they do not significantly change the energy dissipation capacity. Furthermore, unidirectional trajectories with predominant components in the y-direction reduce the elements' capacity to dissipate energy.

**Keywords:** Biaxial bending · Hysteretic behaviour · Energy dissipation · Steel members · Cyclic loading

## 1 Introduction and Background

In the last decades, the state of knowledge and methods used to predict collapse under earthquake loading has evolved significantly [1]. In earthquake events, seismic action excites the buildings in different directions simultaneously. To accurately capture the building's response, either by its orientation towards the seismic action or by its structural irregularities, three-dimensional models are frequently used. Although the response


H. Varum et al. (Eds.): IWEBSE 2023, LNCE 236, pp. 282–292, 2023.
https://doi.org/10.1007/978-3-031-36562-1_22

characteristics of steel members under monotonic and unidirectional lateral cyclic loading are well understood, it is important to study the behaviour of structural elements subjected to biaxial cyclic bending [2].

Studies conducted on RC members [3] concluded that biaxial horizontal cyclic loading can increase the strength and stiffness degradation in comparison to uniaxial loading. However, there are few studies that aimed to investigate the performance of steel members under bidirectional lateral cyclic loading, most of them focused on thin-walled profiles [4].

Since buildings made with European open sections (e.g., IPE and HE profiles) are typically braced only in one direction, this study aims to evaluate the influence of biaxial bending on the behaviour of these members, in terms of resistance, deformation, and energy dissipation capacity. To achieve this objective, steel members with IPE sections are analysed using finite element analysis (FEA) under unidirectional and bidirectional cyclic lateral loading, with different loading paths. The performance indices obtained for members under bidirectional loading are then compared with the ones obtained for members subjected to unidirectional loading conditions.

## 2 Numerical Study

### 2.1 FE Model

Two specimens with different cross-sections (commercial IPE300 and IPE600) were considered in this study, and their dimensions are presented in Table 1. The length (L) of both members is 1350 mm. Both specimens are assumed to be made of structural steel S355 with a yield plateau and an initial yield stress of 288 MPa.

Finite element analysis is carried out using the commercial software ABAQUS [5]. The model consists of a cantilever member that is fixed at its base, with the edge fully restrained to ensure fixed boundary conditions. The member is excited from its top with cyclic lateral loads in the x-direction and y-direction, as shown in Fig. 1. Both cross-section edges were constrained as a rigid body to avoid stress concentration. The member plates were modelled using quadratic shell elements S4R, which are known to capture well the post-buckling behaviour.

A very refined mesh size of b/24 (where b is the flange width) was adopted to ensure accuracy, and the same mesh density was used along the full length of the members to capture any lateral and torsional buckling. Geometrical imperfections are included in the model by considering the manufacturer tolerances on shape and dimensions allowed by the European Standard EN10034:1993 [6] for I and H members. This approach is based on the work of Elkady and Lignos [7] and Mohabeddine *et al.* [8]. To simulate the cyclic behaviour of the material, the combined nonlinear isotropic/kinematic hardening plasticity model based on the Chaboche proposal [9] available in the ABAQUS materials library was adopted.

The finite element model was then validated against experimental tests on different members, and the outcomes are depicted in Fig. 2., as described in Eshaghi *et al.* [10]. The validation of the numerical model was extended to evaluate its energy dissipation capacity. The results of the experimental test conducted by D'Aniello *et al.* [11] on IPE300 were compared to those obtained by the numerical model used in this study.

Figure 3 shows the hysteresis curves obtained for comparison purposes. The difference between the dissipated energy obtained in the experimental tests and the numerical model used is around 4%.

**Table 1.** IPE300 and IPE600 dimensions and Nonlinear Isotropic and Kinematic hardening parameters.

| Column # | *h* (mm) | *b* (mm) | *tw* (mm) | *tf* (mm) |
|---|---|---|---|---|
| IPE300 | 300 | 150 | 7.1 | 10.7 |
| IPE600 | 600 | 220 | 12 | 19 |

For all columns: Length (L) = 1350 mm;
Material: $E = 171.89$ GPa; $\sigma_0 = 288$ N/mm$^2$
Isotropic: $Q_\infty = 78.97$ N/mm$^2$; $b = 2.06$
Kinematic component: $C = 14.127$ N/mm$^2$; $\gamma = 97.87$

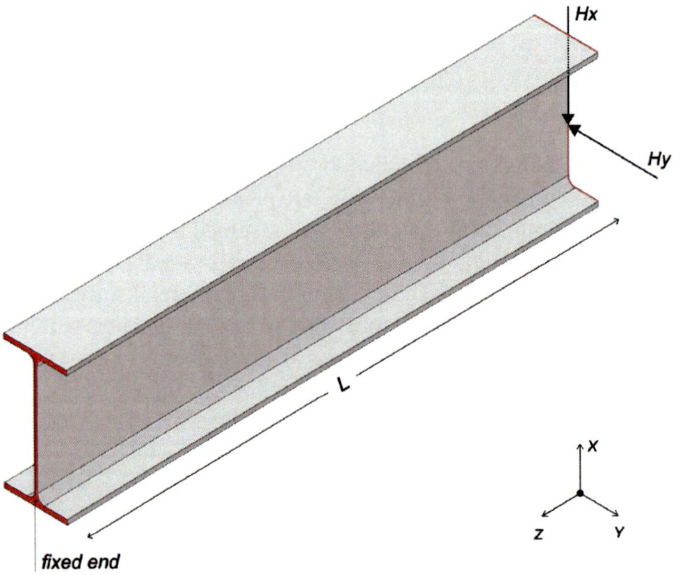


**Fig. 1.** 3D representation of member model

## 2.2 Loading Paths and Protocol

The cyclic lateral displacements were imposed at the top of the member using 11 quasi-static cyclic lateral loading protocols. Eight of these protocols were unidirectional with six of them having components in x and y directions (diagonal loads) and three were bidirectional.

The load protocol used in this study is based on the symmetric cyclic loading protocol provided in ANSI/AISC 341–16 [12]. The loading was displacement-controlled and

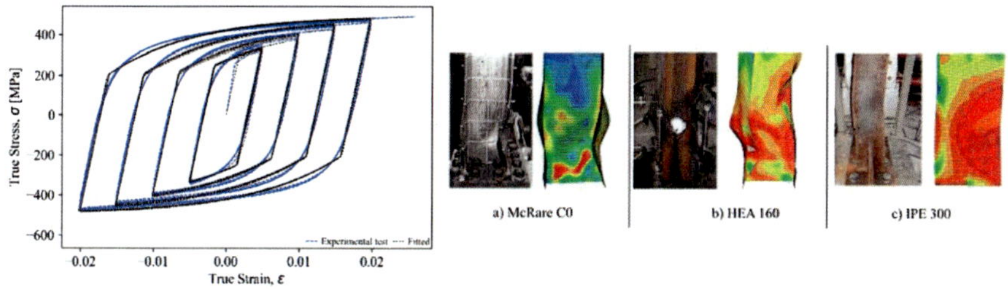


**Fig. 2.** Comparison between the buckling behaviour simulated in the numerical model and from the numerical model obtained by Eshaghi *et al.* [10].

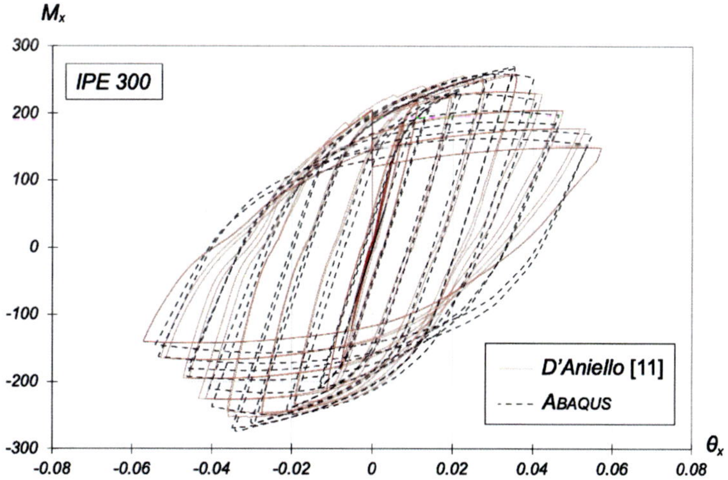


**Fig. 3.** Comparison between the hysteretic behaviour obtained in the experimental tests by D'Aniello et al. [11] and from the numerical model used in this work with ABAQUS software.

applied with increasing amplitudes. The following nominal peak displacement levels (in mm) were considered: 3.75, 5, 7.50, 10, 15, 20, 30, 40, 50, 60, 70, 80, 90 and 100. For the first three displacement peaks (3.75, 5.00 and 7.50) six cycles are imposed. Then, four cycles are performed with a peak of 10 mm, followed by two cycles with peaks of 15 and 20 mm. Afterwards, the load was applied with increments of 10 mm and two cycles are performed at each step. This loading protocol is illustrated in Fig. 4(c).

The schematic representation of the load paths is illustrated in Fig. 4(b). In the unidirectional 0° path, the specimens are only excited in the x-direction. In the unidirectional 90° path, the specimens are only excited in the y-direction. In the unidirectional [0°, 90°] path, six different paths are analysed: 5°, 20°, 30°, 45°, 60 and 75° (all of them with respect to the x-direction). In these cases, the applied displacements have components in both the x and y directions. As for the bidirectional loading, the paths consist of displacement cycles having square, circular and rhombus shapes in the lateral displacement space. For all the 11 cases, the loading protocol used is the one presented earlier, with

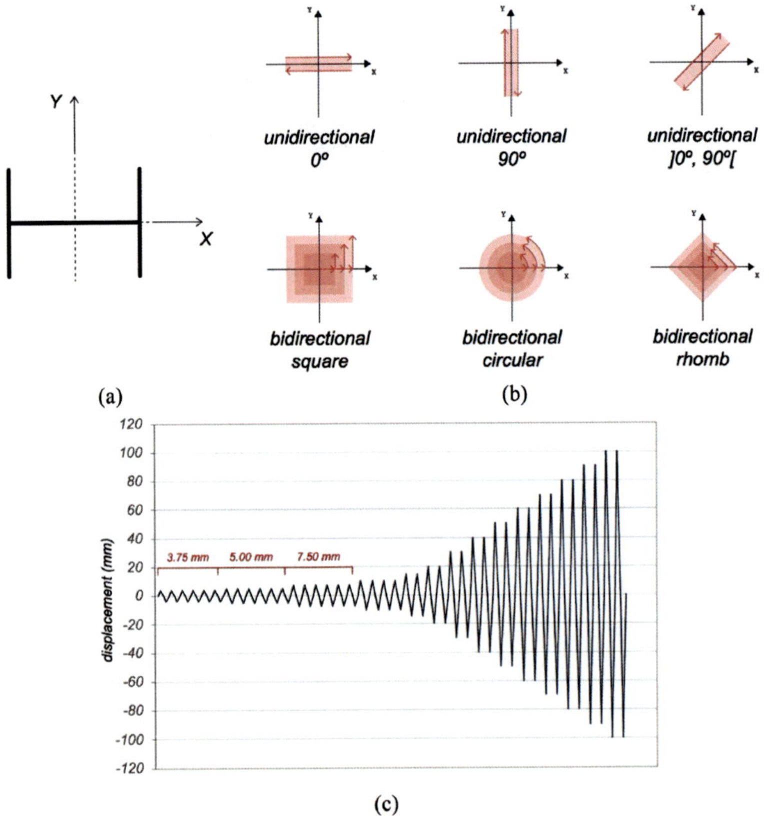


**Fig. 4.** (a) Profile axis, (b) Loading paths and (c) protocols used in the model.

increasing amplitudes of lateral displacement. In the present research, axial loads were not applied to the profiles.

## 3 Global Member Behaviour

Through the analysis of the hysteretic force-rotation curves of all tested members, it is possible to draw some conclusions when comparing the results obtained by members with unidirectional and bidirectional forces.

Figure 5 depicts the hysteretic curves of IPE600 for the uniaxial lateral load path (with 0° and 90°) and all bidirectional lateral load paths (circular, rhombus and square). In this figure, $M_x$ and $M_y$ are the horizontal force components in the x-direction and y-direction, respectively. The shape of the lateral force-deformation loops of IPE300 are similar to those of IPE600 and, for brevity, are not presented.

- The hysteretic behaviour recorded under bidirectional lateral loading (b-d) is quite different from the one recorded under unidirectional lateral loading (a) due to strong coupling effects between the two orthogonal directions.

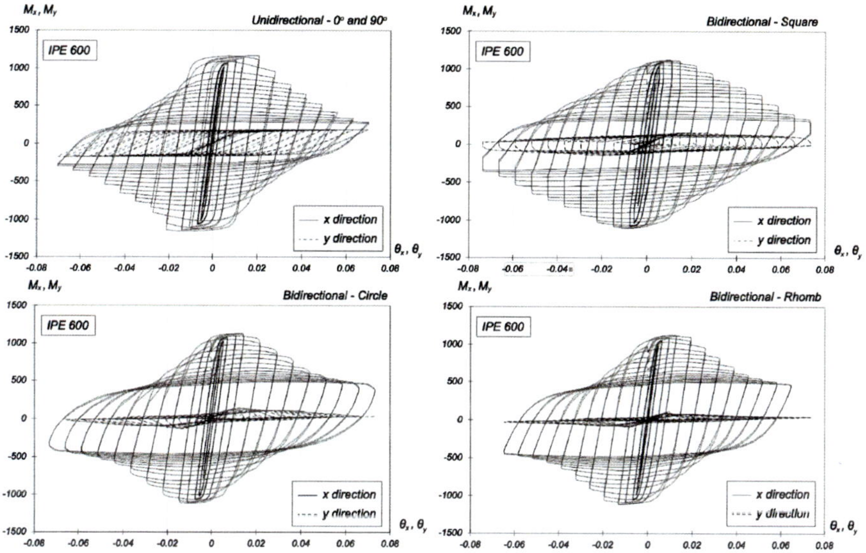


**Fig. 5.** Cyclic hysteresis moment-rotation curves for IPE600 members under (a) unidirectional load path, (b) squared load path, (c) circular load path and (d) rhombus load path.

- With respect to the maximum moment, when comparing the results obtained for each biaxial analysis against the uniaxial analysis in one specific direction (x or y), it can be noticed that lower values were obtained for all biaxial tests than uniaxial ones. This reduction is particularly evident in the weakest direction (y direction), where the biaxial loading induces a reduction of the maximum moment of 20–40%. However, in the stronger direction (x-direction), the observed reductions are small, around 3%.
- The rotation capacity of the profile subjected to unidirectional loads is greater than that obtained by profiles subjected to bidirectional loads. This is especially relevant since the lower rotation capacity is associated with reduction of the seismic performance of the structure.

# 4 Dissipated Energy

## 4.1 Individual Cycle Energy

According to Ucak and Tsopelas [4] and Rodrigues *et al.* [13], the energy dissipated during each displacement cycle can be evaluated by summing the areas enclosed by the hysteresis curves in the x and y directions, as shown in Fig. 6 and presented in Eq. 1.

$$E = \int M_x \theta_x + \int M_y \theta_y \tag{1}$$

The energy dissipated during each individual loading cycle and the accumulated energy dissipation along each analyzed member were calculated. The results for the IPE600 member are presented in Fig. 7 (a–j), where it can be observed that the first cycle

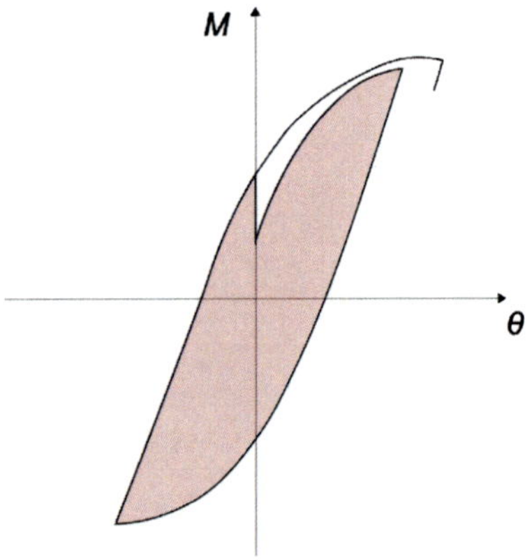


**Fig. 6.** Definition of dissipated energy per cycle

of each displacement peak had the highest energy dissipation compared to subsequent cycles with the same displacement peaks.

For members under unidirectional cyclic loading, the reduction of the energy dissipated in subsequent cycles is approximately 10% of the energy dissipated in the first cycle, with this difference decreasing as the angle of applied load increases. However, for bidirectional load paths, the reduction in energy dissipation is more evident, reaching about 20%. This could be attributed to the induced damage in the first cycle, which reduces the stiffness and resistance of the member, thus reducing its energy dissipation capacity in subsequent cycles.

Additionally, a significant drop in energy dissipation was observed after the occurrence of local buckling of the plates that constitute the profile, resulting in a high level of degradation of the member strength.

### 4.2 Cumulative and Total Energy Dissipation

As mentioned earlier, the cumulative dissipated energy was also calculated. Figure 7 shows the cumulative energy dissipated for each of the analysed cases, in addition to the energy dissipated individually in each cycle (for brevity, only the results for the IPE600 profile will be presented).

Figure 8(a) compares the cumulative dissipated energy per resultant rotation (Eq. 2) for the profiles subjected to unidirectional loads with components in the x-direction at angles 0°, 5°, 20°, 30°, 45°, 60° and 75°. Figure 7(b) compares the cumulative dissipated energy obtained when applying unidirectional loads only with components in the x-direction (0°) and when applying bidirectional loads (circular, square and rhombus). Finally, Fig. 9 presents the total dissipated energy obtained for each of the load paths analysed in this work.

After analyzing the results, several conclusions can be drawn:

- Regarding unidirectional loads, it can be noticed that the cumulative dissipated energy curves obtained for angles of 0°, 5°, 20° and 30° with the x-direction are identical.

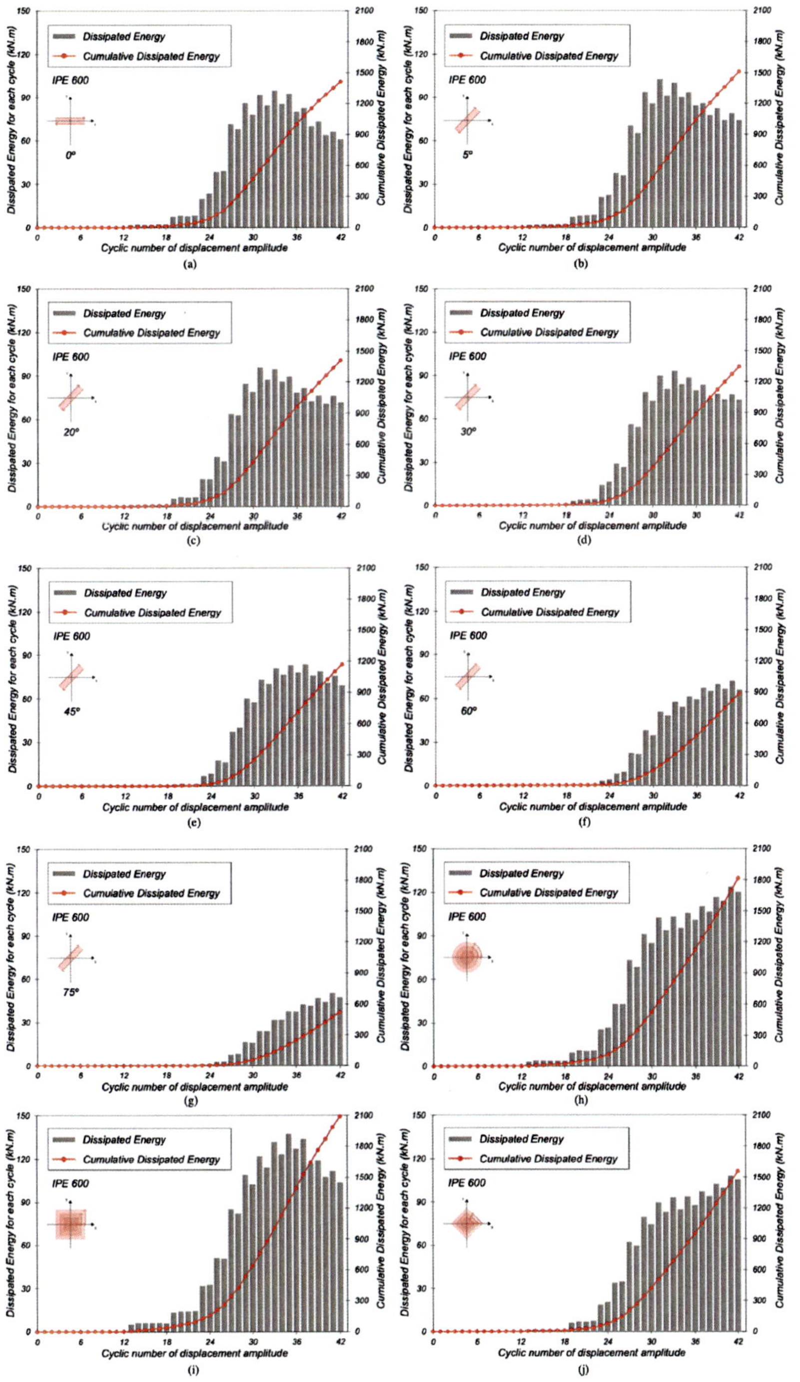


**Fig. 7.** Individual cycle energy and cumulative dissipated energy for IPE600.

- On the other hand, the energy curve referring to the load with larger angles (45°, 60° and 75°) with the x-direction presents the lowest total energy dissipation, since in it, the y-component is much higher, which is expected due to bending of the minor inertia axis.
- Unidirectional load paths with larger x-components and smaller y-components generally have energy dissipation results similar to those obtained for loads only in the x-direction. The biaxial component becomes more relevant in terms of ductility loss when the angle is up to 30 [14].
- Bidirectional loads generally dissipate more energy in biaxial load paths than in the corresponding unidirectional paths. However, when analysing the dissipated energy obtained in the columns under unidirectional load in x-direction, the resulting energy evolution is very similar to that obtained from rhombus load path analyses.
- The rhombus load path dissipates less energy than the other biaxial load paths and dissipates around 25% less than the squared load, which is the load path that dissipates the most energy.

$$\theta_{resultant} = \sqrt{\theta_x^2 + \theta_y^2} \tag{2}$$

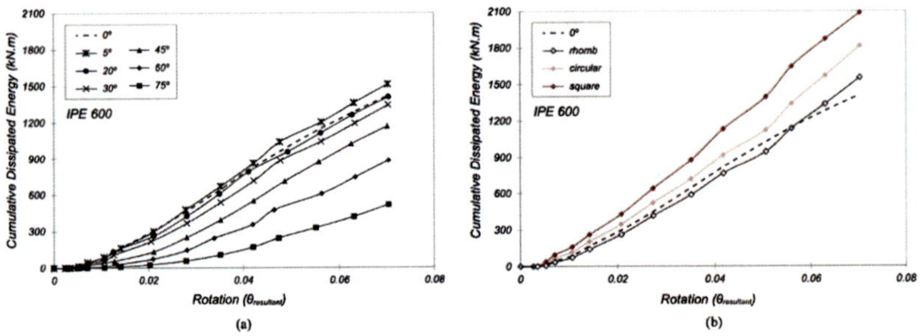


**Fig. 8.** Comparison of cumulative dissipated energy for members with different load paths (unidirectional and bidirectional loads).

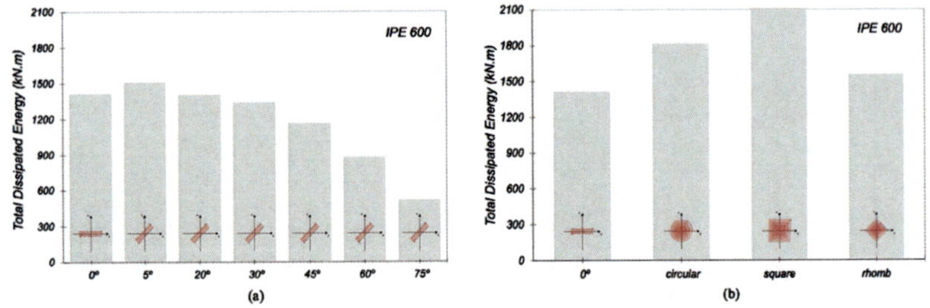


**Fig. 9.** Comparison of total dissipated energy for members with different load paths (unidirectional and bidirectional loads).

## 5 Conclusions

In this numerical study, IPE300 and IPE600 members were analyzed under both unidirectional and bidirectional cyclic loading to investigate their energy dissipation capabilities. The study yielded several important conclusions.

Firstly, the results showed that unidirectional load paths with larger components in the x-direction and smaller components in the y-direction have similar energy dissipation results to those obtained for loads applied only in the x-direction. Therefore, the biaxial component of these load paths can be disregarded without significant effects on energy dissipation capacity, except when the angle is up to 30°, which leads to ductility loss.

Secondly, the study found that unidirectional loads applied at larger angles (45°, 60°, and 75°) in the x-direction have the lowest total energy dissipation, as expected due to the bending of the minor inertia axis.

Thirdly, regarding bidirectional loads, the results showed that dissipated energy in biaxial load paths is generally higher than in corresponding unidirectional paths. Nevertheless, the dissipated energy obtained in the columns under unidirectional load in the x-direction presents a similar energy evolution to that obtained from rhombus load path analyses.

Lastly, the study observed that members subjected to bidirectional loads with load paths of circular, squared, and rhombus shapes exhibit greater energy dissipation than those with components only in the x-direction.

Overall, the findings provide encouragement to extend the methodology adopted in this study to members with other cross-sectional shapes, particularly hollow members. The authors also recognize the importance of evaluating the influence of the level of axial load applied to the profiles, which is currently underway for future publications. The authors are currently working on widening the scope of this numerical investigation and seeking experimental validation for the results obtained.

**Acknowledgments.** This research was supported by UID/ECI/04708/2019 - CONSTRUCT - Instituto de I&D em Estruturas e Construções, funded by national funds through the FCT/MCTES (PIDDAC). The first author acknowledges the financial support provided by "FCT – Fundação para a Ciência e Tecnologia", Portugal, namely through the Ph.D. grant with reference 2022.12364.BD.

## References

1. Lignos, D.G., Krawinkler, H.: Deterioration modeling of steel components in support of collapse prediction of steel moment frames under earthquake loading. J. Struct. Eng. **137**(11), 1291–1302 (2011)
2. Rodrigues, H., Furtado, A., Arêde, A., Varum, H.: Estudo experimental do comportamento sísmico de pilares de betão armado sujeitos à exão biaxial. Revista Portuguesa de Engenharia de Estruturas, Ed. LNEC. Série III. **1**, 91–98 (2016)
3. Rodrigues, H.: Biaxial seismic behaviour of reinforced concrete members, PhD Thesis, Departamento de Engenharia, Universidade de Aveiro, Aveiro (2012)
4. Ucak, A., Tsopelas, P.: Load path effects in circular steel members under bidirectional lateral cyclic loading. J. Struct. Eng. **141**, 1–11 (2014)

5. Smith, M. ABAQUS/Standard User's Manual, Version 6.9. Dassault Systèmes Simulia Corp. (2009)
6. Standards, B.: BS EN 10034:1993 Structural steel I and H sections Tolerances on shape and dimensions (1993)
7. Elkady, A., Lignos, D.G.: Improved seismic design and nonlinear modeling recommendations for wide-flange steel members. J. Struct. Eng. **144**(9), 04018162 (2018)
8. Mohabeddine, A., Koudri, Y.W., Correia, J.A.F.O., Castro, J.M.: Rotation capacity of steel members for the seismic assessment of steel buildings. J. Eng. Struct. **244**, 112760 (2021)
9. Chaboche, J.L.: Time-independent constitutive theories for cyclic plasticity. Int J Plast **2**(2), 149 (1986)
10. Eshaghi, C., Mazarei, A., Barros, R.C., Romão, X., Castro, J.M.: Machine learning approach for predicting flexural behavior variations of I-shaped steel beams due to manufacturing tolerances. COMPDYN, Athens, Greece (2023)
11. D'Aniello, M., Landolfo, R., Piluso, V., Rizzano, G.: Ultimate behavior of steel beams under non-uniform bending. J. Constr. Steel Res. **78**, 144–158 (2012)
12. ANSI/AISC. AISC 341–16: Seismic Provisions for Structural Steel Buildings, AMERICAN I. Chicago (2016)
13. Rodrigues, H., Varum, H., Arêde, A., Costa, A.: A comparative analysis of energy dissipation and equivalent viscous damping of RC members subjected to uniaxial and biaxial loading. Eng. Struct. **35**, 149–164 (2012)
14. Araújo, M., Macedo, L., Castro, J.M.: Evaluation of the rotation capacity limits of steel members defined in EC8-3. J Constr Steel Res. **135**, 11–29 (2017)

# Experimental Study of RC Frames with Window and Door Openings Under Cyclic Loading

Nisar Ali Khan[1,2], Lin Zhou[3], Fabio Di Trapani[4](✉), Cristoforo Demartino[5,6], and Giorgio Monti[7]

[1] Department of Architecture, University Roma Tre, Largo Giovanni Battista Marzi, 10, 00153 Roma, Italy

[2] Department of Civil Engineering, Faculty of Engineering and Technology, International Islamic University, Sector H-10, Islamabad 44000, Pakistan

[3] College of Civil Engineering and Architecture, Zhejiang University, Hangzhou, China
linzhou1990@zju.edu.cn

[4] Dipartimento di Ingegneria Strutturale Edile e Geotecnica, Politecnico di Torino, Corso Duca degli Abruzzi 24, 10129 Turin, Italy

[5] Zhejiang University- University of Illinois at Urbana Champaign Institute, Zhejiang University, Haining, Zhejiang, China

[6] Department of Civil and Environmental Engineering, University of Illinois at Urbana-Champaign, Urbana, IL 61801, USA

[7] Dipartimento di Ingegneria Strutturale e Geotecnica, Sapienza Università di Roma, Rome, Italy

**Abstract.** This paper aims to experimentally investigate the effect of different types of openings on the cyclic response of masonry infilled reinforced concrete frames. Six 2/3-scale square specimens, consisting of bare frame, fully infilled walls, and walls with door or window openings arranged with hollow clay bricks were tested under quasi-static lateral cyclic loading up to large drifts. The experimental responses were analyzed in terms of strength, stiffness, ductility, energy dissipation, equivalent damping and damage mechanisms and compared against those observed in the reference bare frame and fully infilled frame. Results indicated that the typology of the openings significantly alter the resisting mechanisms, although without substantial modification to the lateral resistance capacity. Furthermore, the limit state thresholds were shown to be achieved at substantially different inter-storey drifts, suggesting differing damage metrics.

**Keywords:** Infilled RC Frames · Openings · Masonry · Experimental testing · Energy dissipation · Cyclic response

## 1 Introduction

The vital role played by masonry infills in determining the seismic performance of reinforced concrete (RC) frame structures has been the subject of extensive research through post-earthquake damage observations. Reflecting this, numerous researchers have conducted comprehensive experimental studies in this area over the last 50 years. The field


H. Varum et al. (Eds.): IWEBSE 2023, LNCE 236, pp. 293–303, 2023.
https://doi.org/10.1007/978-3-031-36562-1_23

of RC frames with solid infills has seen numerous valuable experimental campaigns, including those by [1–10]. These studies have made a significant contribution to the understanding of the behavior of RC frames with solid infills under seismic loading. However, tests conducted on infilled frames with openings are considerably fewer than those carried out on solid ones [11]. Among these tests, [12–14] investigated one-storey steel-infilled frames, while most of the experimental studies concentrated on RC infilled frames [15–25]. Moreover, [26–28] tested full-scale reinforced concrete multi-storey frames that included both solid infills and infills with openings. Based on the experimental test results, various authors have proposed empirical relationships to modify the equivalent strut approach to account for the reduction in strength and stiffness caused by the presence of openings. However, most existing empirical relationships only consider the opening area as an input parameter and do not account for the effect of opening typology on the seismic response of infilled frames. This simplified assumption is due to the limited knowledge of underlying damage mechanisms. Thus, further research is needed to develop more comprehensive models that take into consideration these factors. To achieve this, a more thorough understanding of the damage mechanisms related to infilled frames is crucial.

The study involved comparing the cyclic responses of infilled frames with openings to those of the bare frame and the specimens with solid infills, based on strength, stiffness, ductility, energy dissipation, equivalent damping and damage mechanisms. The aim was to gain a better understanding of how the mechanical response of infilled frames with window and door openings is modified, focusing on overall response parameters and damage mechanisms that are activated based on the typology of the openings within the infill.

## 2 Experimental Program

### 2.1 Specimen Details

In the experimental program, eight 2/3 scaled specimens were tested using a cyclic loading protocol. The specimens tested in this study included a bare frame, a fully infilled frame, and six infilled frames with openings. The latter were classified into three types: central window openings (CW), side window openings (SW), and central door openings (CD). Two replicas of each type, identified as F1 and F2, were arranged. Geometric details of specimens shown in Fig. 1.

The infills consisted of two wythes of hollow clay masonry bricks (Fig. 2). The mortar courses were approximately 10 mm thick. The RC frame used in all tests had geometrical and reinforcement details shown in Fig. 2.

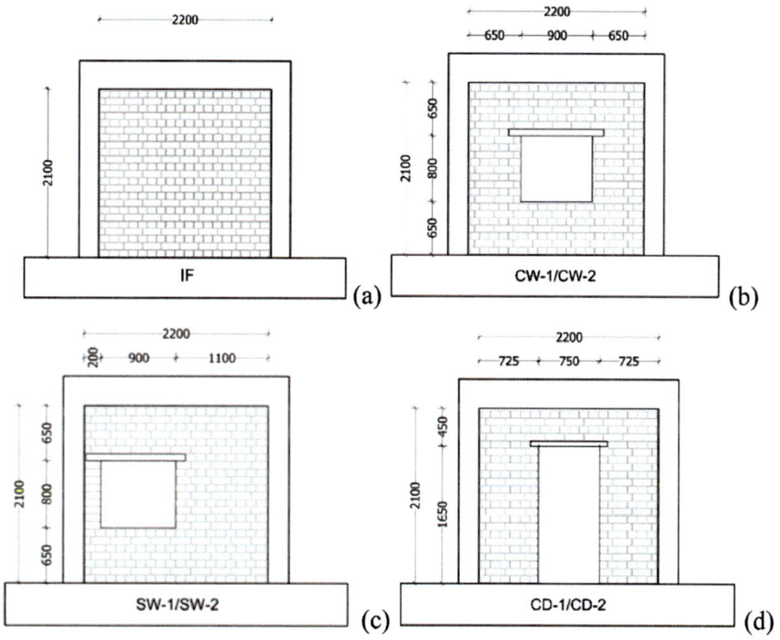


**Fig. 1.** Geometric details of specimens: (a) IF; (b) CW; (c) SW; (d) CD. Dimensions in mm.

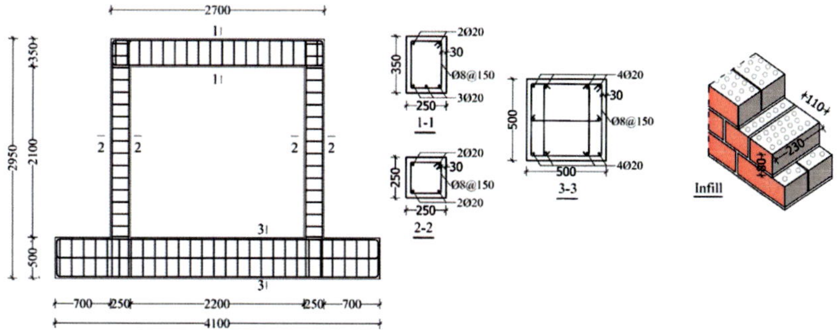


**Fig. 2.** Reinforcement details of the frame and arrangement of the masonry (dimensions in mm).

## 2.2 Material Properties

Comprehensive material testing was conducted, which included compressive tests on 150 × 150 × 150 mm concrete cubes and 50 × 50 × 50 mm mortar cubes, tensile tests on steel rebars, and compressive tests on bricks with a 20% opening percentage. The compressive tests on the bricks were carried out in the direction parallel to the holes. Table 1 present the results of these tests.

**Table 1.** Mechanical properties of materials.

| Specimen | Concrete (Compr. Strength) | Steel (Tensile strength) | | Brick (Compr. Strength) | Mortar (Compr. Strength) |
|---|---|---|---|---|---|
| | $f_c$ (MPa) | $f_y$ (MPa) | $f_t$ (MPa) | $f_b$ (MPa) | $f_{mr}$ (MPa) |
| #1 | 13.78 | 441 | 638 | 10.5 | 5.0 |
| #2 | 13.95 | 438 | 618 | 11.6 | 2.6 |
| #3 | 13.33 | 431 | 668 | 12 | 3.4 |
| #4 | 13.87 | 420 | 202 | 8.7 | 3.4 |
| #5 | 13.65 | 491 | 616 | 13.2 | 4.4 |
| #6 | 12.62 | 428 | 585 | 12.2 | 1.6 |
| #7 | | | | 26.9 | 3.8 |
| #8 | | | | 14.4 | 3.6 |
| #9 | | | | 14.6 | 2.3 |
| #10 | | | | 20.2 | 2.2 |
| #11 | | | | | 2.5 |
| #12 | | | | | 4.2 |
| Mean | 13.53 | 441.50 | 554.50 | 14.43 | 3.25 |
| Std. Dev | 0.50 | 25.37 | 174.85 | 5.35 | 1.02 |
| COV | 4% | 6% | 32% | 37% | 31% |

## 2.3 Masonry Properties

Masonry prisms measuring 900 × 750 × 230 mm were subjected to compression tests along the direction of the holes in the bricks, which is perpendicular to the mortar bed joints. The average compressive strength of the masonry specimens ($f_{mm}$) was found to be 3.36 MPa, and the average elastic Young's modulus ($E_m$) was 3730 MPa. Diagonal shear tests were also conducted on masonry wall specimens with dimensions of 620 × 620 × 230 mm, an average shear resistance in the absence of compressive loads ($f_{vm}$ = 0.356 MPa) and an average shear modulus ($G_m$ = 318.8 MPa).

## 2.4 Test Setup and Instrumentation

All specimens were tested at the structural laboratory of Zhejiang Technical University in Hangzhou, China. A pictures of a specimen in the testing apparatus is shown in Fig. 3. The horizontal lateral load was applied to the mid-height of the beam using a double-action hydraulic actuator connected to the reaction wall and hinged to the steel beam. The hydraulic jack had a stroke of ± 250 mm and a nominal load capacity of ±2000 kN. To transfer the horizontal force to the frame during reverse loading cycles, a system comprising two steel plates and eight prestressed rebars was employed on the steel beam. Furthermore, the vertical load was constant and equally distributed between the two columns (195 kN per column) via two independent hydraulic jacks, corresponding to an axial load ratio of 0.2. The distribution of the vertical load to the columns was facilitated through two steel plates that enabled replication of the effects of gravity loads.

**Fig. 3.** Test setup and instrumentation.

Each specimen underwent a cyclic displacement protocol consisting of 30 cycles, with each cycle repeated three times for every target displacement level. The testing procedure for displacements is beginning at ±10 mm and increasing by increments of 10 mm after completing each set of three cycles until reaching a maximum displacement level of 100 mm. This resulted in interstorey drift ranges from 0.47% to 4.7%. During all tests, measurements were taken for lateral force versus top displacement (mid-height of the RC frame beam) response while following the same testing protocol.

## 3 Experimental Results

The study compared and analyzed the results in terms of strength, stiffness, ductility, energy dissipation, equivalent damping and damage mechanisms. Specifically, the investigation focused on how opening typology influenced the response of the specimens, with a particular emphasis on comparing them to bare frame and fully infilled frame cases. The cyclic responses of all specimens are depicted in Fig. 4, while the damage patterns in correspondence of the peak load are shown in Fig. 5. Detailed interpretation of these outcomes is provided in subsequent sections for better understanding.

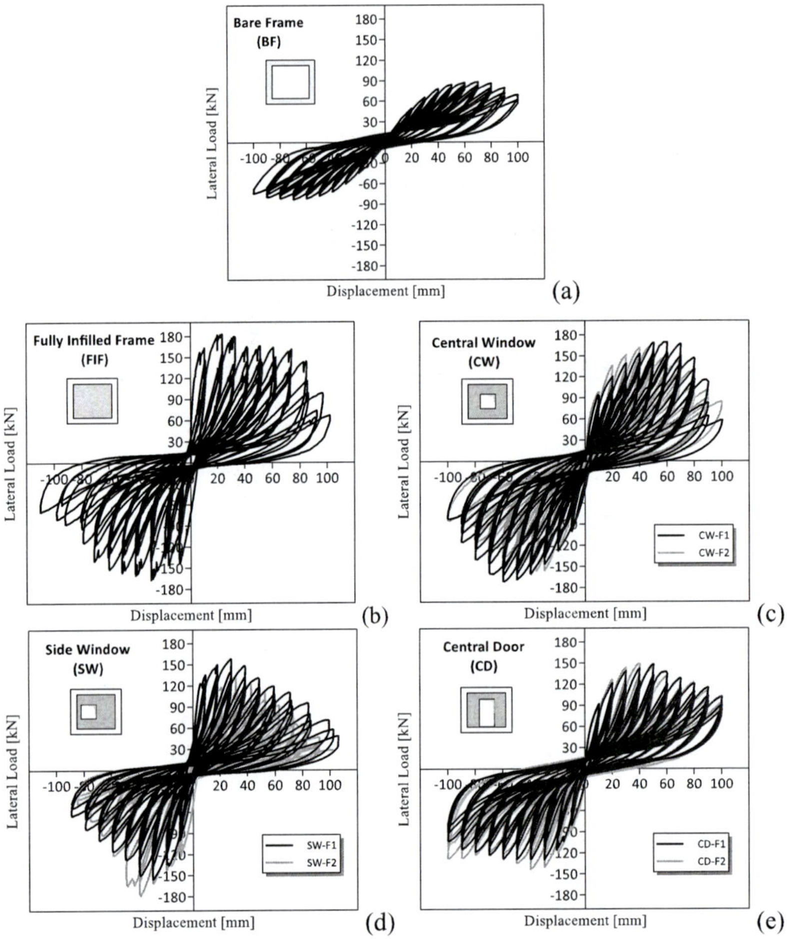


**Fig. 4.** Cyclic responses of the specimens: (a) Bare frame (BF); (b) Fully infilled frame (FIF); (c) Central window (CW); (d) Side window (SW); (e) Central door (CD).

### 3.1 Strength, Stiffness, Ductility

Figure 6 shows the backbone curves of the different specimens, displaying their strength and stiffness more clearly. The initial stiffness ($K_0$) was determined by calculating the slope of the line connecting the origin to the point where the first change in slope occurs, indicating the initial cracking stage. Table 2 summarizes the initial stiffness values for each specimen, along with the ratios of these values for infilled frames compared to bare frames ($K_0/K_{B0}$) and for infilled frames with openings compared to fully infilled frames ($K_0/K_{FI,0}$). The fully infilled frame had a significantly higher initial stiffness (about 21 times) than the bare frame, while the presence of openings reduced this initial stiffness by about 70% compared to the fully infilled frame. Despite variations in opening typology, the initial stiffness of infilled frames with openings remained similar, averaging 27% of the fully infilled frame's initial stiffness (as shown in Table 2).

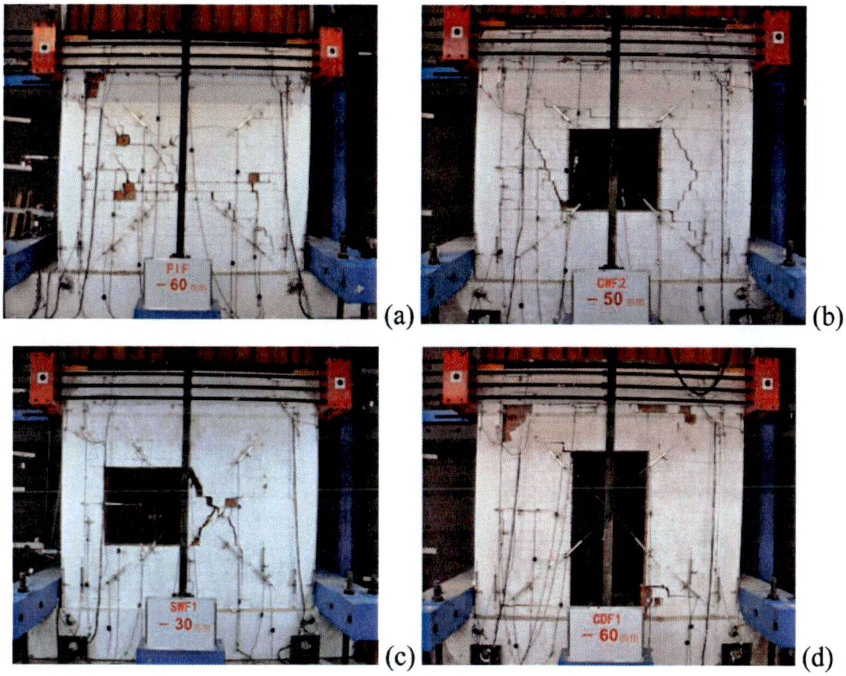


**Fig. 5.** Damage patterns of the specimens at the peak load: (a) Fully infilled frame (FIF); (b) Central window (CW); (c) Side window (SW); (d) Central door (CD).

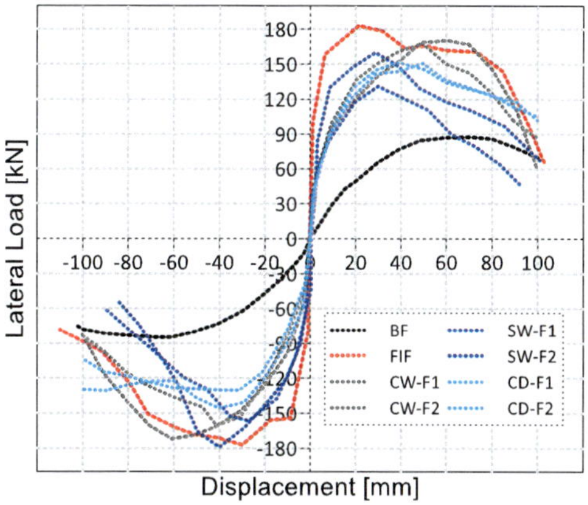


**Fig. 6.** Backbone curves: positive and negative envelopes.

Table 3 reports the peak resistances ($R_{peak}$), displacements ($d_{peak}$), interstorey drifts ($d_{r,peak}$); ultimate resistances ($R_{ult}$), displacements ($d_{ult}$) and interstorey drifts ($d_{r,ult}$) of the specimens. Several considerations can be drawn from the experimental data. Firstly, infilled frames with openings may develop alternative mechanisms to resist loading and achieve peak resistances comparable to fully infilled frames. Secondly, the peak

**Table 2.** Initial stiffness and initial stiffness ratios of the specimens.

| Specimen | $K_0$ | $K_0 / K_{B,0}$ | $K_0 / K_{FI}$ |
|---|---|---|---|
| | (kN/mm) | (–) | (–) |
| BF | 3.3 | – | – |
| FIF | 70.8 | 21.25 | 1.00 |
| CW-F1/2 | 17.6 | 5.28 | 0.25 |
| SW-F1/2 | 21.0 | 6.30 | 0.30 |
| DW-F1/2 | 19.6 | 5.87 | 0.28 |

and ultimate drifts of infilled frames with openings vary significantly depending on the opening typology. Central openings shift the peak and ultimate drifts relative to fully infilled frames, with central door opening specimens showing a ductile behavior.

**Table 3.** Peak and ultimate resistances and displacements of the specimens.

| Specimen | $R_{peak}$ | $R_{peak}/R_{B,peak}$ | $d_{peak}$ | $d_{r,peak}$ | $d_{peak}/d_{B,peak}$ | $R_{ult}$ | $d_{ult}$ | $d_{r,ult}$ | $d_{ult}/d_{B,ult}$ |
|---|---|---|---|---|---|---|---|---|---|
| | (kN) | (–) | (mm) | (%) | (–) | (kN) | (mm) | (%) | (–) |
| BF | 86 | 1.00 | 60.0 | 2.86% | 1.00 | 73.1 | 100 | 4.76% | 1.00 |
| FIF | 178 | 2.07 | 31.0 | 1.48% | 0.52 | 151.3 | 73 | 3.48% | 0.73 |
| CW-F1/2 | 163 | 1.90 | 49.0 | 2.33% | 0.82 | 138.6 | 76 | 3.62% | 0.76 |
| SW-F1/2 | 152 | 1.77 | 28.0 | 1.33% | 0.47 | 129.2 | 51 | 2.43% | 0.51 |
| DW-F1/2 | 142 | 1.65 | 40.0 | 1.90% | 0.67 | 120.7 | 90 | 4.29% | 0.90 |

### 3.2 Energy Dissipation and Equivalent Damping

The assessment of energy dissipation involved the evaluation of various energy-related parameters, including the energy dissipation per cycle ($W_d$), the cumulative energy dissipation ($\Sigma W_d$), and the average energy dissipation per unit length ($W_d/2\delta$). The latter was determined by dividing the energy dissipation per cycle by the total peak-to-peak displacement variation for each cycle ($2\delta$). The results are presented in Fig. 7. The infilled frame specimens exhibited significantly higher energy dissipation than the bare frame at each cycle (Fig. 7a), regardless of the presence of openings. Each displacement level consists of three cycles and as the number of cycles increases, the energy dissipation in cycles with the same displacement decreases, indicating a degradation of the mechanical behavior (Fig. 4, Fig. 7a). Different energy dissipation trends were observed across the various cases. The fully infilled frame demonstrated the highest average dissipation up to 1.5% interstorey drift, but decreased rapidly up to 0.5% drift (Fig. 7b). Beyond 2.0% drift, $W_d/2\delta$ stabilized at a constant value. Infilled frame specimens with central window (CW) and door openings (CD) displayed a more stable trend. For these specimens, the average energy dissipation per unit length was initially significantly lower than the FI specimen, but it remained constant from 1 to 3.5% drift. In this range, the CW specimen exhibited approximately double the energy dissipation capacity of the CD specimen. A different response was recognized from the specimen with the eccentric window (SW). The latter had a similar energy dissipation capacity to the CW specimen up to 1.5% drift (**Fig. 15b**). Beyond this point, the energy dissipation constantly decreased. This is also evident in terms of cumulative energy dissipation (Fig. 7c). It is also noteworthy that beyond 1.5% drift, the CW specimen dissipated more energy than the FIF (Fig. 7b). It can be generally noted that specimens with higher stiffness (FI), and therefore subjected to earlier damage, exhibited higher energy dissipation at lower drifts, which was reduced when the displacement demand increased. Conversely, specimens with lower initial stiffness (CW and CD) exhibited a stable energy dissipation capacity, which was maintained even in response to high drift demands.

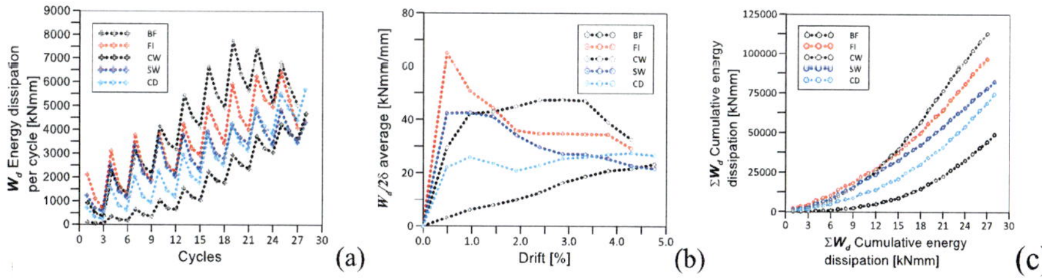


**Fig. 7.** Energy dissipation: (a) Energy dissipation per cycle; (b) Average energy dissipation per unit length. (c) Cumulative energy dissipation.

The equivalent viscous damping ($\xi_{eq}$) at the different cycles was also evaluated as Eq. (1):

$$\xi_{eq} = \frac{W_d}{2\pi(|W_e^+| + |W_e^-|)} \tag{1}$$

where $W_d$ is the dissipated energy per cycle, and $W_e^+$ and $W_e^-$ are the elastic energy at the positive and negative peak displacements, respectively. Figure 8 show that the infilled frame behaved differently from the bare frame: while the bare frame had a steady increase between 2% and 6% drifts (Fig. 8b), the infilled frame showed a trend characterized by a decrease up to 2%–2.5% drifts, followed by an increase at higher drift levels. For the FI, CW and SW specimens, the average values of $\xi_{eq}$ were relatively similar, primarily falling between 3.8% and 6%, while the BF and CD specimens had lower values in the range of 2.5–4.5%.

## 4 Conclusions

In this study, eigth 2/3 scale infilled frames constructed with hollow clay brick masonry and one bare frame were experimentally investigated under quasi-static cyclic lateral loading. The specimens included solid infills as well as those with central door and window openings. The paper analyzed the effect of opening typology on strength, stiffness,

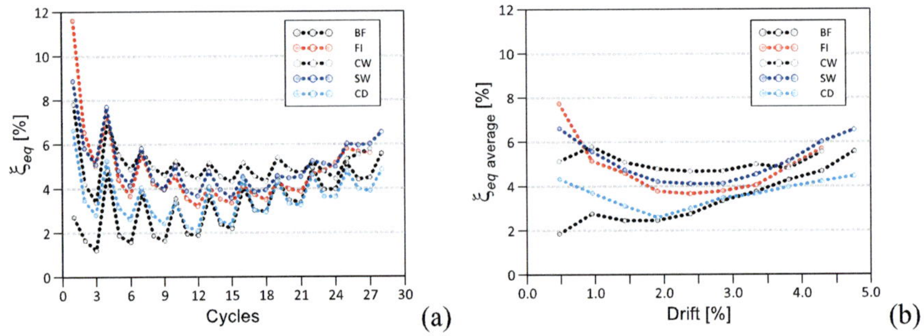


**Fig. 8.** Equivalent viscous damping: (a) equivalent viscous damping per cycle; (b) average equivalent viscous damping at the different drifts.

ductility, energy dissipation, equivalent damping and damage mechanisms. Based on the comprehensive findings, the following conclusions can be drawn:

- The existence of openings in masonry infills substantially alters the overall response compared to fully infilled frames. Nonetheless, a strong interaction between the infill and frame persists due to the emergence of alternative resisting mechanisms.
- The tested infilled frames with openings exhibited a lower initial stiffness (-70%) compared to fully infilled frames. However, the initial stiffness of infilled frames with openings was still six times greater than that of bare frames.
- The peak resistance of the tested infilled frames with the considered opening ratios did not exhibit significant reductions when compared to the fully infilled frame case.
- The presence of openings substantially affected both the peak and ultimate drifts. Specimens with central openings, CW and DW, displayed a shift in peak drifts of + 58% and +29%, respectively, whereas their ultimate drifts were similar to that of the fully infilled frame.
- All infilled frame specimens, displayed a substantially greater energy dissipation capacity compared to the bare frame. Infilled frames with central openings exhibited a consistent dissipation capacity, even when subjected to significant drift.

## References

1. Mehrabi, A.B., Shing, P.B., Schuller, M.P., Noland, L.: Experimental evaluation of masonry-infilled RC frames. J. Struct. Eng. (ASCE) **122**(3), 228–237 (1996)
2. Calvi, G.M., Bolognini, D.: Seismic response of reinforced concrete frames infilled with weakly reinforced masonry panels. J. Earthq. Eng. **5**(02), 153–185 (2001)
3. Al-Chaar, G., Issa, M., Sweeney, S.: Behavior of masonry-infilled nonductile reinforced concrete frames. J. Struct. Eng. **128**, 1055–1063 (2002)
4. Papia, M., Cavaleri, L., Fossetti, M.: Infilled frames: developments in the evaluation of the stiffening effect of infills. Struct. Eng. Mech. **16**(6), 675–693 (2003)
5. Colangelo, F.: Pseudo-dynamic seismic response of reinforced concrete frames infilled with non-structural brick masonry. Earthq. Eng. Struct. Dyn. **34**, 1219–1241 (2005)
6. da Porto, F., Guidi, M., Dalla Benetta, N., Verlato, F.: Combined in-plane/out-of-plane experimental behaviour of reinforced and strengthened infill masonry walls. In: 12th Canadian Masonry Symposium, Vancouver, Canada, 2–5 June 2013 (2013)

7. Cavaleri, L., Di Trapani, F.: Cyclic response of masonry infilled RC frames: experimental results and simplified modeling. Soil Dyn. Earthq. Eng. **65**, 224–242 (2014)
8. Bergami, A., Nuti, C.: Experimental tests and global modeling of masonry infilled frames. Earthq. Struct. **9**(2), 281–303 (2015)
9. Verderame, G.M., Ricci, P., Del Gaudio, C., De Risi, M.T.: Experimental tests on masonry infilled gravity- and seismic-load designed RC frames. In: 16th International Brick and Block Masonry Conference, IBMAC 2016, pp. 1349–1358 (2016)
10. Morandi, P., Hak, S., Magenes, G.: In-plane experimental response of strong masonry infills. Eng. Struct. **156**, 503–521 (2018)
11. Mohammadi, M., Nikfar, F.: Strength and stiffness of masonry-infilled frames with central openings based on experimental results. J. Struct. Eng. **139**, 974–984 (2013)
12. Dawe, J.L., Seah, C.K.: Behaviour of masonry infilled steel frames. Can. J. Civ. Eng. **16**(6), 865–876 (1989)
13. Mosalam, K.M., White, R.N., Gergely, P.: Static response of infilled frames using quasi-static experimentation. J. Struct. Eng. **123**(11), 1462–1469 (1997)
14. Tasnimi, A.A., Mohebkhah, A.: Investigation on the behavior of brick-infilled steel frames with openings, experimental and analytical approaches. Eng. Struct. **33**(3), 968–980 (2011)
15. Kakaletsis, D.J., Karayannis, C.G.: Experimental investigation of infilled R/C frames with eccentric openings. Struct. Eng. Mech. **26**(3), 231–250 (2007)
16. Kakaletsis, D.J., Karayannis, C.G.: Influence of masonry strength and openings on infilled R/C frames under cycling loading. J. Earthq. Eng. **12**(2), 197–221 (2008)
17. Kakaletsis, D.J., Karayannis, C.G.: Experimental investigation of infilled reinforced concrete frames with openings. ACI Struct. J. **106**(2), 132–141 (2009)
18. Blackard, B., Willam, K., Mettupalayam, S.: Experimental observations of masonry infilled reinforced concrete frames with openings. ACI Symp. Paper **265**, 122–199 (2009)
19. Sigmund, V., Penava, D.: Influence of openings, with and without confinement, on cyclic response of infilled R-C Frames-an experimental study. J. Earthq. Eng. **18**(1), 113–146 (2014)
20. Zhai, C., Kong, J., Wang, X., Chen, Z.: Experimental and finite element analytical investigation of seismic behavior of full-scale masonry infilled RC frames. J. Earthq. Eng. **20**(7), 1171–1198 (2016)
21. Mansouri, A., Marefat, M.S., Khanmohammadi, M.: Experimental evaluation of seismic performance of low-shear strength masonry infills with openings in reinforced concrete frames with deficient seismic details. Struct. Des. Tall Special Build. **23**(15), 1190–1210 (2014)
22. Morandi, P., Hak, S., Magenes, G.: Performance-based interpretation of in-plane cyclic tests on RC frames with strong masonry infills. Eng. Struct. **156**, 503–521 (2018)
23. Basha, S.H., Surendran, S., Kaushik, H.B.: Empirical models for lateral stiffness and strength of masonry-infilled RC frames considering the influence of openings. J. Struct. Eng. **146**(4), 04020021 (2020)
24. Di Trapani, F., Giordano, L., Mancini, G.: Progressive collapse response of reinforced concrete frame structures with masonry infills. J. Eng. Mech. **146**(31), 04020002 (2020)
25. Di Trapani, F., Cavaleri, L., Ferrotto, M.F.: FE modeling of partially steel-jacketed (PSJ) RC columns using CDP model. Comput. Concr. **22**(2), 143–152 (2018)
26. Carvalho, E.C., Coelho, E.: Seismic assessment, strengthening and repair of structures. radECOEST2-ICONS report no. 2, European Commission - training and mobility of researchers programme (2001)
27. Al-Chaar, G., Lamb, G.E., Issa, M.: Effect of openings on structural performance of unreinforced masonry infilled frames. ACI Symp. Paper **211**, 247–262 (2003)
28. Stavridis, A., Koutromanos, I., Shing, P.B.: Shake-table tests of a three-story reinforced concrete frame with masonry infill walls. Earthq. Eng. Struct. Dyn. **41**, 1089–1108 (2011)

# Pore Pressure Evaluation by Seismic Loads on the Ground Using an Energy Formulation

Rubén Galindo[1(✉)], Humberto Varum[2], Antonio Lara[1], Hernán Patiño[3], and Daniel Panique[1]

[1] Universidad Politécnica de Madrid, Madrid, Spain
rubenangel.galindo@upm.es
[2] University of Porto, Porto, Portugal
[3] Colombian Institute of Geotechnical Researches, Bogotá, Colombia

**Abstract.** This research proposes an energy formulation to estimate the excess pore pressure of a granular nature soil, or more generally, cohesive. The development of the formulation is validated and calibrated with a complete experimental laboratory campaign using the cyclic simple shear test. Good predictions are obtained for the tested cohesive soil using different combinations of stress states applied to the sample, with different levels of static load prior to the cyclic. In addition, the developed formulation is applied to a seismic load on a granular soil well reported in the literature, to show the pore pressure generation law obtained by simulation and its comparison with the values obtained in the laboratory.

**Keywords:** Liquefaction · Excess pore pressure · Model calibration · Energy formulation · Seismic soil response

## 1 Introduction

The effect of dynamic loads (particularly seismic actions) on the response of the ground is conditioned by the presence of water in its pores. In particular, the saturated condition is the most dangerous, since any tendency to change the volume of the soil is coerced by the presence of water in the pores, generating excess pore pressure. The deformability of a soil is given by the repositioning of its particles and the structure, so that when an external load acts, the volume of voids is reduced; however, the instantaneous circulation of pore water is not possible in the process of applying dynamic loads, developing an increase in pore pressure. Under certain conditions, this pore pressure can cancel the contact between particles, producing the well-known phenomenon of liquefaction.

Many investigations and efforts have been made to develop constitutive models that allow the adequate evaluation of the deformations of the ground when a dynamic load acts. These models have been developed with a higher level of sophistication to incorporate the different response phenomena and introduce the influence of water in a coupled way. Among these models, some initial references to each type of approach can be highlighted: (a) models of nested surfaces [1–3]; (b) endochronic models [4, 5]; (c) models of the boundary surface [6–8]; (d) hypoplastic models [9, 10]; (e) generalized


H. Varum et al. (Eds.): IWEBSE 2023, LNCE 236, pp. 304–318, 2023.
https://doi.org/10.1007/978-3-031-36562-1_24

plasticity theory [11]; (f) hysteresis model [12]. These models have grown in complexity, requiring many parameters and a highly specialized calibration stage. In addition, the incorporation of these models in advanced numerical software is limited and they have been incorporated for specific cases.

The classical dynamic calculation that has been used for applications in civil engineering has focused on the linear equivalent method [13] where the elastic and damping parameters are evaluated according to the stress level. However, practical applications for saturated conditions also require evaluating the reduction in geotechnical resistance due to increased pore pressures under dynamic loading.

The formulation of excess pore pressure received a lot of push in the 1970s with equations that, in general, were made to depend on the number of load cycles [14, 15] and which has remained an active topic of research to the present [16, 17]. Most of the empirical models depend on the number of cycles either directly or indirectly, which creates a theoretical and numerical difficulty for the correct modeling of cyclical phenomena. The consideration of a discrete domain in terms of the generation of cycles supposes having to detect the achievement of response stress cycles on each soil element, which in multiaxial states is not immediate.

In this regard, in recent years, energy methods have been considered to assess the generation of excess pore pressure [18–20]. These methods are the most consistent and promising since they can be applied to obtain the generation of pore pressure regardless of the complexity of the acting stresses, since the increase in pore pressure is evaluated incrementally as the stress state varies. Thus, any constitutive stress-strain model can be used for practical calculation, evaluating the stress path at each point, and coupling the energy formulation to update the effective pressure in the ground.

In this article, an energy approach is proposed for the evaluation of the pore pressure generated, from an intensive dynamic campaign in a cohesive material, which therefore generalizes the geotechnical investigation of this type of formulations beyond granular soils. In addition, as an example, this formulation is also used to evaluate the effect of a seismic load applied on a certain soil reported in the scientific literature.

## 2 Description of the Tested Soil

The material used in this research was obtained through two boreholes carried out at depths between 30 and 52 m with respect to sea level in the Port of PRAT in Barcelona. This material is a soil of alluvial origin that corresponds to the Holocene era classified as silty clays and clayey silts of low to medium plasticity. The soil is typically plastic, contractive and normally consolidated or with a low degree of consolidation.

The main index properties of the material used are the following: natural moisture has a mean value of 29.24% with a standard deviation of 3.13; the bulk density with a mean value of 1967.8 kg/m$^3$ and a standard deviation of 38.7; the average fine content has a value of 98.66% and a standard deviation of 2.62; the specific gravity has a mean value equal to 2.78 and a standard deviation of 0.02; the plasticity index 18.15% and a standard deviation of 1.64.

## 2.1 Experimental Program

The experimental data are based on shear simple tests: 16 tests monotonic and 78 cyclic. The samples were consolidated at the field consolidation stress before applying the different combinations of static and cyclic shear stresses as shown in Fig. 1. This figure indicates on the vertical axis the cyclic shear stress normalized by the consolidation stress ($\sigma'_{0v}$), and on the horizontal axis the initial or static shear stress also normalized by the consolidation stress. If the most critical combinations is analyzed in Fig. 1, it can indicate that for a null initial shear stress ($\tau_0/\sigma'_{0v} = 0$), the maximum cyclic shear stress ($\tau_c = 25\%$) occurs. If, on the other hand, the cyclic shear stress is zero, the maximum initial shear stress corresponds to the static shear strength ($\tau_0/\sigma'_{0v} \approx 30\%$).

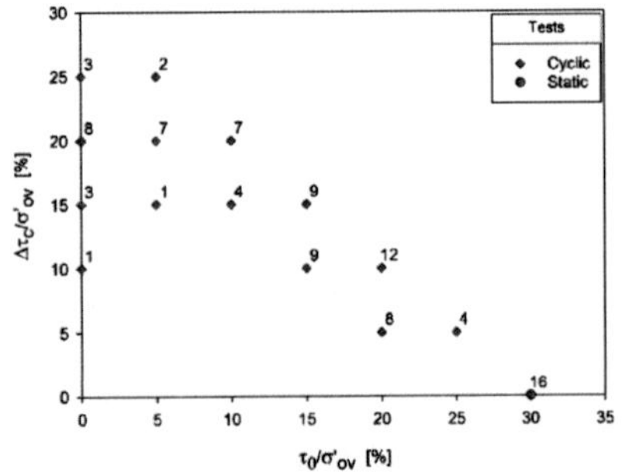


**Fig. 1.** Combination of stresses from the simple static and cyclic shear tests carried out.

The samples tested were always of the same size, with a diameter of 7 cm and 1.91 cm high. The end of the tests was in case of reaching a permanent shear strain ($\gamma_p$) of 15%, a cyclic strain ($\gamma_c$) of 15% or number of cycles ($N$) of 1300.

## 2.2 Laboratory Results

### Monotonic Simple Shear Tests

The typical behavior of normalized shear strength and pore pressure with respect to vertical consolidation stress is shown representatively with 4 tests at different consolidation stresses in Fig. 2. The maximum undrained shear strength ($\tau_m$) was developed from a very large shear strain, varying between 12% and 22%.

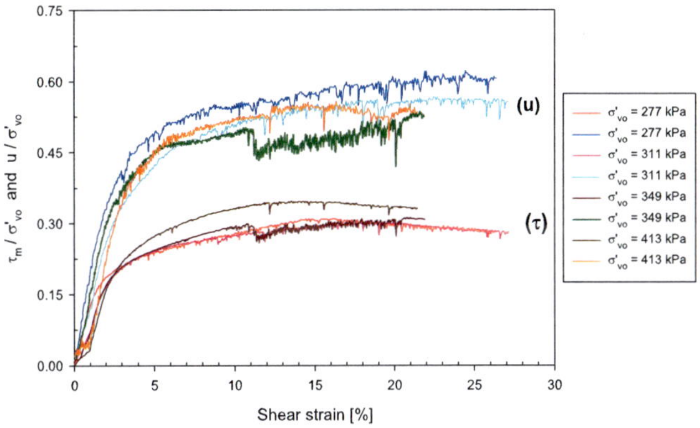


**Fig. 2.** Variation of normalized monotonic shear stress and pore pressure as a function of shear strain.

The stress-strain behavior was typically plastic. The pore water pressure generated during the failure stage was positive in all cases. This indicates a contractive behavior that shows the presence of a deposit that is little over-consolidated or normally consolidated. It is usual to evaluate the approximate maximum shear stress developed by soils using expressions such as $\tau_m = T \cdot \sigma'_{0v}$, where $\tau_m$ is the undrained shear strength in monotonic test, $T$ is a dimensionless constant and $\sigma'_{0v}$ is the in-situ effective vertical stress.

In the case of the cohesive soil, the maximum shear stress was developed for very large shear strains, which could be unsuitable for certain types of structure supported by the soil. It is, therefore, proposed that $T$ should be evaluated as a function of the monotonic shear strain ($\tau_m$), Fig. 3. Thus, the constant $T$ is replaced by a variable as a function of the shear strain and the in-situ effective vertical stress:

$$\tau_m = \lfloor 0.07 \cdot ln(\gamma_m) + 0.43 \rfloor \cdot \sigma'_{ov} \tag{1}$$

Figure 3 shows that the constant $T$ equal to 0.3 used to estimate the resistance of the clay soil corresponds to a deformation equal to 15%.

With the information gathered from the laboratory tests, it is also possible to obtain the shear strength lines for different values of $\tau_m$. The shear strength parameters, $c'$ and $\phi'$, as a function of the shear strain (therefore reference is made to the mobilized strength). To obtain the mobilized shear strength parameters in terms of effective stress, it is possible to use empirical functions Eq. (2) and (3). The suggested empirical functions are as follows:

$$c' = 9 \cdot ln(\gamma_m) + 40 \tag{2}$$

$$\phi' = 6.2 \cdot ln(\gamma_m) + 36.5 \tag{3}$$

where $\phi'$ is the mobilized friction angle, $c'$ is the apparent cohesion and $\gamma_m$ is the monotonic shear strain.

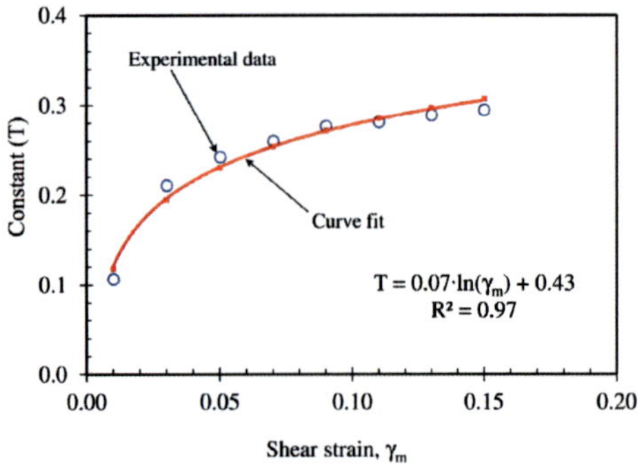


**Fig. 3.** Dimensionless constant $T$ to evaluate the maximum shear stress developed by soils from the in-situ effective vertical stress.

## Cyclic Simple Shear Tests

The cyclic simple shear tests were analyzed considering the relationship of the increase in pore pressure ($u$) with the number of cycles, as well as the relationship stress – strain shown. Table 1 shows for some tests, the combination of static and cyclic shear stresses for each graph as well as a summary of the main results of the tests analyzed in Fig. 4 and 5.

**Table 1.** Combination of stresses used and summary of the results of the tests analyzed in Fig. 3 and 4.

| Reference | | $\tau_0/\sigma'_{0v}$ | $\tau_c/\sigma'_{0v}$ | $N$ | $\gamma_p$ | $\gamma_c$ | $u/\sigma'_{0v}$ |
|---|---|---|---|---|---|---|---|
| Figure 4 | Figure 5 | [ %] | [ %] | [cycles] | [%] | [%] | [ %] |
| a) | a) | 0 | 25 | 15 | 8.9 | 17.8 | 72 |
| a) | – | 0 | 20 | 72 | 10.8 | 15.1 | 76 |
| a) | e) | 0 | 15 | 500 | 11.4 | 15.4 | 84 |
| b) | – | 5 | 25 | 15 | 7.3 | 12.9 | 71 |
| b) | c) | 5 | 20 | 55 | 9 | 11.6 | 80 |
| c) | b) | 10 | 20 | 28 | 15.5 | 7.7 | 75 |
| c) | d) | 10 | 15 | 87 | 17.8 | 6.5 | 77 |
| d) | – | 15 | 15 | 1300 | 15.5 | 2 | 92 |
| d) | – | 15 | 10 | 1300 | 10.4 | 0.6 | 64 |
| e) | f) | 20 | 10 | 1300 | 13.6 | 0.5 | 58 |
| e) | – | 20 | 5 | 1300 | 4.1 | 0.1 | 42 |
| f) | – | 25 | 5 | 1300 | 9.6 | 0.15 | 41 |

The behavior of the normalized pore pressure generation with the number of cycles is shown in Fig. 4. It was observed that, for initial shear stress values of 0, 5 and 10%; the behavior of the pore pressure generation decreases as the cyclic shear stress increases ($\tau_c / \sigma'_{0v}$). On the other hand, for initial shear stresses of 15 and 20%; the trend changes and the generation of pore pressure tends to be higher when the cyclic shear stress increases. It is worth mentioning that in this situation the tests were limited to number of cycles of 1300.

When analyzing the number of cycles with the increase in pore pressure at the end of the test, it turns out that the value of the pore pressure when the end of the test is produced in a few cycles tends to be greater than the pore pressure for tests with many cycles.

The lower the initial shear stress, the end of the test occurs with a lower number of cycles, as can be seen in Fig. 4a. While for large initial shear stresses the number of cycles tends to be greater as can be seen in Fig. 4f. On the other hand, large values of cyclic shear stress (20 and 25%) imply the "failure" with less than 100 cycles as can be

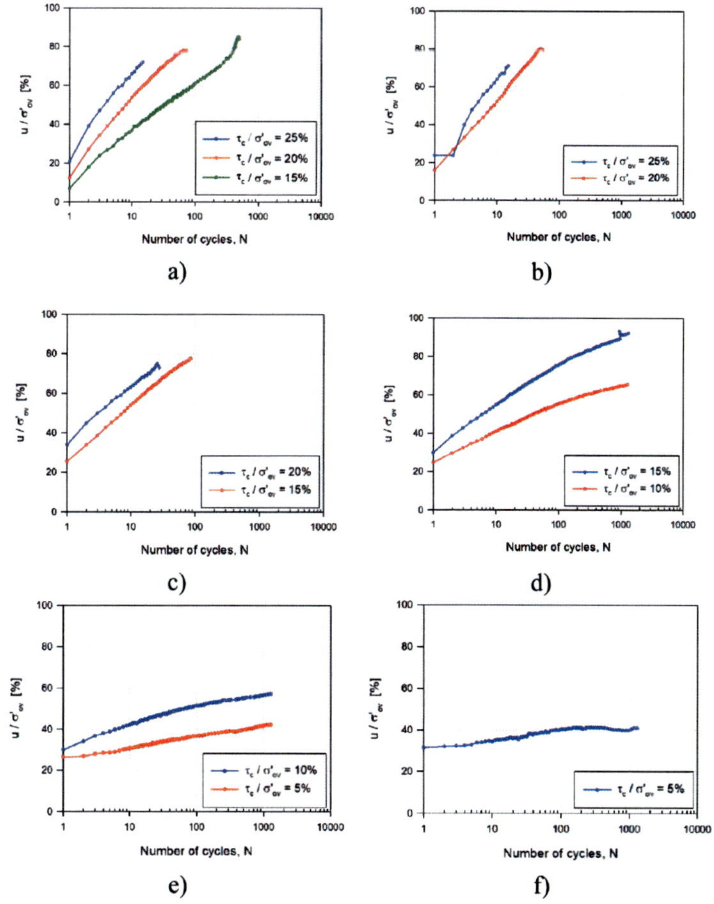


**Fig. 4.** Relationship of the increase in pore pressure with the number of cycles. a) $\tau_0 / \sigma'_{0v} = 0$; b) $\tau_0 / \sigma'_{0v} = 5\%$; c) $\tau_0 / \sigma'_{0v} = 10\%$; d) $\tau_0 / \sigma'_{0v} = 15\%$; e) $\tau_0 / \sigma'_{0v} = 20\%$; f) $\tau_0 / \sigma'_{0v} = 25\%$.

seen in Fig. 4a y 4b. If the cyclic shear stress values are lower (10 and 5%), the "failure" is produced for many cycles (more than 1000 cycles) as observed in Fig. 4e.

The behavior of stress – strain shows that the total shear strain has a cyclic component ($\gamma_c$) as well as a permanent one ($\gamma_p$). Figure 5 shows only 6 graphs of behavior as a function of the number of cycles ($N$). The trend can be indicated for initial shear stress values of 0, 5 and 10%; thus, the charts with higher cyclic shear stress have the lower permanent strain ($\gamma_p$). However, for values of initial shear stress of 15 and 20%, the trend changes and the greater permanent strains are accumulated with the cycles. For low cyclic shear stresses, soil failure occurs for a greater number of cycles as shown in Fig. 5d, 5e and 5f; while for high cyclic shear stresses of the order of 25 or 20%, the number of cycles is lower as shown in Fig. 5a, 5b and 5c.

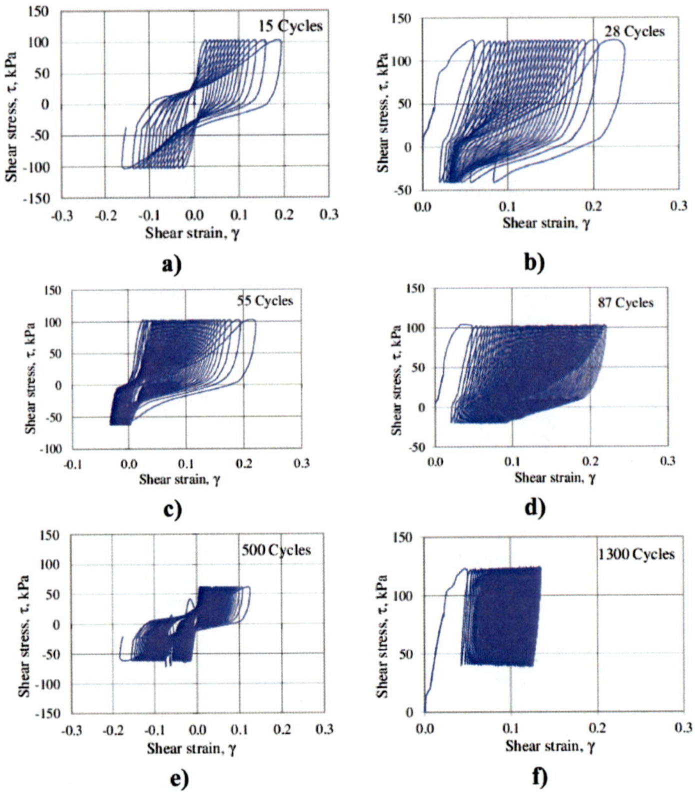


**Fig. 5.** Stress-strain relationship with the number of cycles. a)) $\tau_0/\sigma'_{0v} = 0$ and $\tau_c/\sigma'_{0v} = 25\%$; b) $\tau_0/\sigma'_{0v} = 10\%$ and $\tau_c/\sigma'_{0v} = 20\%$; c) $\tau_0/\sigma'_{0v} = 5\%$ and $\tau_c/\sigma'_{0v} = 20\%$; d) $\tau_0/\sigma'_{0v} = 10\%$ and $\tau_c/\sigma'_{0v} = 15\%$; e) $\tau_0/\sigma'_{0v} = 0$ and $\tau_c/\sigma'_{0v} = 15\%$; f) $\tau_0/\sigma'_{0v} = 20\%$ and $\tau_c/\sigma'_{0v} = 10\%$.

## 3 Energy Formulation

The pore pressure ratio ($r_u$) is defined as the quotient between the excess pore pressure generated and the effective initial confinement stress of the soil ($\sigma'_{0v}$), in such a way that this ratio can vary from zero to one, being unity when complete transfer of effective stress to the water occurs. The GMP energy base model used by Green et al. [18] allows establishing an empirical equation that relates $r_u$ to the energy dissipated per unit volume of the soil, so that:

$$r_u = \sqrt{\frac{W_s}{PEC}} \tag{4}$$

where $W_s$ is the energy dissipated per unit volume of the soil divided by the effective initial confining pressure and PEC is a calibration parameter called "pseudo energy capacity". This formulation was developed using test data on non-plastic sand-silt mix materials and for stress states without pore pressure generated by initial static load.

Thus, a generalization of this energy formulation is proposed, which is applicable to any type of soil, including cohesive soils where the effective stress does not cancel out, and which may be subjected to states of prior stress that generates initial pore pressure.

Based on data from the cohesive soil described in the previous section, the following empirically based energy formulation is proposed to evaluate the pore pressure generated in any soil:

$$r_u = r_o + \frac{(r_{umax} - r_o)}{1 + \sqrt{\frac{W_{s0}}{W_s}}} \tag{5}$$

where $r_0$ is the pore pressure generation rate generated by the applied static stress before the cyclic phase and $r_{umax}$ is the maximum pore pressure rate that can be generated in the soil for the loading conditions. For its part, $W_s$ is the energy dissipated per unit volume of the soil divided by the initial effective confining pressure and $W_{s0}$ is a parameter of the model that refers to the rate of decrease in the generation of pore pressure called the "energy growth parameter", in such a way that its evolution can be defined in the $r_u$ – $(W_s)^{1/2}$ curve.

### 3.1 Evaluation of the Generated Energy

For general load states, energy increments can be related to stress conditions and strain increments according to the following equation:

$$dW_s = \frac{\left(\sigma'_y d\varepsilon_y + 2\sigma'_x d\varepsilon_x + \tau_{yx} d\gamma_{yx} + \tau_{xy} d\gamma_{xy}\right)}{\sigma'_{v0}} \tag{6}$$

where $dW_s$ is the increase in normalized dissipated energy with respect to the initial mean effective stress; $\sigma'_y$ is the effective vertical stress; $d\varepsilon_y$ is the vertical strain increment; $\sigma'_x$ is the effective horizontal stress; $d\varepsilon_x$ is the increment of horizontal strain; $\tau_{yx}$ is the shear stress along the horizontal plane; $d\gamma_{yx}$ is the shear strain increment for stress

$\tau_{yx}$; $\tau_{xy}$ is the shear stress along the vertical plane; $d\gamma_{xy}$ is the shear strain increment for stress $\tau_{xy}$; and $\sigma'_{0v}$ is the initial effective vertical stress.

In the case of the undrained simple shear test, used in the laboratory for the dynamic analysis of the reference soil, Eq. (6) can be expressed for its numerical evaluation as:

$$W_s = \frac{1}{2\sigma'_{v0}} \sum_{i=1}^{n-1} (\tau_{i+1} + \tau_i)(\gamma_{i+1} - \gamma_i) \tag{7}$$

where $\tau_i$ and $\gamma_i$ are the shear stress and strain applied in step $i$ of the discretization, respectively.

### 3.2 Energy Growth Parameter ($W_{s0}$)

The physical interpretation of the parameter $(W_{s0})^{½}$ corresponds to a point in the central zone of evolution of the pore pressure ratio marked in Fig. 6, whose evaluation will allow us to adequately define all the growth of pore pressure up to its maximum value $r_{umax}$. For practical purposes, the point obtained from the following geometric construction can be adopted: the projection, on the abscissa axis, of the intersection of the tangent line at the origin to the pore pressure generation curve at the $r_u$ – $(W_s)^{½}$ plane with the asymptote corresponding to the maximum generation of pore pressure $r_{umax}$.

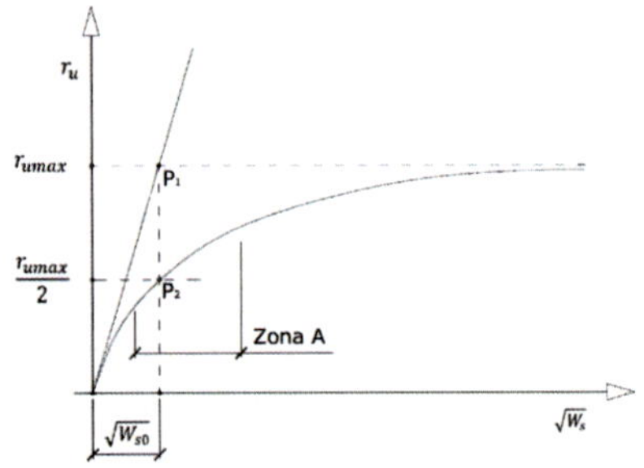


**Fig. 6.** Geometric determination of the parameter $(W_s)^{½}$.

As $(W_s)^{½}$ is a valid parameter for any combination of static and cyclic shear stresses in the simple shear test, then it will be valid to calibrate it in zero initial stress tests. Furthermore, these tests are the most used and also the most convenient because they allow the maximum generation of pore pressure in the soil to be developed, allowing the search for the adjustment parameter more precise. In this case, it can be shown mathematically that the geometric construction described above is equivalent to the pore pressure generation point of 50% of the maximum, writing $W_{s0} = W_{s,0.5}r_{umax}$.

### 3.3 Maximum Value of the Pore Pressure Ratio ($r_{umax}$)

The value of $r_u$ that is reached in a soil depends on the evolution of the dynamic load and the response of the soil with time, that is, on the energy accumulated in the time of the dynamic load. The maximum value of $r_u$ is limited by the $r_{umax}$ parameter.

The value of $r_{umax}$ that a soil without cohesion can develop is equal to one, which, as is well known, corresponds to the state of liquefaction. However, in a cohesive soil, the cohesive component of the strength limits the possibility of soil liquefaction and separation of particles. To analyze this parameter, it is convenient to remember for the simple shear test, the following relationship between the static shear stress with respect to the effective confinement stress [21]:

$$\frac{\tau_{xy}}{\sigma'_y} = m \tan \Psi \tag{8}$$

where $\Psi$ is the angle between the major principal stress and the vertical plane and $m$ is a constant function of the tested material. In addition, it should be noted that was experimentally demonstrate [22, 23] that in the Critical State, where the constant shear stress ($\tau_{xy} = \tau_{stat}$) is reached without volume change of the sample and therefore a maximum pore pressure is developed, values of $\Psi$ close to 45° are produced. Therefore, $\tau_{stat}/\sigma'_y = \beta$ of constant value, y then the value of the maximum pore pressure ratio generated in the static simple shear test is:

$$r_{umax}^{stat} = 1 - \frac{\tau_{stat}}{\beta \cdot \sigma'_{v0}} \tag{9}$$

Therefore, in a simple static shear test, the parameter $\beta$ can be obtained from the maximum pore pressure generated for the static failure stress.

Since a cohesive soil has an apparent cohesion ($c'$), it is immediate from Eq. (9) to deduce the maximum ratio of pore pressure developed under dynamic shear stress:

$$r_{umax} = 1 - \frac{c'}{\beta \cdot \sigma'_{v0}} \tag{10}$$

In our case, the apparent cohesion value corresponds to strain equal the 15% at which we define failure in the simple shear test. If there is a static shear stress component ($\tau_O$), the maximum value $r_{umax}$ of Eq. (10) must also be limited as indicated in Eq. (9) so that in general a virtual cohesion of value ($c' + \tau_O$) can be used.

## 4 Simulation of Generated Pore Pressure

In this section, the energy formulation proposed in Eq. (5) is used to simulate the generation of pore pressure of the tested cohesive soil and validate it with the data obtained experimentally.

Expression (10) allows the evaluation of the maximum pore pressure ratio. That equation requires the parameters $\beta$ and c′. Using Eq. (9), $\beta$ can be obtained for the 16 monotonic tests carried out; while using Eq. (2), an apparent cohesion of 25.25 kPa can be obtained for 15% strain. Table 2 shows the results obtained, where the $r_{umax}$ values have little deviation from the mean value of 0.90.

**Table 2.** $\beta$ parameter calibration.

| Sample Ident | $\sigma'_{v0}$ $\sigma'_{v0}$ (kPa) | $\tau_m$ (kPa) | $u_{max}$ (kPa) | $\beta$ (Eq. (9)) | $r_{umax}$ (Eq. (10)) |
|---|---|---|---|---|---|
| SA-1-M1 | 277 | 86 | 160 | 0,73 | 0,88 |
| SA-1-M2 | 283 | 80 | 159 | 0,65 | 0,86 |
| SA-1-M4 | 311 | 94 | 167 | 0,65 | 0,88 |
| SA-1-M6 | 349 | 109 | 163 | 0,58 | 0,88 |
| SA-1-M7 | 366 | 112 | 215 | 0,74 | 0,91 |
| SA-1-M8 | 389 | 131 | 210 | 0,73 | 0,91 |
| SA-1-M9 | 401 | 145 | 242 | 0,91 | 0,93 |
| SA-1-M10 | 413 | 143 | 230 | 0,78 | 0,92 |
| SA-2-M1 | 294 | 98 | 147 | 0,67 | 0,87 |
| SA-2-M2 | 315 | 100 | 205 | 0,91 | 0,91 |
| SA-2-M3 | 328 | 117 | 190 | 0,85 | 0,91 |
| SA-2-M4 | 343 | 140 | 240 | 1,36 | 0,95 |
| SA-2-M5 | 347 | 102 | 190 | 0,65 | 0,89 |
| SA-2-M6 | 364 | 110 | 195 | 0,65 | 0,89 |
| SA-2-M7 | 373 | 101 | 182 | 0,53 | 0,87 |
| SA-2-M8 | 384 | 148 | 175 | 0,71 | 0,91 |

To obtain the energy growth parameter $W_{s0}$, the cyclic tests without static initial stress that correspond to cyclic ratios ($\tau_c/\sigma'_{v0}$) of 20 and 25% are used, which generate sufficiently high pore pressures. Table 3 shows, for these tests, the values of the parameter $W_{s0}$, which is adjusted by identifying the stress-strain evaluation cycle of each test in which the $r_{umax}$ of 50% (0.45) is reached. As can be seen in the Table, the value of this parameter is in all cases very close to the average value of 0.04, corresponding to an accumulated energy of 45%.

**Table 3.** $W_{s0}$ parameter calibration.

| Sample Ident | $\sigma'_{v0}$ $\sigma'_{v0}$ (kPa) | $\tau_0/\sigma'_{v0}$ $\tau_0/\sigma'_{v0}$ | $\tau_c/\sigma'_{v0}$ $\tau_c/\sigma'_{v0}$ | $u_{max}/\sigma'_{v0}$ | $N$ | $W_{s0}$ |
|---|---|---|---|---|---|---|
| SA-1-M6 | 349 | 0.00 | 0.20 | 0.85 | 146 | 0.035 |
| SA-1-M6 | 349 | 0.00 | 0.25 | 0.72 | 14 | 0.043 |
| SA-1-M7 | 366 | 0.00 | 0.20 | 0.8 | 79 | 0.037 |
| SA-1-M9 | 401 | 0.00 | 0.20 | 0.85 | 34 | 0.038 |
| SA-1-M10 | 413 | 0.00 | 0.20 | 0.78 | 72 | 0.039 |

*(continued)*

**Table 3.** (*continued*)

| Sample Ident | $\sigma'_{v0}$ $\sigma'_{v0}$ (kPa) | $\tau_0/\sigma'_{v0}$ $\tau_0/\sigma'_{v0}$ | $\tau_c/\sigma'_{v0}$ $\tau_c/\sigma'_{v0}$ | $u_{max}/\sigma'_{v0}$ | $N$ | $W_{s0}$ |
|---|---|---|---|---|---|---|
| SA-2-M5 | 347 | 0.00 | 0.20 | 0.76 | 23 | 0.04 |
| SA-2-M5 | 347 | 0.00 | 0.25 | 0.72 | 9 | 0.042 |
| SA-2-M6 | 364 | 0.00 | 0.25 | 0.75 | 27 | 0.042 |
| SA-2-M7 | 373 | 0.00 | 0.20 | 0.75 | 29 | 0.041 |
| SA-2-M8 | 384 | 0.00 | 0.20 | 0.8 | 83 | 0.038 |

The formulation developed for its validation is applied to three different cases as indicated in Fig. 7: (a) case of cyclic stress without static initial stress; (b) case of cyclic stress with a 10% static initial stress ratio; (c) cyclic stress case with a 20% static initial stress ratio. The pore pressure results for the simulations show adequate prediction for different static and cyclic stress combinations.

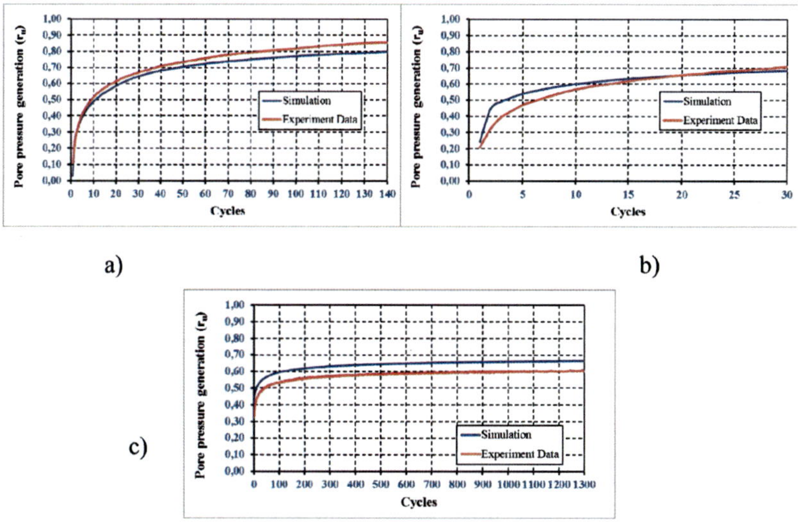


**Fig. 7.** Pore pressure simulations: (a) initial confinement stress $\sigma'_{v0} = 349$ kPa, zero static shear stress and cyclical shear stress of 70 kPa); (b) initial confinement stress $\sigma'_{v0} = 349$ kPa, static shear stress of 35 kPa and cyclic shear stress of 70 kPa; (c) initial confinement stress $\sigma'_{v0} =$ 349 kPa, static shear stress of 70 kPa and cyclic shear stress of 35 kPa.

## 5 Application to Seismic Load

Once the energy model has been formulated, calibrated and validated with dynamic laboratory tests, its application to seismic loads is intended, for which the shear stress applied to the ground follows an irregular evolution over time.

To verify its applicability for seismic loads, an example case reported in the bibliography [23] for Nevada sand (uniform sand) and tested in cyclic simple shear test has been used. In this soil the evolution of stress, strain and pore pressure is known for a transient loading history corresponding to the Palm Spring earthquake recorded in 1986 (magnitude equal to 6). The cyclic shear stress applied to the soil over time is plotted in Fig. 8a, and the response obtained in the cyclic simple shear test is shown in Fig. 8b.

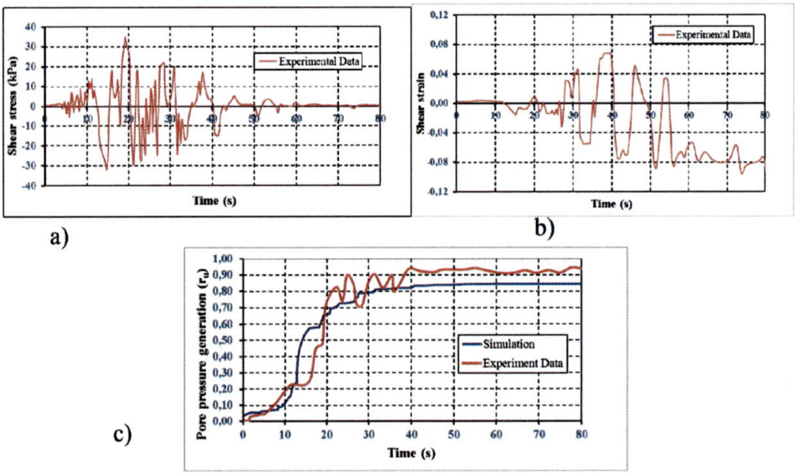


**Fig. 8.** Nevada sand subjected to seismic load: (a) Shear stress-time; (b) Shear strain-time; (c) Comparison of the simulation and the laboratory values of the excess of the pore pressure.

When applying the energy model developed in this research, $r_{umax}$ equal to 1 should be considered, as it is a granular material that can undergo liquefaction and there is no initial static stress. The value of the energy growth parameter $W_{s0}$ can be evaluated from the response time of the excess pore pressure in the cyclic simple shear test, for which 50% of the $r_{umax}$ is developed. In this case, a value of $W_{s0}$ parameter of 0.0025 is obtained.

Figure 8c shows the comparison of the pore pressure of the laboratory and the simulation using the proposed energy formulation. It can be observed that the global values of pore pressure development are quite adjusted to the experimental values, and that the initial evolution of growth can be adequately captured as well as the area of abrupt growth that occurs between 10 and 20 s.

## 6 Conclusions

In this research, an energy-based mathematical model has been presented to evaluate the generation of excess pore pressures in the soil against the action of dynamic loads. The parameters that allow a general model, valid for any type of soil, have been identified, and they have been calibrated using a complete dynamic laboratory campaign with cyclic simple shear tests on a cohesive soil. The validation with tests with different combinations of static and cyclic stresses has been satisfactory. In addition, the excess pressure generated in a soil reported by the technical literature under seismic loading has been evaluated, obtaining results that show an adequate prediction.

**Acknowledgment.** Cooperation between the University of Porto and the Universidad Politécnica de Madrid is supported by the mobility program of the Universidad Politécnica de Madrid in collaboration with Santander Bank.

## References

1. Iwan, W.D.: On a class of models for the yielding behavior of continuous and composite systems. J. Appl. Mech. **34**, 612–617 (1967)
2. Mroz, Z.: On the description of anisotropic workhardening. J. Mech. Phys. Solids **15**(3), 163–175 (1967)
3. Prevost, J.H.: Undrained shear tests on clays. J. Geotech. Geoenviron. Eng. **105**(1), 49–64 (1979)
4. Valanis, K.C.: A theory of viscoplasticity without a yield surface, Part I: general theory. Arch. Mech. Stosow **23**, 517–534 (1971)
5. Bazant, Z., Krizek, R.: Saturated sand as inelastic two-phase medium. ASCE J. Eng. Mech. Div. **101**(4), 317–332 (1975)
6. Dafalias, Y.F., Herrmann, L.R.: Bounding surface formulation of soil plasticity. In: Pande, G.N., Zienkiewicz, O.C. (eds.) Chapter 10 in Soil Mechanics-Transient and Cyclic Loads, pp. 253–282. John Wiley and Sons Ltd. (1982)
7. Whittle, A.J.: Evaluation of a constitutive model for overconsolidated clays. Geotecnique **43**(2), 289–313 (1993)
8. Pestana, J.M., Biscontin, G.: A simplified model describing the cyclic behavior of lightly overconsolidated clays in simple shear. Geotechnical Engineering Report 2001; No UCB/GT/2000–03 (2001)
9. Kolymbas, D., Herle, I., Wolffersdorff, P.A.: Hypoplastic constitutive equation with back stress. Int. J. Numer. Anal. Methods Geomechan. **19**(6), 415–446 (1995)
10. Gudehus, G.: A comprehensive constitutive equation for granular materials. Soils Found. **36**(1), 1–12 (1996)
11. Pastor, M., Zienkiewicz, O.C., Chan, A.H.C.: Generalized plasticity and the modeling of soil behavior. Int. J. Numer. Anal. Methods Geomech. **14**, 151–190 (1990)
12. Galindo, R.A., Lara, A., Melentijevic, S.: Hysteresis model for dynamic load under large strains. Int. J. Geomech. **19**(6) (2019)
13. Seed, H.B., Idriss, I.M.: Soil moduli and damping factors for dynamic response analyses/ Earthquake Engineering Research Center. Report No. EERC 70–10. University of California. Berkeley, California (1970)
14. Lee, K.L., Albaisa, A.: Earthquake induced settlement in saturated sands. J. Geotech. Eng. **100**(4), 387–406 (1974)
15. Booker, J.R., Rahman, M., Seed, H.B.: GADFLEA: a computer program for the analysis of pore pressure generation and dissipation during cyclic or earthquake loading. California University, Berkeley (USA). Earthquake Engineering Research Center (1976)
16. Sun, J., Yuan, X.: A simplified formula for estiming realtime pore water pressure of anisotropically consolidated saturated sands under random earthquake loads. In: Proceeding GeoShanghai, Geotechnical Special Publication No 150, ASCE/GEO Institute, Reston, pp. 444–451 (2006)

17. Khashila, M., Hussien, M., Karray, M., Chekired, M.: The use of pore pressure build-up as damage metric in computation of equivalent number of uniform strain cycles. Can. Geotech. J. **55**(4), 538–550 (2017)
18. Green, R.A., Mitchell, J.K., Polito, C.P.: An energy-based pore pressure generation model for cohesionless soils. In: Smith, D.W., Carter, J.P. (eds.) Proceeding John Booker Memorial Symposium. Developments in Theoretical Geomechanics, Balkema, Rotterdam, Netherlands, pp. 383–390 (2000)
19. Polito, C.P., Green, R.A., Lee, J.: Pore pressure generation models for sands and silty soils subjected to cyclic loading. J. Geotech. Geoenviron. Eng. **134**(10), 1490–1500 (2008)
20. Polito, C.P., Moldenhauer, H.: Energy Dissipation and Pore Pressure Generation in Stress- and Strain-Controlled Cyclic Triaxial Tests. Engineering Faculty Publications, Patents, Presentations. 82 (2019)
21. Oda, M., Konishi, J.: Rotation of principal stresses in granular material during simple shear. Soils Found. **14**(4), 39–53 (1974)
22. Ochiai, H.: The behavior of sands in direct shear tests. Geotech. Test. J. **15**(4), 93–100 (1975)
23. Kwan, W.S., Sideras, S., El Mohtar, C.S., Kramer, S.: Pore pressure generation under different transient loading histories. 10NCEE, U.S. National Conference on Earthquake Engineering, Frontiers of Earthquake Engineering, Alaska, 21–25 July 2014 (2014)

# Author Index




H. Varum et al. (Eds.): IWEBSE 2023, LNCE 236, pp. 319–320, 2023.
https://doi.org/10.1007/978-3-031-36562-1